Springer Monographs in Mathematics

For further volumes:
www.springer.com/series/3733

Lavinia Corina Ciungu

Non-commutative Multiple-Valued Logic Algebras

 Springer

Lavinia Corina Ciungu
Department of Mathematics
University of Iowa
Iowa City, IA, USA

ISSN 1439-7382   ISSN 2196-9922 (electronic)
Springer Monographs in Mathematics
ISBN 978-3-319-03299-3   ISBN 978-3-319-01589-7 (eBook)
DOI 10.1007/978-3-319-01589-7
Springer Cham Heidelberg New York Dordrecht London

Mathematics Subject Classification: 03G10, 06F35, 06D35, 03G25, 03B50, 06F20, 06D50, 08A30, 08B26, 08A72, 03B52, 03G12, 28E10

Printed on acid-free paper

Springer is part of Springer Science+Business Media (www.springer.com)

Dedicated to my precious son David Edward

Introduction

In 1920 Łukasiewicz introduced his three valued logic ([223]), the first model of multiple-valued logic. The n-valued propositional logic for $n > 3$ was constructed in 1922 and the $\aleph_0$-valued Łukasiewicz-Tarski logic in 1930 ([224]). The first completeness theorem for $\aleph_0$-valued Łukasiewicz-Tarski logic was given by Wajsberg in 1935. As a direct generalization of two-valued calculus, Post introduced in 1921 an n-valued propositional calculus distinct from that of Łukasiewicz ([239]).

In the early 1940s Gr.C. Moisil was the first to develop the theory of n-valued Łukasiewicz algebras with the intention of algebraizing Łukasiewicz's logic ([226, 227]), but an example of A. Rose from 1956 established that for $n \geq 5$ the Łukasiewicz implication can no longer be defined on a Łukasiewicz algebra. Consequently, the structures introduced by Moisil are models for Łukasiewicz logic only for $n = 3$ and $n = 4$. These algebras are now called *Łukasiewicz-Moisil algebras* or *LM algebras* for short ([14]).

The loss of implication has led to another type of logic, today called *Moisil logic*, distinct from the Łukasiewicz system. The logic corresponding to n-valued Łukasiewicz-Moisil algebras was created by Moisil in 1964. The fundamental concept of Moisil logic is *nuancing*. During 1954–1973 Moisil introduced the θ-valued LM algebras without negation, applied multiple-valued logics to switching theory and studied algebraic properties of LM algebras (representation, ideals, residuation) ([228]). Moisil's works have been continued by many mathematicians ([149, 151]). A. Iorgulescu introduced and studied θ-valued LM algebras with negation ([170]), while V. Boicescu defined and investigated n-valued LM algebras without negation ([13]).

Today these multiple-valued logics have been developed into fuzzy logics, which connect quantum mechanics, mathematical logic, probability theory, algebra and soft computing.

In 1958 Chang defined *MV-algebras* ([38]) as the algebraic counterpart of $\aleph_0$-valued Łukasiewicz logic and he gave another completeness proof of this logic ([39]).

An *MV-algebra* is an algebra $(A, \oplus, ^-, 0)$ with a binary operation $\oplus$, a unary operation $^-$ and a constant 0 satisfying the following equations:

(MV_1) $(x \oplus y) \oplus z = x \oplus (y \oplus z)$;

(MV_2) $x \oplus y = y \oplus x$;

(MV_3) $x \oplus 0 = x$;

(MV_4) $(x^-)^- = x$;

(MV_5) $x \oplus 0^- = 0^-$;

(MV_6) $(x^- \oplus y)^- \oplus y = (y^- \oplus x)^- \oplus x$.

Studies on MV-algebras have been developed in [5–8, 22, 77, 81, 87, 89, 91, 120, 139, 146, 147, 153, 213, 214, 217–219, 247].

Starting from the systems of positive implicational calculus, weak systems of positive implicational calculus and BCI and BCK systems, in 1966 Y. Imai and K. Iséki introduced the BCK-algebras ([168]).

In 1977 R. Grigolia introduced MV_n-algebras to model the n-valued Łukasiewicz logic ([157]) and it was proved that there is a connection between n-valued Łukasiewicz algebras and MV_n-algebras ([171–173, 191, 216]).

One of the most famous results in the theory of MV-algebras was Mundici's theorem from 1986 which states that the category of MV-algebras is equivalent to the category of Abelian ℓ-groups with strong unit ([229]).

The non-commutative generalizations of MV-algebras called *pseudo-MV alge-bras* were introduced by G. Georgescu and A. Iorgulescu in [135] and [137] and they can be regarded as algebraic semantics for a non-commutative generalization of a multiple-valued reasoning ([215]). The pseudo-MV algebras were introduced independently by J. Rachůnek ([241]) under the name of *generalized MV-algebras*.

A. Dvurečenskij proved in [97] that any pseudo-MV algebra is isomorphic with some interval in an ℓ-group with strong unit, that is, the category of pseudo-MV algebras is equivalent to the category of unital ℓ-groups.

Residuation is a fundamental concept of ordered structures and categories and Ward and Dilworth were the first to introduce the concept of a *residuated lattice* as a generalization of ideal lattices of rings ([262]). The theory of residuated lattices was used to develop algebraic counterparts of fuzzy logics ([256]) and substructural logics ([234]).

A residuated lattice is defined as an algebra $\mathcal{A} = (A, \wedge, \vee, \odot, \to, \rightsquigarrow, e)$ of type $(2, 2, 2, 2, 2, 0)$ satisfying the following conditions:

(A_1) $(A, \wedge, \vee)$ is a lattice;

(A_2) $(A, \odot, e)$ is a monoid;

(A_3) $x \odot y \leq z$ iff $x \leq y \to z$ iff $y \leq x \rightsquigarrow z$ for any $x, y, z \in A$ (*pseudo-residuation*).

A residuated lattice with a constant 0 (which can denote any element) is called a *pointed residuated lattice* or *full Lambek algebra* (*FL-algebra*, for short). If $x \leq e$ for all $x \in A$, then $\mathcal{A}$ is called an *integral residuated lattice*. An FL-algebra $\mathcal{A}$ which satisfies the condition $0 \leq x \leq e$ for all $x \in A$ is called FL_w-*algebra* or *bounded in-tegral residuated lattice* ([129]). In this case we put $e = 1$, so that an FL_w-algebra will be denoted $(A, \wedge, \vee, \odot, \to, \rightsquigarrow, 0, 1)$. Clearly, if $\mathcal{A}$ is an FL_w-algebra, then $(A, \wedge, \vee, 0, 1)$ is a bounded lattice.

In order to formalize the multiple-valued logics induced by continuous t-norms on the real unit interval $[0, 1]$, P. Hájek introduced in 1998 a very general multiple-valued logic, called *Basic Logic* (or BL) ([158]). Basic Logic turns out to be a common ingredient in three important multiple-valued logics: $\aleph_0$-valued Łukasiewicz logic, Gödel logic and Product logic. The Lindenbaum-Tarski algebras for Basic Logic are called *BL-algebras* ([23, 82, 220–222, 255–257]). Apart from their logical interest, BL-algebras have important algebraic properties and they have been intensively studied from an algebraic point of view.

The well-known result that a t-norm on [0, 1] has residuum if and only if the t-norm is left-continuous makes clear that BL is not the most general t-norm based logic. In fact, a weaker logic than BL, called *Monoidal t-norm based logic* (MTL, for short) was defined in [117] and proved in [197] to be the logic of left-continuous t-norms and their residua. The algebraic counterpart of this logic is MTL-algebra, also introduced in [117].

G. Georgescu and A. Iorgulescu introduced in [136] the *pseudo-BL algebras* as a natural generalization of BL-algebras in the non-commutative case. A pseudo-BL algebra is an FL_w-algebra which satisfies the conditions:

(A_4) $(x \to y) \odot x = x \odot (x \rightsquigarrow y) = x \wedge y$ (*pseudo-divisibility*);
(A_5) $(x \to y) \vee (y \to x) = (x \rightsquigarrow y) \vee (y \rightsquigarrow x) = 1$ (*pseudo-prelinearity*).

Properties of pseudo-BL algebras were deeply investigated by A. Di Nola, G. Georgescu and A. Iorgulescu in [85] and [86]. Some classes of pseudo-BL algebras were investigated in [143] and the corresponding propositional logic was established by Hájek in [158] and [159].

A more general structure than the pseudo-BL algebra is the *weak pseudo-BL algebra* or *pseudo-MTL algebra* introduced by P. Flondor, G. Georgescu and A. Iorgulescu in [122]. Pseudo-MTL algebras are FL_w-algebras satisfying condition (A_5) and they include as a particular case the *weak BL-algebras* which is an alternative name for MTL-algebras.

Properties of pseudo-MTL algebras are also studied in [46, 144, 181].

An FL_w-algebra which satisfies condition (A_4) is called a *divisible residuated lattice* or *bounded $R\ell$-monoid*. Properties of divisible residuated lattices were studied by A. Dvurečenskij, J. Rachůnek and J. Kühr ([105, 111, 205, 240]).

Pseudo-BCK algebras were introduced in 2001 by G. Georgescu and A. Iorgulescu ([138]) as non-commutative generalizations of BCK-algebras. Properties of pseudo-BCK algebras and their connection with other fuzzy structures were established by A. Iorgulescu in [179–182].

For a guide through the pseudo-BCK algebras realm we refer the reader to the monograph [186].

Another generalization of pseudo-BL algebras was given in [148], where *pseudo-hoops* were defined and studied. Pseudo-hoops were originally introduced by Bosbach in [15] and [16] under the name of *complementary semigroups*. It was proved that a pseudo-hoop has the pseudo-divisibility condition and it is a meet-semilattice, so a bounded $R\ell$-monoid can be viewed as a bounded pseudo-hoop together with the join-semilattice property. In other words, a bounded pseudo-hoop is a meet-semilattice ordered residuated, integral and divisible monoid.

Other topics in multiple-valued logic algebras have been studied in [34, 36, 92, 132, 141, 150, 248].

The notion of a *state* is an analogue of a probability measure and it has a very important role in the theory of quantum structures ([108]). The basic idea of states is an averaging of events (elements) of a given algebraic structure. Since in the case of Łukasiewicz ∞-valued logic the set of events has the structure of an MV-algebra, the theory of probability on this logic is based on the notion of a state defined on an MV-algebra. Besides mathematical logic, Riečan and Neubrunn studied MV-algebras as fields of events in generalized probability theory ([250]). Therefore, the study of states on MV-algebras is a very active field of research ([40, 83, 84, 119, 133, 246]) which arises from the general problem of investigating probabilities defined for logical systems.

States on an MV-algebra $(A, \oplus, ^-, 0)$ were first introduced by D. Mundici in [230] as functions $s : A \longrightarrow [0, 1]$ satisfying the conditions:

$s(1) = 1$ *(normality)*;
$s(x \oplus y) = s(x) + s(y)$ if $x \odot y = 0$ *(additivity)*,

where $x \odot y = (x^- \oplus y^-)^-$.

They are analogous to finitely additive probability measures on Boolean algebras and play a crucial role in MV-algebraic probability theory ([249]).

States on other commutative and non-commutative algebraic structures have been defined and investigated by many authors ([20, 21, 102, 133, 134, 140, 142, 258, 259]).

The aim of this book is to present new results regarding non-commutative multiple-valued logic algebras and some of their applications. Almost all the results are based on the author's recent papers ([42–75]).

The book consists of nine chapters.

The Chap. 1 is devoted to pseudo-BCK algebras. After presenting the basic definitions and properties, we prove new properties of pseudo-BCK algebras with pseudo-product and pseudo-BCK algebras with pseudo-double negation. Examples of proper pseudo-BCK algebras, good pseudo-BCK algebras and pseudo-BCK lattices are given, and the orthogonal elements in a pseudo-BCK algebra are characterized. Finally, we define the maximal and normal deductive systems of a pseudo-BCK algebra with pseudo-product and we study their properties.

In Chap. 2 we recall the basic properties of pseudo-hoops, we introduce the notions of join-center and cancellative-center of pseudo-hoops and we define and study algebras on subintervals of pseudo-hoops. Additionally, new properties of a pseudo-hoop are proved.

Chapter 3 is devoted to residuated lattices. We investigate the properties of the Boolean center of an FL_w-algebra and we define and study the directly indecomposable FL_w-algebras. One of the main results consists of proving that any linearly ordered FL_w-algebra is directly indecomposable. Finally, we define and study FL_w-algebras of fractions relative to a meet-closed system.

In Chap. 4 we present some specific properties of other non-commutative multiple-valued logic algebras: pseudo-MTL algebras, bounded $R\ell$-monoids, pseudo-BL algebras and pseudo-MV algebras. As main results, we extend to the case of pseudo-MTL algebras some results regarding prime filters proved for pseudo-BL algebras. The Glivenko property for a good pseudo-BCK algebra is defined and it is shown that a good pseudo-hoop has the Glivenko property.

Chapter 5 deals with special classes of non-commutative residuated structures: local, perfect and Archimedean structures. The local bounded pseudo-BCK(pP) algebras are characterized in terms of primary deductive systems, while the perfect pseudo-BCK(pP) algebras are characterized in terms of perfect deductive systems. One of the main results consists of proving that the radical of a bounded pseudo-BCK(pP) algebra is a normal deductive system. We also prove that any linearly ordered pseudo-BCK(pP) algebra and any locally finite pseudo-BCK(pP) algebra are local. Other results state that any local FL_w-algebra and any locally finite FL_w-algebra are directly indecomposable. The classes of Archimedean and hyperarchimedean FL_w-algebras are introduced and it is proved that any locally finite FL_w-algebra is hyperarchimedean and any hyperarchimedean FL_w-algebra is Archimedean.

Chapter 6 is devoted to the presentation of states on multiple-valued logic algebras. We introduce the notion of states on pseudo-BCK algebras and we study their properties. One of the main results consists of proving that any Bosbach state on a good pseudo-BCK algebra is a Riečan state, however the converse turns out not to be true. We also prove that every Riečan state on a good pseudo-BCK algebra with pseudo-double negation is a Bosbach state. In contrast to the case of pseudo-BL algebras, we show that there exist linearly ordered pseudo-BCK algebras having no Bosbach states and that there exist pseudo-BCK algebras having normal filters which are maximal, but having no Bosbach states.

Some specific properties of states on FL_w-algebras, pseudo-MTL algebras, bounded $R\ell$-monoids and subinterval algebras of pseudo-hoops are proved.

A special section is dedicated to the existence of states on the residuated structures, showing that every perfect FL_w-algebra admits at least a Bosbach state and every perfect pseudo-BL algebra has a unique state-morphism.

Finally, we introduce the notion of a local state on a perfect pseudo-MTL algebra and we prove that every local state can be extended to a Riečan state.

In Chap. 7 we generalize measures on BCK algebras introduced by A. Dvurečenskij in [94] and [108] to pseudo-BCK algebras that are not necessarily bounded. In particular, we show that if A is a downwards-directed pseudo-BCK algebra and m a measure on it, then the quotient over the kernel of m can be embedded into the negative cone of an Abelian, Archimedean ℓ-group as its subalgebra. This result will enable us to characterize nonzero measure-morphisms on downwards-directed pseudo-BCK algebras as measures whose kernel is a maximal filter. We study state-measures on pseudo-BCK algebras with strong unit and we show how to characterize state-measure-morphisms as extremal state-measures or as state-measures whose kernel is a maximal filter. In particular, we show that for unital pseudo-BCK algebras that are downwards-directed, the quotient over the kernel can be embedded into the negative cone of an Abelian, Archimedean ℓ-group with strong unit. We generalize to pseudo-BCK algebras the identity between de Finetti maps and Bosbach states, following the results proved by Kühr and Mundici in [211] who showed that de Finetti's coherence principle, which has its origins in Dutch bookmaking, has a strong relationship with MV-states on MV-algebras. We also generalize this for state-measures on unital pseudo-BCK algebras that are downwards-directed.

Chapter 8 is devoted to generalized states on residuated structures. The study of these generalized states is motivated by their interpretation as a new type of semantics for non-commutative fuzzy logics. Usually, the truth degree of sentences in a fuzzy logic is a number in the interval $[0, 1]$ or, more generally, an element of an FL_w-algebra. Similarly, for generalized states, the probability of sentences is evaluated in an arbitrary FL_w-algebra.

We define the generalized states of type I and type II and generalized state-morphisms and we study the relationship between them. We prove that any perfect FL_w-algebra admits strong type I and type II states. Some conditions are given for a generalized state of type I on a linearly ordered bounded $R\ell$-monoid to be a state operator. The notion of a strong perfect FL_w-algebra is introduced and it is proved that any strong perfect FL_w-algebra admits a generalized state-morphism. The notion of a generalized Riečan state is also introduced and the main results are proved based on the Glivenko property defined for the non-commutative case. The main results consist of proving that any order-preserving type I state is a generalized Riečan state and in some particular conditions the two states coincide. We introduce the notion of a generalized local state on a perfect pseudo-MTL algebra A and we prove that, if A is relatively free of zero divisors, then every generalized local state can be extended to a generalized Riečan state.

Chapter 9 deals with residuated structures with internal states. We define the notions of state operator, strong state operator, state-morphism operator, weak state-morphism operator and we study their properties. We prove that every strong state pseudo-hoop is a state pseudo-hoop and any state operator on an idempotent pseudo-hoop is a weak state-morphism operator. It is proved that for an idempotent pseudo-hoop A a state operator on $\text{Reg}(A)$ can be extended to a state operator on A. One of the main results of this chapter consists of proving that every perfect pseudo-hoop admits a nontrivial state operator. Other results compare the state operators with states and generalized states on a pseudo-hoop. Some conditions are given for a state operator to be a generalized state and for a generalized state to be a state operator.

We hope that this book will be useful to graduate students and researchers in the area of algebras of multiple-valued logics.

I wish to firstly thank my adviser George Georgescu for guiding many of my steps in this field.

This manuscript owes a lot to Afrodita Iorgulescu for her careful reading and remarks.

I am also in debt to Anatolij Dvurečenskij for his suggestions and fruitful collaborations.

On a personal note, I am very grateful to my parents for all their support and encouragement over the years.

Last but not least I wish to thank my husband for his wonderful companionship.

Iowa City, USA   Lavinia Corina Ciungu
May 2013

Contents

Chapter 1
Pseudo-BCK Algebras

BCK algebras were originally introduced by K. Isèki in [194] with a binary operation $*$ modeling the set-theoretical difference and with a constant element 0, that is, a least element. Another motivation is from classical and non-classical propositional calculi modeling logical implications. Such algebras contain as a special subfamily the family of MV-algebras where some important fuzzy structures can be studied. For more about BCK algebras, see [167, 174–179, 182–187, 189, 192, 193, 225].

Pseudo-BCK algebras were introduced by G. Georgescu and A. Iorgulescu in [138] as algebras with "two differences", a left- and right-difference, instead of one $*$ and with a constant element 0 as the least element. In [112], a special subclass of pseudo-BCK algebras, called Łukasiewicz pseudo-BCK algebras, was introduced and it was shown that each such algebra is always a subalgebra of the positive cone of some ℓ-group (not necessarily Abelian). The class of Łukasiewicz pseudo-BCK algebras is a variety whereas the class of pseudo-BCK algebras is not; it is only a quasivariety because it is not closed under homomorphic images. Nowadays pseudo-BCK algebras are used in a dual form, with two implications, $\rightarrow$ and $\rightsquigarrow$ and with one constant element 1, that is the greatest element. Thus such pseudo-BCK algebras are in the "negative cone" and are also called "left-ones". Further properties of pseudo-BCK algebras and their connection with other fuzzy structures were established by A. Iorgulescu in [179–182]. For a guide through the pseudo-BCK algebras realm, see the monograph [186]. Studies on pseudo-BCK algebras were also developed in [107, 163, 190, 206, 208–210].

In this chapter we prove new properties of pseudo-BCK algebras with pseudo-product and pseudo-BCK algebras with pseudo-double negation and we show that every pseudo-BCK algebra can be extended to a good one. Examples of proper pseudo-BCK algebras, good pseudo-BCK algebras and pseudo-BCK lattices are given and the orthogonal elements in a pseudo-BCK algebra are characterized. Finally, we define the maximal and normal deductive systems of a pseudo-BCK algebra with pseudo-product and we study their properties.

L.C. Ciungu, *Non-commutative Multiple-Valued Logic Algebras*,
Springer Monographs in Mathematics, DOI 10.1007/978-3-319-01589-7_1,
© Springer International Publishing Switzerland 2014

1.1 Definitions and Properties

Definition 1.1 A *pseudo-BCK algebra* (more precisely, *reversed left-pseudo-BCK algebra*) is a structure $\mathcal{A} = (A, \leq, \rightarrow, \rightsquigarrow, 1)$ where $\leq$ is a binary relation on A, $\rightarrow$ and $\rightsquigarrow$ are binary operations on A and 1 is an element of A satisfying, for all $x, y, z \in A$, the axioms:

$(psBCK_1)$ $x \rightarrow y \leq (y \rightarrow z) \rightsquigarrow (x \rightarrow z)$, $x \rightsquigarrow y \leq (y \rightsquigarrow z) \rightarrow (x \rightsquigarrow z)$;
$(psBCK_2)$ $x \leq (x \rightarrow y) \rightsquigarrow y$, $x \leq (x \rightsquigarrow y) \rightarrow y$;
$(psBCK_3)$ $x \leq x$;
$(psBCK_4)$ $x \leq 1$;
$(psBCK_5)$ if $x \leq y$ and $y \leq x$, then $x = y$;
$(psBCK_6)$ $x \leq y$ iff $x \rightarrow y = 1$ iff $x \rightsquigarrow y = 1$.

A pseudo-BCK algebra $\mathcal{A} = (A, \leq, \rightarrow, \rightsquigarrow, 1)$ is *commutative* if $\rightarrow = \rightsquigarrow$. Any commutative pseudo-BCK algebra is a BCK-algebra.

In the sequel we will refer to the pseudo-BCK algebra $(A, \leq, \rightarrow, \rightsquigarrow, 1)$ by its universe A.

Proposition 1.1 *The structure* $(A, \leq, \rightarrow, \rightsquigarrow, 1)$ *is a pseudo-BCK algebra iff the algebra* $(A, \rightarrow, \rightsquigarrow, 1)$ *of type* $(2, 2, 0)$ *satisfies the following identities and quasi-identity:*

$(psBCK_1')$ $(x \rightarrow y) \rightsquigarrow [(y \rightarrow z) \rightsquigarrow (x \rightarrow z)] = 1$;
$(psBCK_2')$ $(x \rightsquigarrow y) \rightarrow [(y \rightsquigarrow z) \rightarrow (x \rightsquigarrow z)] = 1$;
$(psBCK_3')$ $1 \rightarrow x = x$;
$(psBCK_4')$ $1 \rightsquigarrow x = x$;
$(psBCK_5')$ $x \rightarrow 1 = 1$;
$(psBCK_6')$ $(x \rightarrow y = 1$ *and* $y \rightarrow x = 1)$ *implies* $x = y$.

Proof Obviously, any pseudo-BCK algebra satisfies $(psBCK_1')$–$(psBCK_6')$.

Conversely, assume that an algebra $(A, \rightarrow, \rightsquigarrow, 1)$ satisfies $(psBCK_1')$–$(psBCK_6')$. Applying $(psBCK_3')$ and $(psBCK_1')$ we get:

$$x \rightsquigarrow \left[(x \rightarrow y) \rightsquigarrow y\right] = (1 \rightarrow x) \rightsquigarrow \left[(x \rightarrow y) \rightsquigarrow (1 \rightarrow y)\right] = 1.$$

Similarly, by $(psBCK_4')$ and $(psBCK_2')$ we have:

$$x \rightarrow \left[(x \rightsquigarrow y) \rightarrow y\right] = (1 \rightsquigarrow x) \rightarrow \left[(x \rightsquigarrow y) \rightarrow (1 \rightsquigarrow y)\right] = 1.$$

Applying $(psBCK_3')$ and $(psBCK_2')$ we have:

$$x \rightarrow x = 1 \rightarrow (x \rightarrow x) = (1 \rightsquigarrow 1) \rightarrow \left[(1 \rightsquigarrow x) \rightarrow (1 \rightsquigarrow x)\right] = 1.$$

Similarly, by $(psBCK_4')$ and $(psBCK_1')$ we get:

$$x \rightsquigarrow x = 1 \rightsquigarrow (x \rightsquigarrow x) = (1 \rightarrow 1) \rightsquigarrow \left[(1 \rightarrow x) \rightsquigarrow (1 \rightarrow x)\right] = 1.$$

Moreover, if $x \rightarrow y = 1$ then $x \rightsquigarrow y = x \rightsquigarrow [(x \rightarrow y) \rightsquigarrow y] = 1$ and similarly, if $x \rightsquigarrow y = 1$ then $x \rightarrow y = x \rightarrow [(x \rightsquigarrow y) \rightarrow y] = 1$.

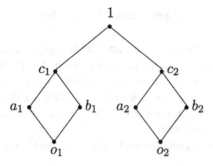

Fig. 1.1 Example of proper pseudo-BCK algebra

It follows that $x \rightarrow y = 1$ iff $x \rightsquigarrow y = 1$.

We deduce that the relation $\leq$ defined by $x \leq y$ iff $x \rightarrow y = 1$ is a partial order on A which makes $(A, \leq, \rightarrow, \rightsquigarrow, 1)$ a pseudo-BCK algebra.   □

In the sequel, we shall use either $(A, \leq, \rightarrow, \rightsquigarrow, 1)$ or $(A, \rightarrow, \rightsquigarrow, 1)$ for a pseudo-BCK algebra.

Example 1.1 Consider $A = \{o_1, a_1, b_1, c_1, o_2, a_2, b_2, c_2, 1\}$ with $o_1 < a_1, b_1 < c_1 < 1$ and a_1, b_1 incomparable, $o_2 < a_2, b_2 < c_2 < 1$ and a_2, b_2 incomparable. Assume that any element of the set $\{o_1, a_1, b_1, c_1\}$ is incomparable with any element of the set $\{o_2, a_2, b_2, c_2\}$ (see Fig. 1.1).

Consider the operations $\rightarrow$, $\rightsquigarrow$ given by the following tables:

$\rightarrow$	o_1	a_1	b_1	c_1	o_2	a_2	b_2	c_2	1
o_1	1	1	1	1	o_2	a_2	b_2	c_2	1
a_1	o_1	1	b_1	1	o_2	a_2	b_2	c_2	1
b_1	a_1	a_1	1	1	o_2	a_2	b_2	c_2	1
c_1	o_1	a_1	b_1	1	o_2	a_2	b_2	c_2	1
o_2	o_1	a_1	b_1	c_1	1	1	1	1	1
a_2	o_1	a_1	b_1	c_1	o_2	1	b_2	1	1
b_2	o_1	a_1	b_1	c_1	c_2	c_2	1	1	1
c_2	o_1	a_1	b_1	c_1	o_2	c_2	b_2	1	1
1	o_1	a_1	b_1	c_1	o_2	a_2	b_2	c_2	1

$\rightsquigarrow$	o_1	a_1	b_1	c_1	o_2	a_2	b_2	c_2	1
o_1	1	1	1	1	o_2	a_2	b_2	c_2	1
a_1	b_1	1	b_1	1	o_2	a_2	b_2	c_2	1
b_1	o_1	a_1	1	1	o_2	a_2	b_2	c_2	1
c_1	o_1	a_1	b_1	1	o_2	a_2	b_2	c_2	1
o_2	o_1	a_1	b_1	c_1	1	1	1	1	1
a_2	o_1	a_1	b_1	c_1	b_2	1	b_2	1	1
b_2	o_1	a_1	b_1	c_1	b_2	c_2	1	1	1
c_2	o_1	a_1	b_1	c_1	b_2	c_2	b_2	1	1
1	o_1	a_1	b_1	c_1	o_2	a_2	b_2	c_2	1

Then $(A, \leq, \rightarrow, \rightsquigarrow, 1)$ is a proper pseudo-BCK algebra.

We recall the definition of an ℓ-group. The language of lattice-ordered groups (ℓ-groups) involves both the group operations and the binary lattice operations.

By a *lattice-ordered group* (ℓ-group) we will mean an ordered group $(G, \leq)$ such that $(G, \leq)$ is a lattice. The ℓ-group G is called an ℓu-*group* if there exists an element $u > 0$ such that for any $x \in G$ there is an $n \in \mathbb{N}$ such that $x \leq nu$. The element u is called a *strong unit*.

For details regarding ℓ-groups we refer the reader to [2, 12, 76].

Example 1.2 Let $(G, \vee, \wedge, +, -, 0)$ be an ℓ-group.
On the negative cone $G^- = \{g \in G \mid g \leq 0\}$ we define:

$$g \rightarrow h := h - (g \vee h) = (h - g) \wedge 0,$$
$$g \rightsquigarrow h := -(g \vee h) + h = (-g + h) \wedge 0.$$

Then $(G^-, \leq, \rightarrow, \rightsquigarrow, 0)$ is a pseudo-BCK algebra.

Remark 1.1 (Definition of union) Let $(A_i, \leq, \rightarrow_i, \rightsquigarrow_i, 1_i)_{i \in I}$ be a collection of pseudo-BCK algebras such that:

(i) $1_i = 1$ for all $i \in I$,
(ii) $A_i \cap A_j = \{1\}$ for all $i, j \in I, i \neq j$.

Let $A = \bigcup_{i \in I} A_i$ and define:

$$x \rightarrow y := \begin{cases} x \rightarrow_i y & \text{if } x, y \in A_i, i \in I \\ y & \text{otherwise,} \end{cases}$$

$$x \rightsquigarrow y := \begin{cases} x \rightsquigarrow_i y & \text{if } x, y \in A_i, i \in I \\ y & \text{otherwise.} \end{cases}$$

Then $(A, \leq, \rightarrow, \rightsquigarrow, 1)$ is a pseudo-BCK algebra called the *union* of the pseudo-BCK algebras $(A_i, \leq, \rightarrow_i, \rightsquigarrow_i, 1_i)_{i \in I}$.

Note that the notion of union defined above is not related to the notion of ordinal sum defined in Chap. 2.

Proposition 1.2 *In any pseudo-BCK algebra A the following properties hold*:

$(psbck\text{-}c_1)$ $x \leq y$ *implies* $y \rightarrow z \leq x \rightarrow z$ *and* $y \rightsquigarrow z \leq x \rightsquigarrow z$;
$(psbck\text{-}c_2)$ $x \leq y, y \leq z$ *implies* $x \leq z$;
$(psbck\text{-}c_3)$ $x \rightarrow (y \rightsquigarrow z) = y \rightsquigarrow (x \rightarrow z), x \rightsquigarrow (y \rightarrow z) = y \rightarrow (x \rightsquigarrow z)$;
$(psbck\text{-}c_4)$ $z \leq y \rightarrow x$ *iff* $y \leq z \rightsquigarrow x$;
$(psbck\text{-}c_5)$ $z \rightarrow x \leq (y \rightarrow z) \rightarrow (y \rightarrow x), z \rightsquigarrow x \leq (y \rightsquigarrow z) \rightsquigarrow (y \rightsquigarrow x)$;
$(psbck\text{-}c_6)$ $x \leq y \rightarrow x, x \leq y \rightsquigarrow x$;

($psbck$-c_7) $1 \to x = x = 1 \rightsquigarrow x$;
($psbck$-c_8) $x \to x = x \rightsquigarrow x = 1$;
($psbck$-c_9) $x \to 1 = x \rightsquigarrow 1 = 1$;
($psbck$-c_{10}) $x \le y$ implies $z \to x \le z \to y$ and $z \rightsquigarrow x \le z \rightsquigarrow y$;
($psbck$-c_{11}) $[(y \to x) \rightsquigarrow x] \to x = y \to x$, $[(y \rightsquigarrow x) \to x] \rightsquigarrow x = y \rightsquigarrow x$.

Proof

($psbck$-c_1) Since $x \le y$, applying ($psBCK_6$), ($psBCK_1$) and ($psBCK_4$) we get $1 = x \to y \le (y \to z) \rightsquigarrow (x \to z)$, so $(y \to z) \rightsquigarrow (x \to z) = 1$ for all $z \in A$.
Applying ($psBCK_6$) again we get $y \to z \le x \to z$.
Similarly, $y \rightsquigarrow z \le x \rightsquigarrow z$.

($psbck$-c_2) By ($psbck$-c_1), $x \le y$ implies $y \to z \le x \to z$. Since $y \le z$ we have $y \to z = 1$, so $x \to z = 1$. Applying ($psBCK_6$) we get $x \le z$.

($psbck$-c_3) Applying ($psBCK_1$) we have $y \to x \le (x \to z) \rightsquigarrow (y \to z)$ and by ($psbck$-c_1) we get $[(x \to z) \rightsquigarrow (y \to z)] \rightsquigarrow u \le (y \to x) \rightsquigarrow u$ for any $u \in A$.
From this inequality, replacing z with $u \rightsquigarrow z$, x with $x \rightsquigarrow z$ and u with $(u \rightsquigarrow x) \rightsquigarrow [y \to (u \rightsquigarrow z)]$ we get

$$\left[[(x \rightsquigarrow z) \to (u \rightsquigarrow z)] \rightsquigarrow [y \to (u \rightsquigarrow z)]\right] \rightsquigarrow \left[(u \rightsquigarrow x) \rightsquigarrow [y \to (u \rightsquigarrow z)]\right]$$
$$\le \left[y \to (x \rightsquigarrow z)\right] \rightsquigarrow \left[(u \rightsquigarrow x) \rightsquigarrow [y \to (u \rightsquigarrow z)]\right].$$

By ($psBCK_1$) we have $u \rightsquigarrow x \le (x \rightsquigarrow z) \to (u \rightsquigarrow z)$ and applying ($psbck$-c_1) it follows that the left-hand side of the above inequality is equal to 1.
Thus the right-hand side is also equal to 1, so $y \to (x \rightsquigarrow z) \le (u \rightsquigarrow x) \rightsquigarrow [y \to (u \rightsquigarrow z)]$.
Replacing x with $y \to z$ and u with x we get

$$y \to \left[(y \to z) \rightsquigarrow z\right] \le \left[x \rightsquigarrow (y \to z)\right] \rightsquigarrow \left[y \to (x \rightsquigarrow z)\right].$$

But, by ($psBCK_2$) we have $y \le (y \to z) \rightsquigarrow z$, so $y \to [(y \to z) \rightsquigarrow z] = 1$.
It follows that $[x \rightsquigarrow (y \to z)] \rightsquigarrow [y \to (x \rightsquigarrow z)] = 1$.
Therefore $x \rightsquigarrow (y \to z) \le y \to (x \rightsquigarrow z)$.
On the other hand, by ($psBCK_2$) we have $x \le (x \rightsquigarrow z) \to z$ and applying ($psbck$-c_1) we get $[(x \rightsquigarrow z) \to z] \rightsquigarrow (y \to z) \le x \rightsquigarrow (y \to z)$.
By ($psBCK_1$) we have $y \to x \le (x \to z) \rightsquigarrow (y \to z)$ and replacing x with $x \rightsquigarrow z$ we get $y \to (x \rightsquigarrow z) \le [(x \rightsquigarrow z) \to z] \rightsquigarrow (y \to z) \le x \rightsquigarrow (y \to z)$.
We conclude that $x \to (y \rightsquigarrow z) = y \rightsquigarrow (x \to z)$.
Similarly, $x \rightsquigarrow (y \to z) = y \to (x \rightsquigarrow z)$.

($psbck$-c_4) From $z \le y \to x$, by ($psBCK_2$) and ($psbck$-c_1) we have

$$y \le (y \to x) \rightsquigarrow x \le z \rightsquigarrow x.$$

Similarly, from $y \le z \rightsquigarrow x$ we get $z \le (z \rightsquigarrow x) \to x \le y \to x$.

(*psbck-c5*) Applying (*psBCK*$_1$) we have $y \to z \leq (z \to x) \rightsquigarrow (y \to x)$ and according to (*psbck-c$_1$*) we get

$$\left[(z \to x) \rightsquigarrow (y \to x)\right] \to (y \to x) \leq (y \to z) \to (y \to x).$$

By (*psBCK$_2$*) it follows that $z \to x \leq [(z \to x) \rightsquigarrow (y \to x)] \to (y \to x)$, and applying (*psbck-c$_2$*) we conclude that $z \to x \leq (y \to z) \to (y \to x)$.

Similarly, from $y \rightsquigarrow z \leq (z \rightsquigarrow x) \to (y \rightsquigarrow x)$ we get $z \rightsquigarrow x \leq (y \rightsquigarrow z) \rightsquigarrow (y \rightsquigarrow x)$.

(*psbck-c6*) Since $y \leq 1 = x \to x$, it follows by (*psbck-c$_4$*) that $x \leq y \rightsquigarrow x$.

Similarly, from $y \leq 1 = x \rightsquigarrow x$ we get $x \leq y \to x$.

(*psbck-c7*) By (*psbck-c$_6$*) we have $x \leq 1 \to x$ and $x \leq 1 \rightsquigarrow x$.

By (*psBCK$_2$*) we get $1 \leq (1 \to x) \rightsquigarrow x$ and $1 \leq (1 \rightsquigarrow x) \to x$.

It follows that $(1 \to x) \rightsquigarrow x = 1$ and $(1 \rightsquigarrow x) \to x = 1$, so $1 \to x \leq x$ and $1 \rightsquigarrow x \leq x$. Thus $1 \to x = x = 1 \rightsquigarrow x$.

(*psbck-c8*) and (*psbck-c9*) are consequences of the axiom (*psBCK$_6$*).

(*psbck-c10*) Applying (*psbck-c$_7$*), (*psBCK$_6$*) and (*psBCK$_1$*) we have:

$$z \to y = 1 \rightsquigarrow (z \to y) = (x \to y) \rightsquigarrow (z \to y) \geq z \to x \quad \text{and}$$

$$z \rightsquigarrow y = 1 \to (z \rightsquigarrow y) = (x \rightsquigarrow y) \to (z \rightsquigarrow y) \geq z \rightsquigarrow x.$$

(*psbck-c11*) By (*psBCK$_2$*) we have $y \leq (y \to x) \rightsquigarrow x$ and $y \leq (y \rightsquigarrow x) \to x$.

Applying (*psbck-c$_1$*) we get

$$\left[(y \to x) \rightsquigarrow x\right] \to x \leq y \to x \quad \text{and} \quad \left[(y \rightsquigarrow x) \to x\right] \rightsquigarrow x \leq y \rightsquigarrow x.$$

On the other hand, by (*psBCK$_2$*) we have:

$$y \to x \leq \left[(y \to x) \rightsquigarrow x\right] \to x \quad \text{and} \quad y \rightsquigarrow x \leq \left[(y \rightsquigarrow x) \to x\right] \rightsquigarrow x.$$

We conclude that

$$\left[(y \to x) \rightsquigarrow x\right] \to x = y \to x \quad \text{and} \quad \left[(y \rightsquigarrow x) \to x\right] \rightsquigarrow x = y \rightsquigarrow x. \qquad \square$$

Proposition 1.3 Let $(A, \leq, \to, \rightsquigarrow, 1)$ be a pseudo-BCK algebra.

If $\bigvee_{i \in I} x_i$ exists, then so does $\bigwedge_{i \in I}(x_i \to y)$ and $\bigwedge_{i \in I}(x_i \rightsquigarrow y)$ and we have:

(*psbck-c12*) $(\bigvee_{i \in I} x_i) \to y = \bigwedge_{i \in I}(x_i \to y)$, $(\bigvee_{i \in I} x_i) \rightsquigarrow y = \bigwedge_{i \in I}(x_i \rightsquigarrow y)$.

Proof If we let $x = \bigvee_{i \in I} x_i$, it follows that $x_i \leq x$ and applying (*psbck-c$_1$*) we have $x \to y \leq x_i \to y$ for all $i \in I$. Let z be a lower bound of $\{x_i \to y \mid i \in I\}$. Then, by (*psbck-c$_4$*), $z \leq x_i \to y$ implies $x_i \leq z \rightsquigarrow y$ for all $i \in I$, so $x \leq z \rightsquigarrow y$. Applying (*psbck-c$_4$*) again, we get $z \leq x \to y$.

Thus $x \to y$ is the g.l.b. of $\{x_i \to y \mid i \in I\}$.

We conclude that $\bigwedge_{i \in I}(x_i \to y)$ exists and $(\bigvee_{i \in I} x_i) \to y = \bigwedge_{i \in I}(x_i \to y)$.

Similarly, $\bigwedge_{i \in I}(x_i \rightsquigarrow y)$ exists and $(\bigvee_{i \in I} x_i) \rightsquigarrow y = \bigwedge_{i \in I}(x_i \rightsquigarrow y)$. $\qquad \square$

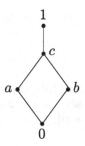

Fig. 1.2 Example of bounded pseudo-BCK algebra

Definition 1.2 If there is an element 0 of a pseudo-BCK algebra $(A, \leq, \rightarrow, \rightsquigarrow, 1)$, such that $0 \leq x$ (i.e. $0 \rightarrow x = 0 \rightsquigarrow x = 1$), for all $x \in A$, then 0 is called the *zero* of A. A pseudo-BCK algebra with zero is called a *bounded pseudo-BCK algebra* and it is denoted by $(A, \leq, \rightarrow, \rightsquigarrow, 0, 1)$.

Example 1.3 Consider $A = \{0, a, b, c, 1\}$ with $0 < a, b < c < 1$ and a, b incomparable (see Fig. 1.2).

Consider the operations $\rightarrow$, $\rightsquigarrow$ given by the following tables:

$\rightarrow$	0	a	b	c	1
0	1	1	1	1	1
a	0	1	b	1	1
b	a	a	1	1	1
c	0	a	b	1	1
1	0	a	b	c	1

$\rightsquigarrow$	0	a	b	c	1
0	1	1	1	1	1
a	b	1	b	1	1
b	0	a	1	1	1
c	0	a	b	1	1
1	0	a	b	c	1

Then $(A, \leq, \rightarrow, \rightsquigarrow, 0, 1)$ is a bounded pseudo-BCK algebra. (As we will see later, A is even a pseudo-BCK lattice.)

Let $(A, \leq, \rightarrow, \rightsquigarrow, 0, 1)$ be a bounded pseudo-BCK algebra. We define two negations $^-$ and $^\sim$: for all $x \in A$,

$$x^- := x \rightarrow 0, \qquad x^\sim := x \rightsquigarrow 0.$$

In the sequel we will use the following notation:

$$x^{--} = \left(x^-\right)^-; \qquad x^{\sim\sim} = \left(x^\sim\right)^\sim; \qquad x^{-\sim} = \left(x^-\right)^\sim; \qquad x^{\sim-} = \left(x^\sim\right)^-.$$

Example 1.4 Let $(G, \vee, \wedge, +, -, 0)$ be an ℓ-group with a strong unit $u \geq 0$. On the interval $[-u, 0]$ we define:

$$x \rightarrow y := (y - x) \wedge 0, \qquad x \rightsquigarrow y := (-x + y) \wedge 0.$$

Then $([-u, 0], \leq, \rightarrow, \rightsquigarrow, -u, 0)$ is a bounded pseudo-BCK algebra with $x^- = -u - x$ and $x^\sim = -x - u$. In a similar way, $((-u, 0], \leq, \rightarrow, \rightsquigarrow, 0)$ is a pseudo-BCK algebra that is not bounded.

Example 1.5 Let $(G, \vee, \wedge, +, -, 0)$ be an ℓ-group with a strong unit $u \geq 0$. On the interval $[0, u]$ we define:

$$x \to y := (u - x + y) \wedge u, \qquad x \rightsquigarrow y := (y - x + u) \wedge u.$$

Then $([0, u], \leq, \to, \rightsquigarrow, 0, u)$ is a bounded pseudo-BCK algebra with $x^- = u - x$ and $x^\sim = -x + u$. If on $[0, u]$ we set $\to_1 = \rightsquigarrow$ and $\rightsquigarrow_1 = \to$, then $([0, u], \leq, \to_1, \rightsquigarrow_1, 0, u)$ is isomorphic with $([-u, 0], \leq, \to, \rightsquigarrow, -u, 0)$ under the isomorphism $x \mapsto x - u, x \in [0, u]$.

Proposition 1.4 *In a bounded pseudo-BCK algebra the following hold*:

($psbck$-c_{13}) $1^- = 0 = 1^\sim, 0^- = 1 = 0^\sim$;

($psbck$-c_{14}) $x \leq x^{-\sim}, x \leq x^{\sim-}$;

($psbck$-c_{15}) $x \to y \leq y^- \rightsquigarrow x^-, x \rightsquigarrow y \leq y^\sim \to x^\sim$;

($psbck$-c_{16}) $x \leq y$ *implies* $y^- \leq x^-$ *and* $y^\sim \leq x^\sim$;

($psbck$-c_{17}) $x \to y^\sim = y \rightsquigarrow x^-$ *and* $x \rightsquigarrow y^- = y \to x^\sim$;

($psbck$-c_{18}) $x^{-\sim-} = x^-, x^{\sim-\sim} = x^\sim$;

($psbck$-c_{19}) $x \to y^{-\sim} = y^- \rightsquigarrow x^- = x^{-\sim} \to y^{-\sim}$ *and* $x \rightsquigarrow y^{\sim-} = y^\sim \to x^\sim = x^{\sim-} \rightsquigarrow y^{\sim-}$;

($psbck$-c_{20}) $x \to y^\sim = y^{\sim-} \rightsquigarrow x^- = x^{-\sim} \to y^\sim$ *and* $x \rightsquigarrow y^- = y^{-\sim} \to x^\sim = x^{\sim-} \rightsquigarrow y^-$;

($psbck$-c_{21}) $(x \to y^{\sim-})^{\sim-} = x \to y^{\sim-}$ *and* $(x \rightsquigarrow y^{-\sim})^{-\sim} = x \rightsquigarrow y^{-\sim}$.

Proof

($psbck$-c_{13}) Since $0 \leq 0$, by ($psBCK_6$) we get $0 \to 0 = 1$ and $0 \rightsquigarrow 0 = 1$, that is, $0^- = 1$ and $0^\sim = 1$.

Taking $x = 1$ and $y = 0$ in ($psBCK_2$) we have $1 \leq (1 \to 0) \rightsquigarrow 0$, hence $(1 \to 0) \rightsquigarrow 0 = 1$. Thus by ($psBCK_6$) we get $1 \to 0 \leq 0$, so $1 \to 0 = 0$, i.e. $1^- = 0$. Similarly, $1^\sim = 0$.

($psbck$-c_{14}) This follows by taking $y = 0$ in ($psBCK_2$).

($psbck$-c_{15}) Applying ($psBCK_1$) for $z = 0$ we get:

$$x \to y \leq (y \to 0) \rightsquigarrow (x \to 0) = y^- \rightsquigarrow x^- \quad \text{and}$$

$$x \rightsquigarrow y \leq (y \rightsquigarrow 0) \to (x \rightsquigarrow 0) = y^\sim \to x^\sim.$$

($psbck$-c_{16}) From $x \leq y$, applying ($psbck$-c_1) we get $y \to 0 \leq x \to 0$, so $y^- \leq x^-$. Similarly, $y^\sim \leq x^\sim$.

($psbck$-c_{17}) By ($psbck$-c_{15}), ($psbck$-c_{14}) and ($psbck$-c_1) we get:

$$x \to y^\sim \leq y^{\sim-} \rightsquigarrow x^- \leq y \rightsquigarrow x^- \quad \text{and} \quad x \rightsquigarrow y^- \leq y^{-\sim} \to x^\sim \leq y \to x^\sim.$$

In the above inequalities we change x and y obtaining:

$$y \to x^\sim \leq x \rightsquigarrow y^- \quad \text{and} \quad y \rightsquigarrow x^- \leq x \to y^\sim.$$

Thus $x \to y^\sim = y \rightsquigarrow x^-$ and $x \rightsquigarrow y^- = y \to x^\sim$.

(*psbck-c$_{18}$*) By (*psbck-c$_{14}$*) and (*psbck-c$_{16}$*) we get $x^{\sim-\sim} \leq x^{\sim}$ and $x^{-\sim-} \leq x^{-}$.
By (*psbck-c$_{14}$*), replacing x with $x^{\sim}$ and x^{-} we get $x^{\sim} \leq x^{\sim-\sim}$ and $x^{-} \leq x^{-\sim-}$, respectively. Thus $x^{\sim-\sim} = x^{\sim}$ and $x^{-\sim-} = x^{-}$.

(*psbck-c$_{19}$*) By (*psbck-c$_{17}$*) we have: $y \rightsquigarrow x^{-} = x \rightarrow y^{\sim}$.
Replacing y with y^{-} we get: $y^{-} \rightsquigarrow x^{-} = x \rightarrow y^{-\sim}$.
Replacing x by $x^{-\sim}$ in the last equality we get: $y^{-} \rightsquigarrow x^{-\sim-} = x^{-\sim} \rightarrow y^{-\sim}$.
Hence applying (*psbck-c$_{18}$*) it follows that: $y^{-} \rightsquigarrow x^{-} = x^{-\sim} \rightarrow y^{-\sim}$.
Thus $x \rightarrow y^{-\sim} = y^{-} \rightsquigarrow x^{-} = x^{-\sim} \rightarrow y^{-\sim}$.
Similarly, $x \rightsquigarrow y^{\sim-} = y^{\sim} \rightarrow x^{\sim} = x^{\sim-} \rightsquigarrow y^{\sim-}$.

(*psbck-c$_{20}$*) The assertions follow by replacing in (*psbck-c$_{19}$*) y with $y^{\sim}$ and y with y^{-}, respectively and applying (*psbck-c$_{18}$*).

(*psbck-c$_{21}$*) Applying (*psbck-c$_{3}$*) and (*psbck-c$_{19}$*) we have:

$$1 = \left(x \rightarrow y^{\sim-}\right) \rightsquigarrow \left(x \rightarrow y^{\sim-}\right) = x \rightarrow \left(\left(x \rightarrow y^{\sim-}\right) \rightsquigarrow y^{\sim-}\right)$$

$$= x \rightarrow \left(\left(x \rightarrow y^{\sim-}\right)^{\sim-} \rightsquigarrow y^{\sim-}\right) = \left(x \rightarrow y^{\sim-}\right)^{\sim-} \rightsquigarrow \left(x \rightarrow y^{\sim-}\right).$$

Hence $\left(x \rightarrow y^{\sim-}\right)^{\sim-} \leq x \rightarrow y^{\sim-}$.
On the other hand, by (*psbck-c$_{14}$*) we have $x \rightarrow y^{\sim-} \leq \left(x \rightarrow y^{\sim-}\right)^{\sim-}$, thus $\left(x \rightarrow y^{\sim-}\right)^{\sim-} = x \rightarrow y^{\sim-}$. Similarly, $\left(x \rightsquigarrow y^{-\sim}\right)^{-\sim} = x \rightsquigarrow y^{-\sim}$. $\square$

We recall some notions and results regarding pseudo-BCK semilattices (see [209]).

Definition 1.3 A *pseudo-BCK join-semilattice* is an algebra $(A, \vee, \rightarrow, \rightsquigarrow, 1)$ such that $(A, \vee)$ is a join-semilattice, $(A, \rightarrow, \rightsquigarrow, 1)$ is a pseudo-BCK algebra and $x \rightarrow y = 1$ iff $x \vee y = y$.

Remark 1.2 It is easy to show that an algebra $(A, \vee, \rightarrow, \rightsquigarrow, 1)$ of type $(2, 2, 2, 0)$ is a pseudo-BCK join-semilattice if and only if $(A, \vee)$ is a join-semilattice and $(A, \rightarrow, \rightsquigarrow, 1)$ satisfies (*psBCK$'_{1}$*)–(*psBCK$'_{5}$*) and the following identities:

(*psBCK$'_{7}$*) $x \vee [(x \rightarrow y) \rightsquigarrow y] = (x \rightarrow y) \rightsquigarrow y$;
(*psBCK$'_{8}$*) $x \rightarrow (x \vee y) = 1$.

Definition 1.4 A *pseudo-BCK meet-semilattice* is an algebra $(A, \wedge, \rightarrow, \rightsquigarrow, 1)$ such that $(A, \wedge)$ is a meet-semilattice, $(A, \rightarrow, \rightsquigarrow, 1)$ is a pseudo-BCK algebra and $x \rightarrow y = 1$ iff $x \wedge y = x$.

Remark 1.3 It is easy to show that an algebra $(A, \wedge, \rightarrow, \rightsquigarrow, 1)$ of type $(2, 2, 2, 0)$ is a pseudo-BCK meet-semilattice if and only if $(A, \wedge)$ is a meet-semilattice and $(A, \rightarrow, \rightsquigarrow, 1)$ satisfies the identities (*psBCK$'_{1}$*)–(*psBCK$'_{5}$*) and the identities:

$(psBCK_7'')$ $x \wedge [(x \to y) \rightsquigarrow y] = x$;
$(psBCK_8'')$ $(x \wedge y) \to y = 1$.

Example 1.6 Given a pseudo-hoop $(A, \odot, \to, \rightsquigarrow, 1)$ (see Chap. 2), then $(A, \wedge, \to, \rightsquigarrow, 1)$ is a pseudo-BCK meet-semilattice, where $x \wedge y = x \odot (x \rightsquigarrow y) = (x \to y) \odot x$.

In the sequel by a *pseudo-BCK semilattice* we mean a pseudo-BCK join-semilattice.

Definition 1.5 Let $(A, \leq, \to, \rightsquigarrow, 1)$ be a pseudo-BCK algebra. If the poset $(A, \leq)$ is a lattice, then we say that A is a *pseudo-BCK lattice*.

A pseudo-BCK lattice is denoted by $(A, \wedge, \vee, \to, \rightsquigarrow, 1)$.

Example 1.7 Consider the bounded pseudo-BCK algebra $(A, \leq, \to, \rightsquigarrow, 0, 1)$ from Example 1.3. Since $(A, \leq)$ is a lattice, it follows that A is a pseudo-BCK lattice.

Let A be a pseudo-BCK algebra. For all $x, y \in A$, define:

$$x \vee_1 y = (x \to y) \rightsquigarrow y, \qquad x \vee_2 y = (x \rightsquigarrow y) \to y.$$

Proposition 1.5 *In any bounded pseudo-BCK algebra A the following hold for all $x, y \in A$:*

(1) $0 \vee_1 x = x = 0 \vee_2 x$;
(2) $x \vee_1 0 = x^{-\sim}, x \vee_2 0 = x^{\sim-}$;
(3) $1 \vee_1 x = x \vee_1 1 = 1 = 1 \vee_2 x = x \vee_2 1$;
(4) $x \leq y$ *implies* $x \vee_1 y = y$ *and* $x \vee_2 y = y$;
(5) $x \vee_1 x = x \vee_2 x = x$.

Proof

(1) $0 \vee_1 x = (0 \to x) \rightsquigarrow x = 1 \rightsquigarrow x = x$ and similarly $0 \vee_2 x = x$.
(2) $x \vee_1 0 = (x \to 0) \rightsquigarrow 0 = x^{-\sim}$ and similarly $x \vee_2 0 = x^{\sim-}$.
(3) We have: $1 \vee_1 x = (1 \to x) \rightsquigarrow x = 1$ and $x \vee_1 1 = (x \to 1) \rightsquigarrow 1 = 1$, so $1 \vee_1 x = x \vee_1 1 = 1$. Similarly, $1 \vee_2 x = x \vee_2 1 = 1$.
(4) $x \vee_1 y = (x \to y) \rightsquigarrow y = 1 \rightsquigarrow y = y$. Similarly, $x \vee_2 y = y$.
(5) This follows from the definitions of $\vee_1$ and $\vee_2$. $\square$

Proposition 1.6 *In any bounded pseudo-BCK algebra A the following hold for all $x, y \in A$:*

(1) $x \vee_1 y^{-\sim} = x^{-\sim} \vee_1 y^{-\sim}$ *and* $x \vee_2 y^{\sim-} = x^{\sim-} \vee_2 y^{\sim-}$;
(2) $x \vee_1 y^{\sim} = x^{-\sim} \vee_1 y^{\sim}$ *and* $x \vee_2 y^{-} = x^{\sim-} \vee_2 y^{-}$;
(3) $(x^{-\sim} \vee_1 y^{-\sim})^{-\sim} = x^{-\sim} \vee_1 y^{-\sim}$ *and* $(x^{\sim-} \vee_2 y^{\sim-})^{\sim-} = x^{\sim-} \vee_2 y^{\sim-}$.

Proof

(1) Applying $(psbck\text{-}c_{19})$ we have:

$$x \vee_1 y^{-\sim} = \left(x \rightarrow y^{-\sim}\right) \rightsquigarrow y^{-\sim} = \left(x^{-\sim} \rightarrow y^{-\sim}\right) \rightsquigarrow y^{-\sim} = x^{-\sim} \vee_1 y^{-\sim};$$

$$x \vee_2 y^{\sim-} = \left(x \rightsquigarrow y^{\sim-}\right) \rightarrow y^{\sim-} = \left(x^{\sim-} \rightsquigarrow y^{\sim-}\right) \rightarrow y^{\sim-} = x^{\sim-} \vee_2 y^{\sim-}.$$

(2) Applying $(psbck\text{-}c_{20})$ we have:

$$x \vee_1 y^{\sim} = \left(x \rightarrow y^{\sim}\right) \rightsquigarrow y^{\sim} = \left(x^{-\sim} \rightarrow y^{\sim}\right) \rightsquigarrow y^{\sim} = x^{-\sim} \vee_1 y^{\sim};$$

$$x \vee_2 y^{-} = \left(x \rightsquigarrow y^{-}\right) \rightarrow y^{-} = \left(x^{\sim-} \rightsquigarrow y^{-}\right) \rightarrow y^{-} = x^{\sim-} \vee_2 y^{-}.$$

(3) Applying $(psbck\text{-}c_{21})$ we have:

$$\left(x^{-\sim} \vee_1 y^{-\sim}\right)^{-\sim} = \left[\left(x^{-\sim} \rightarrow y^{-\sim}\right) \rightsquigarrow y^{-\sim}\right]^{-\sim} = \left(x^{-\sim} \rightarrow y^{-\sim}\right) \rightsquigarrow y^{-\sim}$$
$$= x^{-\sim} \vee_1 y^{-\sim};$$

$$\left(x^{\sim-} \vee_2 y^{\sim-}\right)^{\sim-} = \left[\left(x^{\sim-} \rightsquigarrow y^{\sim-}\right) \rightarrow y^{\sim-}\right]^{\sim-} = \left(x^{\sim-} \rightsquigarrow y^{\sim-}\right) \rightarrow y^{\sim-}$$
$$= x^{\sim-} \vee_2 y^{\sim-}. \qquad \square$$

Proposition 1.7 *In any pseudo-BCK algebra the following hold for all $x, y \in A$:*

$(psbck\text{-}c_{22})$ $(x \vee_1 y) \rightarrow y = x \rightarrow y$ *and* $(x \vee_2 y) \rightsquigarrow y = x \rightsquigarrow y$.

Proof This is a consequence of the property $(psbck\text{-}c_{11})$. $\square$

Lemma 1.1 *Let A be a pseudo-BCK algebra. Then:*

(1) $x \vee_1 y$ $(y \vee_1 x)$ *is an upper bound of* $\{x, y\}$;
(2) $x \vee_2 y$ $(y \vee_2 x)$ *is an upper bound of* $\{x, y\}$

for all $x, y \in A$.

Proof

(1) By $(psBCK_2)$ we have $x \leq (x \rightarrow y) \rightsquigarrow y$.
     Since by $(psbck\text{-}c_6)$, $y \leq (x \rightarrow y) \rightsquigarrow y$, we conclude that $x, y \leq x \vee_1 y$.
     Similarly we get $x, y \leq y \vee_1 x$.
(2) Similar to (1). $\square$

Definition 1.6 Let A be a pseudo-BCK algebra.

(1) If $x \vee_1 y = y \vee_1 x$ for all $x, y \in A$, then A is called $\vee_1$-*commutative*;
(2) If $x \vee_2 y = y \vee_2 x$ for all $x, y \in A$, then A is called $\vee_2$-*commutative*.

Lemma 1.2 *Let A be a pseudo-BCK algebra.*

(1) *If for all* $x, y \in A$, $x \vee_1 y$ $(y \vee_1 x)$ *is the l.u.b. of* $\{x, y\}$, *then* A *is* $\vee_1$-*commutative*;

(2) *If for all* $x, y \in A$, $x \vee_2 y$ $(y \vee_2 x)$ *is the l.u.b. of* $\{x, y\}$, *then* A *is* $\vee_2$-*commutative*.

Proof

(1) Suppose that for all $x, y \in A$, $x \vee_1 y$ $(y \vee_1 x)$ is the l.u.b. of $\{x, y\}$. Then by Lemma 1.1, for all $x, y \in A$ we have $y \vee_1 x \leq x \vee_1 y$ and $x \vee_1 y \leq y \vee_1 x$. Applying $(psBCK_5)$ we get $x \vee_1 y = y \vee_1 x$. Thus A is $\vee_1$-commutative.

(2) Similar to (1). $\qquad\square$

Proposition 1.8 *Let A be a pseudo-BCK algebra.*

(1) *If A is $\vee_1$-commutative, then $x \vee_1 y$ is the l.u.b. of $\{x, y\}$, for all $x, y \in A$;*

(2) *If A is $\vee_2$-commutative, then $x \vee_2 y$ is the l.u.b. of $\{x, y\}$, for all $x, y \in A$.*

Proof

(1) Let $x, y \in A$. According to Lemma 1.1, $x \vee_1 y$ is an upper bound of $\{x, y\}$. Let z be another upper bound of $\{x, y\}$, i.e. $x \leq z$ and $y \leq z$. We will prove that $x \vee_1 y \leq z$. Indeed, applying Proposition 1.5(4) and taking into consideration that A is $\vee_1$-commutative we have:

$$x \vee_1 y \to z = x \vee_1 y \to y \vee_1 z = x \vee_1 y \to z \vee_1 y$$

$$= \big((x \to y) \rightsquigarrow y\big) \to \big((z \to y) \rightsquigarrow y\big).$$

According to $(psBCK_1)$ we have $(b \to c) \rightsquigarrow (a \to c) \geq a \to b$ and replacing a with $z \to y$, b with $x \to y$ and c with y we get:

$$\big((x \to y) \rightsquigarrow y\big) \to \big((z \to y) \rightsquigarrow y\big) \geq (z \to y) \rightsquigarrow (x \to y)$$

$$\geq x \to z \quad \big(\text{by } (psBCK_1)\big).$$

Hence $x \vee_1 y \to z \geq x \to z = 1$ (since $x \leq z$). It follows that $x \vee_1 y \to z = 1$, thus $x \vee_1 y \leq z$. We conclude that $x \vee_1 y$ is the l.u.b. of $\{x, y\}$.

(2) Similar to (1). $\qquad\square$

Theorem 1.1 *If A is a pseudo-BCK algebra, then:*

(1) *A is $\vee_1$-commutative iff it is a join-semilattice with respect to $\vee_1$ (under $\leq$);*

(2) *A is $\vee_2$-commutative iff it is a join-semilattice with respect to $\vee_2$ (under $\leq$).*

Proof This is a consequence of Lemma 1.2 and Proposition 1.8. $\qquad\square$

Corollary 1.1 *Let A be a pseudo-BCK algebra. Then:*

(1) *If A is $\vee_1$-commutative, then $x \vee_1 y \leq x \vee_2 y, y \vee_2 x$ for all $x, y \in A$;*

(2) *If A is $\vee_2$-commutative, then $x \vee_2 y \leq x \vee_1 y$, $y \vee_1 x$ for all $x, y \in A$.*

Proof

(1) According to Lemma 1.1, $x \vee_2 y$, $y \vee_2 x$ are upper bounds of $\{x, y\}$. By Proposition 1.8, $x \vee_1 y$ is the l.u.b. of $\{x, y\}$, thus $x \vee_1 y \leq x \vee_2 y$, $y \vee_2 x$.
(2) Similar to (1).   □

Definition 1.7 A pseudo-BCK algebra is called *sup-commutative* if it is both $\vee_1$-commutative and $\vee_2$-commutative.

Theorem 1.2 *A pseudo-BCK algebra is sup-commutative iff it is a join-semilattice with respect to both $\vee_1$ and $\vee_2$.*

Proof This follows from Theorem 1.1.   □

Corollary 1.2 *If A is a sup-commutative pseudo-BCK algebra, then $x \vee_1 y = x \vee_2 y$ for all $x, y \in A$.*

Proof By Corollary 1.1, $x \vee_1 y \leq x \vee_2 y$ and $x \vee_2 y \leq x \vee_1 y$, hence $x \vee_1 y = x \vee_2 y$.   □

Lemma 1.3 *In a $\vee_1$-commutative ($\vee_2$-commutative) bounded pseudo-BCK algebra A, we have $x^{-\sim} = x$ ($x^{\sim-} = x$, respectively), for all $x \in A$.*

Proof Replacing y with 0 in the identity $x \vee_1 y = y \vee_1 x$, we get $(x \to 0) \rightsquigarrow 0 = (0 \to x) \rightsquigarrow x$, i.e. $x^{-\sim} = x$.
Similarly, replacing y with 0 in $x \vee_2 y = y \vee_2 x$, we get $x^{\sim-} = x$.   □

Corollary 1.3 *Let A be a sup-commutative, bounded pseudo-BCK algebra. Then $x^{-\sim} = x^{\sim-} = x$, for all $x \in A$.*

Proof This follows by replacing y with 0 in the equality $x \vee_1 y = x \vee_2 y$ and applying Lemma 1.3.   □

In a bounded pseudo-BCK algebra A, define, for all $x, y \in A$:

$$x \wedge_1 y := \left(x^- \vee_1 y^- \right)^\sim,$$

$$x \wedge_2 y := \left(x^- \vee_2 y^- \right)^\sim.$$

Lemma 1.4 *Let A be a pseudo-BCK algebra. Then for all $x, y \in A$:*

(1) $x \wedge_1 y$ $(y \wedge_1 x)$ *is a lower bound of* $\{x^{-\sim}, y^{-\sim}\}$;
(2) $x \wedge_2 y$ $(y \wedge_2 x)$ *is a lower bound of* $\{x^{\sim-}, y^{\sim-}\}$.

Proof

(1) By Lemma 1.1 we have $x^-, y^- \leq x^- \vee_1 y^-$, hence $x \wedge_1 y = (x^- \vee_1 y^-)^\sim \leq$ $x^{-\sim}, y^{-\sim}$. Thus $x \wedge_1 y$ is a lower bound of $\{x^{-\sim}, y^{-\sim}\}$.
(2) Similar to (1). □

Proposition 1.9 *Let A be a bounded pseudo-BCK algebra.*

(1) *If A is $\vee_1$-commutative, then $x \wedge_1 y$ $(y \wedge_1 x)$ is the g.l.b. of $\{x, y\}$ and $x \wedge_1 y = y \wedge_1 x$, for all $x, y \in A$;*
(2) *If A is $\vee_2$-commutative, then $x \wedge_2 y$ $(y \wedge_2 x)$ is the g.l.b. of $\{x, y\}$ and $x \wedge_2 y = y \wedge_2 x$, for all $x, y \in A$.*

Proof

(1) By Lemma 1.3, $x^{-\sim} = x$ and $y^{-\sim} = y$. Hence by Lemma 1.4, $x \wedge_1 y$ is a lower bound of $\{x, y\}$. Now let z be another lower bound of $\{x, y\}$, i.e. $z \leq x, y$. It follows that $x^-, y^- \leq z^-$, thus z^- is an upper bound of $\{x^-, y^-\}$. Since A is $\vee_1$-commutative, by Proposition 1.8, $x^- \vee_1 y^-$ is the l.u.b. of $\{x^-, y^-\}$, hence $x^- \vee_1 y^- \leq z^-$. Thus $z = z^{-\sim} \leq (x^- \vee_1 y^-)^\sim = x \wedge_1 y$, i.e. $x \wedge_1 y$ is the g.l.b. of $\{x, y\}$. Since A is $\vee_1$-commutative, we have $x^- \vee_1 y^- = y^- \vee_1 x^-$, hence by definition it follows that $x \wedge_1 y = y \wedge_1 x$, for all $x, y \in A$.
(2) Similar to (1). □

Corollary 1.4 *Let A be a bounded pseudo-BCK algebra.*

(1) *If A is $\vee_1$-commutative, then A is a lattice with respect to $\wedge_1, \vee_1$;*
(2) *If A is $\vee_2$-commutative, then A is a lattice with respect to $\wedge_2, \vee_2$.*

Proof This follows by Propositions 1.8 and 1.9. □

Theorem 1.3 *A bounded sup-commutative pseudo-BCK algebra A is a lattice with respect to both $\vee_1, \wedge_1$ and $\vee_2, \wedge_2$ (under $\leq$) and for all x, y we have:*

$$x \vee_1 y = x \vee_2 y, \qquad x \wedge_1 y = x \wedge_2 y.$$

Proof By Corollary 1.4, A is a lattice with respect to both $\wedge_1, \vee_1$ and $\wedge_2, \vee_2$. By Corollary 1.2, $x \vee_1 y = x \vee_2 y$ for all $x, y \in A$. By Proposition 1.9 we get: $x \wedge_2 y \leq x \wedge_1 y$ and $x \wedge_1 y \leq x \wedge_2 y$, hence $x \wedge_1 y = x \wedge_2 y$ for all $x, y \in A$. □

We recall that a *downwards-directed set* (or a *filtered set*) is a partially ordered set $(A, \leq)$ such that whenever $a, b \in A$, there exists an $x \in A$ such that $x \leq a$ and $x \leq b$.

Dually, an *upwards-directed set* is a partially ordered set $(A, \leq)$ such that whenever $a, b \in A$, there exists an $x \in A$ such that $a \leq x$ and $b \leq x$.

If X is a set, then a *net* in X will be a set $\{x_i \mid i \in I\}$, where $(I, \leq)$ is an upwards-directed set.

We say that a pseudo-BCK algebra A satisfies the *relative cancellation property*, (RCP) for short, if for every $a, b, c \in A$,

$$a, b \leq c \quad \text{and} \quad c \to a = c \to b, c \rightsquigarrow a = c \rightsquigarrow b \quad \text{imply} \quad a = b.$$

We note that a pseudo-BCK algebra A that is sup-commutative and satisfies the (RCP) condition is said to be a *Łukasiewicz pseudo-BCK algebra* (see [112]).

Example 1.8 The pseudo-BCK algebra A from Example 1.3 is downwards-directed with (RCP).

Proposition 1.10 *Any downwards-directed sup-commutative pseudo-BCK algebra has (RCP).*

Proof Consider $a, b, c \in A$ such that $a, b \leq c$ and $c \to a = c \to b, c \rightsquigarrow a = c \rightsquigarrow b$. There exists an $x \in A$ such that $x \leq a, b$.

By $(psbck\text{-}c_1)$, from $a \leq c$ it follows that $c \rightsquigarrow x \leq a \rightsquigarrow x$.

According to Proposition 1.5(4) and $(psbck\text{-}c_3)$ we have:

$$a \rightsquigarrow x = (c \rightsquigarrow x) \vee_1 (a \rightsquigarrow x) = (a \rightsquigarrow x) \vee_1 (c \rightsquigarrow x)$$

$$= \left[(a \rightsquigarrow x) \to (c \rightsquigarrow x) \right] \rightsquigarrow (c \rightsquigarrow x) = \left[c \rightsquigarrow \left[(a \rightsquigarrow x) \to x \right] \right] \rightsquigarrow (c \rightsquigarrow x)$$

$$= \left[c \rightsquigarrow (a \vee_2 x) \right] \rightsquigarrow (c \rightsquigarrow x) = \left[c \rightsquigarrow (x \vee_2 a) \right] \rightsquigarrow (c \rightsquigarrow x)$$

$$= (c \rightsquigarrow a) \rightsquigarrow (c \rightsquigarrow x).$$

Similarly, $b \rightsquigarrow x = (c \rightsquigarrow b) \rightsquigarrow (c \rightsquigarrow x) = (c \rightsquigarrow a) \rightsquigarrow (c \rightsquigarrow x) = a \rightsquigarrow x$.

We have: $a = x \vee_2 a = a \vee_2 x = (a \rightsquigarrow x) \to x = (b \rightsquigarrow x) \to x = b \vee_2 x = x \vee_2 b = b$.

Thus A has (RCP).   □

1.2 Pseudo-BCK Algebras with Pseudo-product

Definition 1.8 A pseudo-BCK algebra with the *(pP) condition* (i.e. with the *pseudo-product* condition) or a *pseudo-BCK(pP) algebra* for short, is a pseudo-BCK algebra $(A, \leq, \to, \rightsquigarrow, 1)$ satisfying the (pP) condition:

(pP) For all $x, y \in A$, $x \odot y$ exists where

$$x \odot y = \min\{z \mid x \leq y \to z\} = \min\{z \mid y \leq x \rightsquigarrow z\}.$$

Example 1.9 Take $A = \{0, a_1, a_2, s, a, b, n, c, d, m, 1\}$ with $0 < a_1 < a_2 < s < a, b < n < c, d < m < 1$ (see Fig. 1.3).

Consider the operations $\to, \rightsquigarrow$ given by the following tables:

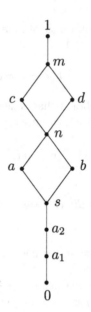

Fig. 1.3 Example of bounded pseudo-BCK(pP) algebra

$\rightarrow$	0	a_1	a_2	s	a	b	n	c	d	m	1
0	1	1	1	1	1	1	1	1	1	1	1
a_1	a_1	1	1	1	1	1	1	1	1	1	1
a_2	a_1	a_1	1	1	1	1	1	1	1	1	1
s	0	a_1	a_2	1	1	1	1	1	1	1	1
a	0	a_1	a_2	m	1	m	1	1	1	1	1
b	0	a_1	a_2	m	m	1	1	1	1	1	1
n	0	a_1	a_2	m	m	m	1	1	1	1	1
c	0	a_1	a_2	m	m	m	m	1	m	1	1
d	0	a_1	a_2	m	m	m	m	m	1	1	1
m	0	a_1	a_2	m	m	m	m	m	m	1	1
1	0	a_1	a_2	s	a	b	n	c	d	m	1

$\rightsquigarrow$	0	a_1	a_2	s	a	b	n	c	d	m	1
0	1	1	1	1	1	1	1	1	1	1	1
a_1	a_2	1	1	1	1	1	1	1	1	1	1
a_2	0	a_1	1	1	1	1	1	1	1	1	1
s	0	a_1	a_2	1	1	1	1	1	1	1	1
a	0	a_1	a_2	m	1	m	1	1	1	1	1
b	0	a_1	a_2	m	m	1	1	1	1	1	1
n	0	a_1	a_2	m	m	m	1	1	1	1	1
c	0	a_1	a_2	m	m	m	m	1	m	1	1
d	0	a_1	a_2	m	m	m	m	m	1	1	1
m	0	a_1	a_2	m	m	m	m	m	m	1	1
1	0	a_1	a_2	s	a	b	n	c	d	m	1

Then $(A, \leq, \rightarrow, \rightsquigarrow, 0, 1)$ is a bounded pseudo-BCK(pP) algebra. The operation $\odot$ is given by the following table:

$\odot$	0	a_1	a_2	s	a	b	n	c	d	m	1
0	0	0	0	0	0	0	0	0	0	0	0
a_1	0	0	0	a_1	a_1	a_1	a_1	a_1	a_1	a_1	a_1
a_2	0	a_1	a_2	a_2	a_2	a_2	a_2	a_2	a_2	a_2	a_2
s	0	a_1	a_2	s	s	s	s	s	s	s	s
a	0	a_1	a_2	s	s	s	s	s	s	s	a
b	0	a_1	a_2	s	s	s	s	s	s	s	b
n	0	a_1	a_2	s	s	s	s	s	s	s	n
c	0	a_1	a_2	s	s	s	s	s	s	s	c
d	0	a_1	a_2	s	s	s	s	s	s	s	d
m	0	a_1	a_2	s	s	s	s	s	s	s	m
1	0	a_1	a_2	s	a	b	n	c	d	m	1

Remark 1.4 Any bounded linearly ordered pseudo-BCK algebra satisfies the (pP) condition. If the pseudo-BCK algebra is not bounded this result is not always valid. Indeed, let $(\mathbb{Q}, \leq, +, -, 0)$ be the additive group of rationals with the usual linear order and take $A = \{x \in \mathbb{Q} \mid -\sqrt{2} < x \leq 0\}$. Then $(A, \rightarrow, 0)$ is a linearly ordered BCK-algebra with $x \rightarrow y = \min\{0, y - x\}$.

We have $\{z \in A \mid (-1) \leq (-1) \rightarrow z = \min\{0, z + 1\}\} = A$.

Thus $(-1) \odot (-1) = \min A$ does not exist in $(A, \rightarrow, 0)$.

Example 1.10

(1) If $(A, \leq, \rightarrow, \rightsquigarrow, 0, 1)$ is the bounded pseudo-BCK algebra from Example 1.3, then:

$$\min\{z \mid b \leq a \rightarrow z\} = \min\{a, b, c, 1\} \quad \text{and}$$

$$\min\{z \mid a \leq b \rightsquigarrow z\} = \min\{a, b, c, 1\}$$

do not exist. Thus $b \odot a$ does not exist, so A is not a pseudo-BCK(pP) algebra.

(2) If $(A, \leq, \rightarrow, \rightsquigarrow, 0, 1)$ is the subreduct of an FL_w-algebra (see Chap. 3), then it is obvious that A is a bounded pseudo-BCK(pP) algebra.

Example 1.11 Consider $A = \{0, a, b, c, 1\}$ with $0 < a < b, c < 1$ and b, c incomparable (see Fig. 1.4).

Consider the operations $\rightarrow, \rightsquigarrow$ given by the following tables:

$\rightarrow$	0	a	b	c	1
0	1	1	1	1	1
a	0	1	1	1	1
b	0	a	1	c	1
c	0	b	b	1	1
1	0	a	b	c	1

$\rightsquigarrow$	0	a	b	c	1
0	1	1	1	1	1
a	0	1	1	1	1
b	0	c	1	c	1
c	0	a	b	1	1
1	0	a	b	c	1

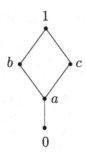

Fig. 1.4 Example of pseudo-BCK algebra without (pP) condition

Then $(A, \leq, \rightarrow, \rightsquigarrow, 0, 1)$ is a bounded pseudo-BCK algebra.

Since $(A, \leq)$ is a lattice, it follows that A is a pseudo-BCK lattice.

We can see that $c \odot b = \min\{z \mid c \leq b \rightarrow z\} = \min\{b, c, 1\}$ does not exist.

Hence A is a pseudo-BCK lattice without the (pP) condition.

Remark 1.5 It is easy to see that from the definition of the (pP) property, in any pseudo-BCK(pP) algebra we have:

$(psbck\text{-}c_{23})$ $x \leq y \rightarrow (x \odot y), x \leq y \rightsquigarrow (y \odot x)$.

Theorem 1.4 *The* (pP) *condition is equivalent to the* (pRP) *(pseudo-residuation property)*:

(pRP) *For all x, y, z the following holds*

$$x \odot y \leq z \quad \textit{iff} \quad x \leq y \rightarrow z \quad \textit{iff} \quad y \leq x \rightsquigarrow z.$$

Proof Assume that the (pP) condition holds.

From $x \odot y \leq z$, applying $(psbck\text{-}c_{10})$ we have $y \rightarrow x \odot y \leq y \rightarrow z$ and by $(psbck\text{-}c_{23})$ we get $x \leq y \rightarrow z$. It is easy to see that $x \leq y \rightarrow z$ implies $x \odot y \leq z$.

Thus $x \odot y \leq z$ iff $x \leq y \rightarrow z$ and similarly, $x \odot y \leq z$ iff $y \leq x \rightsquigarrow z$.

So, (pRP) also holds.

Conversely, suppose that (pRP) is satisfied.

Since $x \odot y \leq x \odot y$, by (pRP) we have $x \leq y \rightarrow x \odot y$, so $x \odot y \in \{z \mid x \leq y \rightarrow z\}$.

But from $x \leq y \rightarrow z$, by (pRP) we have $x \odot y \leq z$, and we conclude that $\min\{z \mid x \leq y \rightarrow z\} = x \odot y$.

One can prove similarly that $\min\{z \mid y \leq x \rightsquigarrow z\} = x \odot y$.

Thus the (pP) condition is also satisfied. □

Theorem 1.5 *Let* $(A, \leq, \rightarrow, \rightsquigarrow, 1)$ *be a pseudo-BCK(pP) algebra. Then the algebra* $(A, \odot, 1)$ *is a monoid, i.e. $\odot$ is associative with identity element 1.*

Proof For an arbitrary element $u \in A$, applying Theorem 1.4, $(psbck\text{-}c_4)$ and $(psbck\text{-}c_3)$ we have:

$$(x \odot y) \odot z \leq u \quad \text{iff} \quad x \odot y \leq z \rightarrow u \quad \text{iff} \quad x \leq y \rightarrow (z \rightarrow u) \quad \text{iff}$$

$$y \leq x \rightsquigarrow (z \rightarrow u) \quad \text{iff} \quad y \leq z \rightarrow (x \rightsquigarrow u) \quad \text{iff}$$

$$y \odot z \leq x \rightsquigarrow u \quad \text{iff} \quad x \odot (y \odot z) \leq u.$$

It follows that $(x \odot y) \odot z = x \odot (y \odot z)$, hence $\odot$ is associative.

By $(psbck\text{-}c_{23})$ we have $y \leq x \rightarrow y \odot x$, so for $y = 1$ we get $1 \leq x \rightarrow 1 \odot x$. Thus $x \rightarrow 1 \odot x = 1$, that is, $x \leq 1 \odot x$. On the other hand, $1 \odot x \leq x$, so $1 \odot x = x$. Similarly, $x \odot 1 = x$, that is, 1 is the identity element.

We conclude that $(A, \odot, 1)$ is a monoid.   □

Remark 1.6 A *partial ordered residuated integral monoid* (*porim*, for short) is a structure $(A, \leq, \odot, \rightarrow, \rightsquigarrow, 1)$, where $(A, \leq)$ is a poset with greatest element 1, $(A, \odot, 1)$ is a monoid and $x \odot y \leq z$ iff $x \leq y \rightarrow z$ iff $y \leq x \rightsquigarrow z$, for all $x, y, z \in A$.

Applying Theorems 1.5, 1.4 and $(psBCK_4)$, it follows that every pseudo-BCK(pP) algebra is a porim.

On the other hand, one can easily prove that every porim is a pseudo-BCK(pP) algebra (see Remark 3.2).

Theorem 1.6 *Pseudo-BCK(pP) algebras are categorically isomorphic to porims.*

Proof This follows from Remark 1.6.   □

Proposition 1.11 *In any pseudo-BCK(pP) algebra the following properties hold:*

$(psbck\text{-}c_{24})$ $x \odot y \leq x, y$;

$(psbck\text{-}c_{25})$ $(x \rightarrow y) \odot x \leq x, y, x \odot (x \rightsquigarrow y) \leq x, y$;

$(psbck\text{-}c_{26})$ $x \leq y$ *implies* $x \odot z \leq y \odot z, z \odot x \leq z \odot y$;

$(psbck\text{-}c_{27})$ $x \rightarrow y \leq (x \odot z) \rightarrow (y \odot z), x \rightsquigarrow y \leq (z \odot x) \rightsquigarrow (z \odot y)$;

$(psbck\text{-}c_{28})$ $x \odot (y \rightarrow z) \leq y \rightarrow (x \odot z), (y \rightsquigarrow z) \odot x \leq y \rightsquigarrow (z \odot x)$;

$(psbck\text{-}c_{29})$ $(y \rightarrow z) \odot (x \rightarrow y) \leq x \rightarrow z, (x \rightsquigarrow y) \odot (y \rightsquigarrow z) \leq x \rightsquigarrow z$;

$(psbck\text{-}c_{30})$ $x \rightarrow (y \rightarrow z) = (x \odot y) \rightarrow z, x \rightsquigarrow (y \rightsquigarrow z) = (y \odot x) \rightsquigarrow z$;

$(psbck\text{-}c_{31})$ $(x \odot z) \rightarrow (y \odot z) \leq x \rightarrow (z \rightarrow y), (z \odot x) \rightsquigarrow (z \odot y) \leq x \rightsquigarrow (z \rightsquigarrow y)$;

$(psbck\text{-}c_{32})$ $x \rightarrow y \leq (x \odot z) \rightarrow (y \odot z) \leq x \rightarrow (z \rightarrow y), x \rightsquigarrow y \leq (z \odot x) \rightsquigarrow (z \odot y) \leq x \rightsquigarrow (z \rightsquigarrow y)$;

$(psbck\text{-}c_{33})$ $(x_{n-1} \rightarrow x_n) \odot (x_{n-2} \rightarrow x_{n-1}) \odot \cdots \odot (x_1 \rightarrow x_2) \leq x_1 \rightarrow x_n$ *and* $(x_1 \rightsquigarrow x_2) \odot (x_2 \rightsquigarrow x_3) \odot \cdots \odot (x_{n-1} \rightsquigarrow x_n) \leq x_1 \rightsquigarrow x_n.$

Proof

$(psbck\text{-}c_{24})$ Applying $(psbck\text{-}c_6)$ we have:

$$x \leq y \rightarrow x, \text{ so } x \odot y \leq x \quad \text{and} \quad y \leq x \rightsquigarrow y, \text{ so } x \odot y \leq y.$$

$(psbck\text{-}c_{25})$ By $(psbck\text{-}c_{24})$ we have $(x \rightarrow y) \odot x \leq x$.

Since $(x \rightarrow y) \odot x = \min\{z \mid x \rightarrow y \leq x \rightarrow z\}$ and taking into consideration that $x \rightarrow y \leq x \rightarrow y$ we get $(x \rightarrow y) \odot x \leq y$. Similarly, $x \odot (x \rightsquigarrow y) \leq x, y$.

($psbck$-c_{26}) Applying ($psbck$-c_{23}) we have $x \leq y \leq z \to y \odot z$, so by (pRP) we get
$x \odot z \leq y \odot z$. Similarly, $z \odot x \leq z \odot y$.

($psbck$-c_{27}) Taking into consideration ($psbck$-c_{25}), ($psbck$-c_{26}) and (pRP) we have
$(x \to y) \odot x \leq y$, so $[(x \to y) \odot x] \odot z \leq y \odot z$.

Thus $x \to y \leq (x \odot z) \to (y \odot z)$ and similarly $x \rightsquigarrow y \leq (z \odot x) \rightsquigarrow (z \odot y)$.

($psbck$-c_{28}) Applying ($psbck$-c_{25}) we get $x \odot (y \to z) \odot y \leq x \odot z$ and by (pRP)
we have $x \odot (y \to z) \leq y \to (x \odot z)$.

Similarly, $(y \rightsquigarrow z) \odot x \leq y \rightsquigarrow (z \odot x)$.

($psbck$-c_{29}) By ($psbck$-c_{25}) and (pRP) we have

$$(y \to z) \odot (x \to y) \odot x \leq (y \to z) \odot y \leq z, \quad \text{so } (y \to z) \odot (x \to y) \leq x \to z.$$

Similarly $(x \rightsquigarrow y) \odot (y \rightsquigarrow z) \leq x \rightsquigarrow z$.

($psbck$-c_{30}) For any u we have:

$$u \leq x \to (y \to z) \quad \text{iff} \quad u \odot x \leq y \to z \quad \text{iff} \quad (u \odot x) \odot y \leq z \quad \text{iff}$$

$$u \odot (x \odot y) \leq z \quad \text{iff} \quad u \leq (x \odot y) \to z, \quad \text{so } x \to (y \to z) = (x \odot y) \to z.$$

Similarly, $x \rightsquigarrow (y \rightsquigarrow z) = (y \odot x) \rightsquigarrow z$.

($psbck$-c_{31}) From $y \odot z \leq y$, applying ($psbck$-c_{10}) and ($psbck$-c_{30}) we get

$$x \odot z \to y \odot z \leq x \odot z \to y = x \to (z \to y).$$

Similarly, $(z \odot x) \rightsquigarrow (z \odot y) \leq x \rightsquigarrow (z \rightsquigarrow y)$.

($psbck$-c_{32}) This is a consequence of properties ($psbck$-c_{27}) and ($psbck$-c_{31}).

($psbck$-c_{33}) This follows from ($psbck$-c_{29}) by induction.   □

Proposition 1.12 *In a bounded pseudo-BCK(pP) algebra the following hold*:

($psbck$-c_{34}) $x \odot 0 = 0 \odot x = 0$;

($psbck$-c_{35}) $y^- \odot (x \to y) \leq x^-$ *and* $(x \rightsquigarrow y) \odot y^\sim \leq x^\sim$;

($psbck$-c_{36}) $x^- \odot x = 0$ *and* $x \odot x^\sim = 0$;

($psbck$-c_{37}) $x \to y^- = (x \odot y)^-$ *and* $x \rightsquigarrow y^\sim = (y \odot x)^\sim$;

($psbck$-c_{38}) $x \leq y^-$ *iff* $x \odot y = 0$ *and* $x \leq y^\sim$ *iff* $y \odot x = 0$;

($psbck$-c_{39}) $x \leq y^-$ *iff* $y \leq x^\sim$;

($psbck$-c_{40}) $x \leq x^\sim \to y$ *and* $x \leq x^- \rightsquigarrow y$.

Proof

($psbck$-c_{34}) From $x \leq 0 \to 0 = 1$ and $x \leq 0 \rightsquigarrow 0 = 1$ we get $x \odot 0 \leq 0$ and $0 \odot x \leq 0$. Thus $x \odot 0 = 0 \odot x = 0$.

($psbck$-c_{35}) This follows from ($psbck$-c_{29}) for $z = 0$.

($psbck$-c_{36}) From $x \to 0 \leq x \to 0$ we get $(x \to 0) \odot x \leq 0$, so $x^- \odot x = 0$.

Similarly, $x \odot x^\sim = 0$.

($psbck$-c_{37}) Applying ($psbck$-c_{30}) we get:

$$x \to y^- = x \to (y \to 0) = x \odot y \to 0 = (x \odot y)^-.$$

Similarly, $x \rightsquigarrow y^\sim = (y \odot x)^\sim$.

($psbck$-c_{38}) Assume $x \leq y^-$. Applying ($psbck$-c_{26}) we get $x \odot y \leq y^- \odot y = 0$, so $x \odot y = 0$. Conversely, if $x \odot y = 0$, by ($psbck$-c_{23}) we get $x \leq y \rightarrow (x \odot y) = y \rightarrow 0 = y^-$. Similarly, $x \leq y^\sim$ iff $y \odot x = 0$.

($psbck$-c_{39}) This follows from ($psbck$-c_{38}).

($psbck$-c_{40}) Since $0 \leq y$, we have $x \odot x^\sim \leq y$, so $x \leq x^\sim \rightarrow y$.
  Similarly, $x \leq x^- \rightsquigarrow y$.   □

Proposition 1.13 *In every bounded pseudo-BCK lattice A we have*:

($psbck$-c_{41}) $(x \vee y)^- = x^- \wedge y^-, (x \vee y)^\sim = x^\sim \wedge y^\sim$;

($psbck$-c_{42}) $(x \wedge y)^- \geq x^- \vee y^-$ and $(x \wedge y)^\sim \geq x^\sim \vee y^\sim$;

($psbck$-c_{43}) $(x \vee y)^{-\sim} \geq x^{-\sim} \vee y^{-\sim}$ and $(x \vee y)^{\sim -} \geq x^{\sim -} \vee y^{\sim -}$.

Proof

($psbck$-c_{41}) According to ($psbck$-c_{12}), for all $x, y, z \in A$ we have:

$$(x \vee y) \rightarrow z = (x \rightarrow z) \wedge (y \rightarrow z) \quad \text{and} \quad (x \vee y) \rightsquigarrow z = (x \rightsquigarrow z) \wedge (y \rightsquigarrow z).$$

Taking $z = 0$ we get $(x \vee y)^- = x^- \wedge y^-$ and $(x \vee y)^\sim = x^\sim \wedge y^\sim$.

($psbck$-c_{42}) By $x \wedge y \leq x$ and $x \wedge y \leq y$ we get $x^- \leq (x \wedge y)^-$ and $y^- \leq (x \wedge y)^-$, respectively. Thus $(x \wedge y)^- \geq x^- \vee y^-$. Similarly, $(x \wedge y)^\sim \geq x^\sim \vee y^\sim$.

($psbck$-c_{43}) Applying ($psbck$-c_{41}) and ($psbck$-c_{42}) we get:

$$(x \vee y)^{-\sim} = \left(x^- \wedge y^-\right)^\sim \geq x^{-\sim} \vee y^{-\sim} \quad \text{and}$$

$$(x \vee y)^{\sim -} = \left(x^\sim \wedge y^\sim\right)^- \geq x^{\sim -} \vee y^{\sim -}.$$ □

Proposition 1.14 *In any pseudo-BCK(pP) lattice the following hold*:

(1) $x \odot (\bigvee_{i \in I} y_i) = \bigvee_{i \in I}(x \odot y_i)$ and $(\bigvee_{i \in I} y_i) \odot x = \bigvee_{i \in I}(y_i \odot x)$;

(2) $y \rightarrow (\bigwedge_{i \in I} x_i) = \bigwedge_{i \in I}(y \rightarrow x_i)$ and $y \rightsquigarrow (\bigwedge_{i \in I} x_i) = \bigwedge_{i \in I}(y \rightsquigarrow x_i)$;

(3) $(\bigvee_{i \in I} x_i) \rightarrow y = \bigwedge_{i \in I}(x_i \rightarrow y)$ and $(\bigvee_{i \in I} x_i) \rightsquigarrow y = \bigwedge_{i \in I}(x_i \rightsquigarrow y)$,

whenever the arbitrary meets and unions exist.

Proof

(1) Since $y_i \leq \bigvee_{i \in I} y_i$ for all $i \in I$, according to ($psbck$-c_{26}) we get $x \odot y_i \leq x \odot (\bigvee_{i \in I} y_i)$ for all $i \in I$. It follows that $\bigvee_{i \in I}(x \odot y_i) \leq x \odot (\bigvee_{i \in I} y_i)$.

On the other hand, $x \odot y_i \leq \bigvee_{i \in I}(x \odot y_i)$ for all $i \in I$, so $y_i \leq x \rightsquigarrow \bigvee_{i \in I}(x \odot y_i)$ for all $i \in I$. If follows that $\bigvee_{i \in I} y_i \leq x \rightsquigarrow \bigvee_{i \in I}(x \odot y_i)$, that is, $x \odot (\bigvee_{i \in I} y_i) \leq \bigvee_{i \in I}(x \odot y_i)$.

Thus $x \odot (\bigvee_{i \in I} y_i) = \bigvee_{i \in I}(x \odot y_i)$.

Similarly, $(\bigvee_{i \in I} y_i) \odot x = \bigvee_{i \in I}(y_i \odot x)$.

(2) For any $u \in A$ we have the following equivalences:

$$u \leq y \rightarrow \left(\bigwedge_{i \in I} x_i \right) \quad \text{iff} \quad u \odot y \leq \bigwedge_{i \in I} x_i \quad \text{iff} \quad u \odot y \leq x_i \text{ for all } i \in I \quad \text{iff}$$

$$u \leq y \rightarrow x_i \text{ for all } i \in I \quad \text{iff} \quad u \leq \bigwedge_{i \in I} (y \rightarrow x_i).$$

Therefore, $y \rightarrow (\bigwedge_{i \in I} x_i) = \bigwedge_{i \in I} (y \rightarrow x_i)$.
Similarly, $y \rightsquigarrow (\bigwedge_{i \in I} x_i) = \bigwedge_{i \in I} (y \rightsquigarrow x_i)$.

(3) For any $u \in A$ we have the following equivalences:

$$u \leq \left(\bigvee_{i \in I} x_i \right) \rightarrow y \quad \text{iff} \quad u \odot \left(\bigvee_{i \in I} x_i \right) \leq y \quad \text{iff} \quad \bigvee_{i \in I} (u \odot x_i) \leq y \quad \text{iff}$$

$$u \odot x_i \leq y \text{ for all } i \in I \quad \text{iff} \quad u \leq x_i \rightarrow y \text{ for all } i \in I \quad \text{iff}$$

$$u \leq \bigwedge_{i \in I} (x_i \rightarrow y).$$

Therefore, $(\bigvee_{i \in I} x_i) \rightarrow y = \bigwedge_{i \in I} (x_i \rightarrow y)$.
Similarly, $(\bigvee_{i \in I} x_i) \rightsquigarrow y = \bigwedge_{i \in I} (x_i \rightsquigarrow y)$.   □

1.3 Pseudo-BCK Algebras with Pseudo-double Negation

Definition 1.9 A bounded pseudo-BCK algebra $(A, \leq, \rightarrow, \rightsquigarrow, 0, 1)$ has the (pDN) (*pseudo-Double Negation*) condition if it satisfies the following condition for all $x \in A$:

(pDN) $x^{-\sim} = x^{\sim -} = x$.

Example 1.12 Let $(G, \vee, \wedge, +, -, 0)$ be a linearly ordered ℓ-group and let $u \in G$, $u < 0$. Define:

$$x \rightarrow y := \begin{cases} 0 & \text{if } x \leq y \\ (u - x) \vee y & \text{if } x > y, \end{cases}$$

$$x \rightsquigarrow y := \begin{cases} 0 & \text{if } x \leq y \\ (-x + u) \vee y & \text{if } x > y. \end{cases}$$

Then $\mathcal{A} = ([u, 0], \rightarrow, \rightsquigarrow, 0 = u, 1 = 0)$ is a pseudo-BCK(pDN) algebra.

Example 1.13 Consider the structure $(A, \leq, \rightarrow, \rightsquigarrow, 0, 1)$ given in Fig. 1.5.

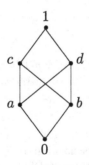

Fig. 1.5 Example of pseudo-BCK(pDN) algebra that is not lattice

The operations $\rightarrow$ and $\rightsquiggle$ on $A = \{0, a, b, c, d, 1\}$ are defined as follows:

$\rightarrow$	0	a	b	c	d	1
0	1	1	1	1	1	1
a	d	1	d	1	1	1
b	c	c	1	1	1	1
c	a	a	d	1	d	1
d	b	c	b	c	1	1
1	0	a	b	c	d	1

$\rightsquiggle$	0	a	b	c	d	1
0	1	1	1	1	1	1
a	c	1	c	1	1	1
b	d	d	1	1	1	1
c	b	d	b	1	d	1
d	a	a	c	c	1	1
1	0	a	b	c	d	1

Then $(A, \leq, \rightarrow, \rightsquiggle, 0, 1)$ is a pseudo-BCK(pDN) algebra that is not lattice.

Example 1.14 Any sup-commutative bounded pseudo-BCK algebra is a pseudo-BCK(pDN) algebra (see Corollary 1.3).

Proposition 1.15 *Let A be a pseudo-BCK(pDN) algebra. Then for all $x, y \in A$ the following hold*:

(*psbck-c_{44}*) $x \rightarrow y = y^- \rightsquiggle x^-, x \rightsquiggle y = y^\sim \rightarrow x^\sim$;
(*psbck-c_{45}*) $x^\sim \rightarrow y = y^- \rightsquiggle x$;
(*psbck-c_{46}*) $(x \rightarrow y^-)^\sim = (y \rightsquiggle x^\sim)^-$.

Proof

(*psbck-c_{44}*) This is a consequence of (*psbck-c_{19}*).
(*psbck-c_{45}*) This follows from (*psbck-c_{44}*) replacing x with $x^\sim$.
(*psbck-c_{46}*) Applying (*psbck-c_{16}*), condition (pDN), (*psbck-c_4*) and (*psbck-c_{44}*), for any $z \in A$ we have:

$$\left(x \rightarrow y^-\right)^\sim \leq z \quad \text{iff} \quad z^- \leq x \rightarrow y^- \quad \text{iff} \quad x \leq z^- \rightsquiggle y^- = y \rightarrow z \quad \text{iff}$$

$$y \leq x \rightsquiggle z = z^\sim \rightarrow x^\sim \quad \text{iff} \quad z^\sim \leq y \rightsquiggle x^\sim \quad \text{iff} \quad \left(y \rightsquiggle x^\sim\right)^- \leq z.$$

Thus $(x \rightarrow y^-)^\sim = (y \rightsquiggle x^\sim)^-$. $\qquad\qquad\square$

Proposition 1.16 *In every pseudo-BCK(pDN) lattice A we have*:

(*psbck-c_{47}*) $(x^- \vee y^-)^\sim = (x^\sim \vee y^\sim)^- = x \wedge y$.

Proof By ($psbck$-c_{41}) we have $(x^- \vee y^-)^\sim = x^{-\sim} \wedge y^{-\sim} = x \wedge y$.

Similarly, $(x^\sim \vee y^\sim)^- = x \wedge y$.   □

Proposition 1.17 *Let A be a pseudo-BCK(pDN) algebra and $x, y \in A$.*
If $x \wedge y$ exists, then $x^- \vee y^-, x^\sim \vee y^\sim$ exist and:

($psbck$-c_{48}) $(x \wedge y)^- = x^- \vee y^-, (x \wedge y)^\sim = x^\sim \vee y^\sim$.

Proof Since $x \wedge y \leq x, y$, we get $x^-, y^- \leq (x \wedge y)^-$. It follows that $(x \wedge y)^-$ is an
upper bound of x^- and y^-. Let u be an arbitrary upper bound of x^- and y^-, that is,
$x^-, y^- \leq u$. Since A satisfies condition (pDN), we get $u^\sim \leq x, y$, so $u^\sim \leq x \wedge y$.
Finally we get $(x \wedge y)^- \leq u$, so $(x \wedge y)^-$ is the least upper bound of x^- and y^-.
Thus $x^- \vee y^-$ exists and $(x \wedge y)^- = x^- \vee y^-$.

Similarly, $x^\sim \vee y^\sim$ exists and $(x \wedge y)^\sim = x^\sim \vee y^\sim$.   □

Corollary 1.5 *In every pseudo-BCK(pDN) lattice A we have:*

($psbck$-c_{49}) $(x^- \wedge y^-)^\sim = (x^\sim \wedge y^\sim)^- = x \vee y$.

Let $(A, \leq, \rightarrow, \rightsquigarrow, 0, 1)$ be a pseudo-BCK(pDN) algebra. Define on A a new operation $\cdot$ as follows:

$$x \cdot y := \left(x \rightarrow y^-\right)^\sim = \left(y \rightsquigarrow x^\sim\right)^- \quad \text{for all } x, y \in A.$$

This operation is well-defined by ($psbck$-c_{46}).

Lemma 1.5 *Let $(A, \leq, \rightarrow, \rightsquigarrow, 0, 1)$ be a pseudo-BCK(pDN) algebra. For any*
$x, y, z \in A$:

$$x \leq y \quad \text{implies} \quad x \cdot z \leq y \cdot z \text{ and } z \cdot x \leq z \cdot y.$$

Proof We have $x \cdot z = (x \rightarrow z^-)^\sim$ and $y \cdot z = (y \rightarrow z^-)^\sim$. From $x \leq y$ and ($psbck$-
c_1) we get $y \rightarrow z^- \leq x \rightarrow z^-$.

Applying ($psbck$-c_{16}) we have $(x \rightarrow z^-)^\sim \leq (y \rightarrow z^-)^\sim$, that is, $x \cdot z \leq y \cdot z$.

Similarly, $x \leq y$ implies $y \rightsquigarrow z^\sim \leq x \rightsquigarrow z^\sim$, so $(x \rightsquigarrow z^\sim)^- \leq (y \rightsquigarrow z^\sim)^-$, that is,
$z \cdot x \leq z \cdot y$.   □

Proposition 1.18 *Let $(A, \leq, \rightarrow, \rightsquigarrow, 0, 1)$ be a pseudo-BCK(pDN) algebra. The fol-*
lowing conditions are equivalent:

(pDN-C_1) $x \cdot y \leq z$ iff $x \leq y \rightarrow z$ iff $y \leq x \rightsquigarrow z$;
(pDN-C_2) $x \cdot y = \min\{z \mid x \leq y \rightarrow z\} = \min\{z \mid y \leq x \rightsquigarrow z\}$;
(pDN-C_3) $y \rightarrow z = \max\{x \mid x \cdot y \leq z\} = \max\{x \mid y \cdot x \leq z\}$.

Proof

(pDN-C_1) $\Rightarrow$ (pDN-C_2) Since $x \cdot y \leq x \cdot y$, by (pDN-C_1) we have $x \leq y \rightarrow (x \cdot y)$.
  If z satisfies $x \leq y \rightarrow z$, then by (pDN-C_1) we get $x \cdot y \leq z$. Thus $x \cdot y = \min\{z \mid$
  $x \leq y \rightarrow z\}$. Similarly, $x \cdot y = \min\{z \mid y \leq x \rightsquigarrow z\}$.

$(pDN\text{-}C_2) \Rightarrow (pDN\text{-}C_1)$ If $x \cdot y \leq z$, then applying $(psbck\text{-}c_{10})$ we have $y \to (x \cdot y) \leq y \to z$. Taking into consideration that, by $(pDN\text{-}C_2)$, $x \leq y \to (x \cdot y)$, we get $x \leq y \to z$.

$(pDN\text{-}C_1) \Rightarrow (pDN\text{-}C_3)$ Since $y \to z \leq y \to z$, by $(pDN\text{-}C_1)$ we get $(y \to z) \cdot y \leq z$. If x satisfies $x \cdot y \leq z$, then by $(pDN\text{-}C_1)$ we have $x \leq y \to z$. Thus $y \to z = \max\{x \mid x \cdot y \leq z\}$.

$(pDN\text{-}C_3) \Rightarrow (pDN\text{-}C_1)$ If $x \cdot y \leq z$, then by $(pDN\text{-}C_3)$ we get $x \leq y \to z$ and applying Lemma 1.5 it follows that $x \cdot y \leq (y \to z) \cdot y$. Since by $(pDN\text{-}C_3)$ we also have $(y \to z) \cdot y \leq z$, we get $x \cdot y \leq z$.

The rest of the proof is similar.   □

Proposition 1.19 *Let* $(A, \leq, \to, \rightsquigarrow, 0, 1)$ *be a pseudo-BCK(pDN) algebra. Then:*

$$x \to y = \left(x \cdot y^{\sim}\right)^{-}, \qquad x \rightsquigarrow y = \left(y^{-} \cdot x\right)^{\sim}.$$

Proof Since $x \cdot y = (x \to y^{-})^{\sim} = (y \rightsquigarrow x^{\sim})^{-}$ and $x^{\sim -} = x = x^{-\sim}$, we get

$$\left(x \cdot y^{\sim}\right)^{-} = \left(x \to y^{\sim -}\right)^{\sim -} = x \to y \quad \text{and} \quad \left(y^{-} \cdot x\right)^{\sim} = \left(x \rightsquigarrow y^{-\sim}\right)^{-\sim} = x \rightsquigarrow y.$$

□

Proposition 1.20 *Let* $(A, \leq, \to, \rightsquigarrow, 0, 1)$ *be a pseudo-BCK(pDN) algebra. Then for all* $x, y, z \in A$:

$$x \cdot y \leq z \quad \text{iff} \quad x \leq y \to z \quad \text{iff} \quad y \leq x \rightsquigarrow z.$$

Proof We will apply the properties $(psbck\text{-}c_4)$ and $(psbck\text{-}c_{44})$:

$$x \cdot y \leq z \quad \text{iff} \quad \left(x \to y^{-}\right)^{\sim} \leq z \quad \text{iff} \quad z^{-} \leq x \to y^{-} \quad \text{iff} \quad x \leq z^{-} \rightsquigarrow y^{-} = y \to z$$

and

$$x \cdot y \leq z \quad \text{iff} \quad \left(y \rightsquigarrow x^{\sim}\right)^{-} \leq z \quad \text{iff} \quad z^{\sim} \leq y \rightsquigarrow x^{\sim} \quad \text{iff} \quad y \leq z^{\sim} \to x^{\sim} = x \rightsquigarrow z.$$

□

Theorem 1.7 *A bounded pseudo-BCK algebra* $(A, \leq, \to, \rightsquigarrow, 0, 1)$ *satisfying condition* (pDN) *has the* (pP) *condition and:*

$$x \odot y = \left(x \to y^{-}\right)^{\sim} = \left(y \rightsquigarrow x^{\sim}\right)^{-}, \qquad x \to y = \left(x \odot y^{\sim}\right)^{-},$$
$$x \rightsquigarrow y = \left(y^{-} \odot x\right)^{\sim}.$$

Proof By Propositions 1.20 and 1.18, it follows that A satisfies the (pP) condition. If we let $x \odot y = \min\{z \mid x \leq y \to z\} = \min\{z \mid y \rightsquigarrow z = x \cdot y\}$, it follows that $x \odot y = x \cdot y = (x \to y^{-})^{\sim} = (y \rightsquigarrow x^{\sim})^{-}$.   □

Theorem 1.8 *Let* $(A, \leq, \rightarrow, \rightsquigarrow, 0, 1)$ *be a bounded pseudo-BCK algebra. The following are equivalent*:

(a) *A has the* (pDN) *condition*;
(b) $x \rightarrow y = y^- \rightsquigarrow x^-$ *and* $x \rightsquigarrow y = y^\sim \rightarrow x^\sim$;
(c) $x^\sim \rightarrow y = y^- \rightsquigarrow x$ *and* $x^- \rightsquigarrow y = y^\sim \rightarrow x$;
(d) $x^- \leq y$ *implies* $y^\sim \leq x$ *and* $x^\sim \leq y$ *implies* $y^- \leq x$.

Proof

(a) $\Rightarrow$ (b) By $(psbck\text{-}c_{19})$ we have:

$$x \rightarrow y = x \rightarrow y^{-\sim} = y^- \rightsquigarrow x^- \quad \text{and} \quad x \rightsquigarrow y = x \rightsquigarrow y^{\sim-} = y^\sim \rightarrow x^\sim.$$

(b) $\Rightarrow$ (c) By $(psbck\text{-}c_{17})$ we have: $x^\sim \rightarrow y^{-\sim} = y^- \rightsquigarrow x^{\sim-}$.
  Applying (b) we get: $x^\sim \rightarrow y = y^- \rightsquigarrow x^{\sim-}$ and $y^- \rightsquigarrow x = x^\sim \rightarrow y^{-\sim}$.
  Thus $x^\sim \rightarrow y = y^- \rightsquigarrow x$. Similarly, $x^- \rightsquigarrow y = y^\sim \rightarrow x$.
(c) $\Rightarrow$ (d) If $x^- \leq y$, then $x^- \rightsquigarrow y = 1$. Applying (c) we get $y^\sim \rightarrow x = 1$, that is, $y^\sim \leq x$. Similarly, $x^\sim \leq y$ implies $y^- \leq x$.
(d) $\Rightarrow$ (a) From $x^- \leq x^-$ and (d) we have $x^{-\sim} \leq x$. Taking into consideration $(psbck\text{-}c_{14})$ we get $x^{-\sim} = x$. Similarly, $x^{\sim-} = x$.

Thus A satisfies the (pDN) condition. $\qquad\qquad\qquad\qquad\qquad\qquad\qquad\qquad\square$

Theorem 1.9 *If* $(A, \leq, \rightarrow, \rightsquigarrow, 0, 1)$ *is a pseudo-BCK(pDN) algebra, then the following are equivalent*:

(a) $(A, \leq)$ *is a meet-semilattice*;
(b) $(A, \leq)$ *is a join-semilattice*;
(c) $(A, \leq)$ *is a lattice*.

Proof

(a) $\Rightarrow$ (b) Consider $x, y \in A$. Since A is a meet-semilattice, $x^- \wedge y^-$ exists. Applying $(psbck\text{-}c_{48})$, it follows that $x^{-\sim} \vee y^{-\sim}$ exists, that is, $x \vee y$ exists.
  Thus A is a join-semilattice.
(b) $\Rightarrow$ (c) Because A is a join-semilattice it follows that $x^- \vee y^-$ exists for all $x, y \in A$. Hence by $(psbck\text{-}c_{47})$, $x \wedge y = (x^- \vee y^-)^\sim$. Thus $x \wedge y$ exists, so A is a lattice.
(c) $\Rightarrow$ (a) This is obvious, since A is a lattice. $\qquad\qquad\qquad\qquad\qquad\qquad\square$

Proposition 1.21 *In every pseudo-BCK(pDN) lattice the following hold*:

(1) $y \rightarrow (\bigwedge_{i \in I} x_i) = \bigwedge_{i \in I} (y \rightarrow x_i)$;
(2) $y \rightsquigarrow (\bigwedge_{i \in I} x_i) = \bigwedge_{i \in I} (y \rightsquigarrow x_i)$.

Proof This is a consequence of Proposition 1.14(2), since every pseudo-BCK(pDN) lattice is a pseudo-BCK(pP) lattice. We give another proof for this result.

Applying (*psbck-c$_{17}$*) and Proposition 1.3 we have:

$$y \rightarrow \left(x_1^- \vee x_2^-\right)^\sim = \left(x_1^- \vee x_2^-\right) \rightsquigarrow y^- = \left(x_1^- \rightsquigarrow y^-\right) \wedge \left(x_2^- \rightsquigarrow y^-\right)$$
$$= \left(y \rightarrow x_1^{-\sim}\right) \wedge \left(y \rightarrow x_2^{-\sim}\right) = (y \rightarrow x_1) \wedge (y \rightarrow x_2).$$

By (*psbck-c$_{47}$*) we have $\left(x_1^- \vee x_2^-\right)^\sim = x_1 \wedge x_2$.

Hence $y \rightarrow (x_1 \wedge x_2) = (y \rightarrow x_1) \wedge (y \rightarrow x_2)$.

By induction we get assertion (1).

(2) Similar to (1). □

Remark 1.7 If the pseudo-BCK lattice A does not have the property (pDN), then the results of Proposition 1.21 do not hold. Indeed, in the pseudo-BCK lattice A from Example 1.3 we have $a \rightarrow (a \wedge b) = a \rightarrow 0 = 0$, while $(a \rightarrow a) \wedge (a \rightarrow b) = 1 \wedge b = b$. Thus $a \rightarrow (a \wedge b) \neq (a \rightarrow a) \wedge (a \rightarrow b)$.

Proposition 1.22 *In every pseudo-BCK(pDN) lattice the following conditions are equivalent:*

(C$_1$) $(x \wedge y) \rightarrow z = (x \rightarrow z) \vee (y \rightarrow z)$ *and* $(x \wedge y) \rightsquigarrow z = (x \rightsquigarrow z) \vee (y \rightsquigarrow z)$;

(C$_2$) $z \rightarrow (x \vee y) = (z \rightarrow x) \vee (z \rightarrow y)$ *and* $z \rightsquigarrow (x \vee y) = (z \rightsquigarrow x) \vee (z \rightsquigarrow y)$.

Proof

(C$_1$) $\Rightarrow$ (C$_2$) By the second identity from (C$_1$) we have:

$$\left(x^- \wedge y^-\right) \rightsquigarrow z^- = \left(x^- \rightsquigarrow z^-\right) \vee \left(y^- \rightsquigarrow z^-\right).$$

Applying (*psbck-c$_{17}$*) and (*psbck-c$_{49}$*) we get:

$$\left(x^- \wedge y^-\right) \rightsquigarrow z^- = z \rightarrow \left(x^- \wedge y^-\right)^\sim = z \rightarrow (x \vee y).$$

By (*psbck-c$_{44}$*) we have:

$$\left(x^- \rightsquigarrow z^-\right) \vee \left(y^- \rightsquigarrow z^-\right) = (z \rightarrow x) \vee (z \rightarrow y).$$

Thus $z \rightarrow (x \vee y) = (z \rightarrow x) \vee (z \rightarrow y)$.

Similarly, from the first identity of (C$_1$) we get the second identity from (C$_2$).

(C$_2$) $\Rightarrow$ (C$_1$) By the second identity from (C$_2$) we get:

$$z^- \rightsquigarrow \left(x^- \vee y^-\right) = \left(z^- \rightsquigarrow x^-\right) \vee \left(z^- \rightsquigarrow y^-\right).$$

Applying (*psbck-c$_{45}$*) and (*psbck-c$_{44}$*) we have:

$$\left(x^- \vee y^-\right)^\sim \rightarrow z = (x \rightarrow z) \vee (y \rightarrow z).$$

By (*psbck-c$_{47}$*), it follows that $(x \wedge y) \rightarrow z = (x \rightarrow z) \vee (y \rightarrow z)$.

Similarly, from the first identity of (C$_2$) we get the second identity from (C$_1$). □

Remark 1.8 The class of pseudo-BCK(pDN) lattices satisfying the conditions (C_1) and (C_2) is not empty. Indeed, one can see that every pseudo-MV algebra satisfies these conditions (see Chap. 4).

Theorem 1.10 *Let A be a pseudo-BCK lattice such that at least one of the following identities holds:*

(C_1^1) $(x \wedge y) \rightarrow z = (x \rightarrow z) \vee (y \rightarrow z)$;
(C_1^2) $(x \wedge y) \rightsquigarrow z = (x \rightsquigarrow z) \vee (y \rightsquigarrow z)$.

Then $(A, \wedge, \vee)$ is distributive.

Proof Let $u = (x \vee y) \wedge (x \vee z)$. Obviously, $x \leq u$ and $y \wedge z \leq u$.

It follows that u is an upper bound of x and $y \wedge z$.

Let v be an arbitrary upper bound of x and $y \wedge z$, that is, $x \leq v$ and $y \wedge z \leq v$. By Proposition 1.3 we get:

$$(x \vee y) \rightarrow v = (x \rightarrow v) \wedge (y \rightarrow v) = y \rightarrow v \quad \text{and}$$

$$(x \vee z) \rightarrow v = (x \rightarrow v) \wedge (z \rightarrow v) = z \rightarrow v.$$

If the identity (C_1^1) is satisfied, then we have:

$$\left[(x \vee y) \rightarrow v \right] \vee \left[(x \vee z) \rightarrow v \right] = (y \rightarrow v) \vee (z \rightarrow v) = (y \wedge z) \rightarrow v = 1 \quad \text{and}$$

$$\left[(x \vee y) \wedge (x \vee z) \right] \rightarrow v = \left[(x \vee y) \rightarrow v \right] \vee \left[(x \vee z) \rightarrow v \right] = 1,$$

that is, $(x \vee y) \wedge (x \vee z) \leq v$, so $u \leq v$.

Thus u is the least upper bound of x and $y \wedge z$.

We conclude that $x \vee (y \wedge z) = (x \vee y) \wedge (x \vee z)$, that is, $(A, \leq)$ is distributive.

Similarly, if (C_1^2) is satisfied, we get the same conclusion. □

Corollary 1.6 *If $(A, \wedge, \vee, \rightarrow, \rightsquigarrow, 0, 1)$ is a pseudo-BCK(pDN) lattice satisfying (C_1) or (C_2), then $(A, \wedge, \vee)$ is distributive.*

Proof If A satisfies C_1, then it satisfies (C_1^1) and (C_1^2), so A is distributive.

If A satisfies (C_2), since by Proposition 1.22 (C_1) is equivalent to (C_2), it follows that A is distributive. □

1.4 Good Pseudo-BCK Algebras

Definition 1.10 A bounded pseudo-BCK algebra A is called *good* if it satisfies the following condition for all $x \in A$:

(good) $x^{-\sim} = x^{\sim-}$.

We remark that any pseudo-BCK(pDN) algebra is good.

Example 1.15 The bounded pseudo-BCK algebras from Examples 1.4 and 1.5 are good pseudo-BCK algebras.

Remark 1.9 Any bounded pseudo-BCK algebra can be extended to a good one. Indeed, consider the bounded pseudo-BCK algebra $\mathcal{A} = (A, \leq, \rightarrow, \rightsquigarrow, 0, 1)$ and an element $0_1 \notin A$. Consider a new pseudo-BCK algebra $\mathcal{A}_1 = (A_1, \leq, \rightarrow_1, \rightsquigarrow_1, 0_1, 1)$, where $A_1 = A \cup \{0_1\}$ and the operations $\rightarrow_1$ and $\rightsquigarrow_1$ are defined as follows:

$$x \rightarrow_1 y := \begin{cases} x \rightarrow y & \text{if } x, y \in A \\ 1 & \text{if } x = 0_1, y \in A_1 \\ 0_1 & \text{if } x \in A, y = 0_1, \end{cases}$$

$$x \rightsquigarrow_1 y := \begin{cases} x \rightsquigarrow y & \text{if } x, y \in A \\ 1 & \text{if } x = 0_1, y \in A_1 \\ 0_1 & \text{if } x \in A, y = 0_1. \end{cases}$$

One can easily check that $\mathcal{A}$ is a subalgebra of $\mathcal{A}_1$ and $\mathcal{A}_1$ is a good pseudo-BCK algebra.

Example 1.16 Consider the pseudo-BCK algebra $\mathcal{A}$ from Example 1.3. Since $(a^-)^\sim = 1$ and $(a^\sim)^- = a$, it follows that $\mathcal{A}$ is not good. $\mathcal{A}$ can be extended to the good pseudo-BCK algebra $\mathcal{A}_1 = (A_1, \leq, \rightarrow, \rightsquigarrow, 0, 1)$, where $A_1 = \{0, a, b, c, d, 1\}$ (in the construction given in Remark 1.9 we replaced c by d, b by c, a by b, 0 by a and 0_1 by 0, so $0 < a < b, c < d < 1$ and b, c are incomparable). The operations $\rightarrow$ and $\rightsquigarrow$ are defined as follows:

$\rightarrow$	0	a	b	c	d	1
0	1	1	1	1	1	1
a	0	1	1	1	1	1
b	0	a	1	c	1	1
c	0	b	b	1	1	1
d	0	a	b	c	1	1
1	0	a	b	c	d	1

$\rightsquigarrow$	0	a	b	c	d	1
0	1	1	1	1	1	1
a	0	1	1	1	1	1
b	0	c	1	c	1	1
c	0	a	b	1	1	1
d	0	a	b	c	1	1
1	0	a	b	c	d	1

One can easily check that $\mathcal{A}_1$ is a good pseudo-BCK algebra.

Moreover, we can see that

$$\min\{z \mid c \leq b \rightarrow z\} = \min\{b, c, d, 1\} \quad \text{and} \quad \min\{z \mid b \leq c \rightsquigarrow z\} = \min\{b, c, d, 1\}$$

do not exist. Thus $c \odot b$ does not exist, so $\mathcal{A}_1$ does not satisfy the (pP) condition.

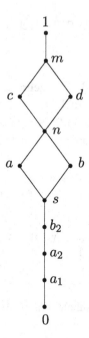

Fig. 1.6 Example of good pseudo-BCK(pP) lattice

Example 1.17 Consider the pseudo-BCK lattice $\mathcal{A}$ from Example 1.9. Since $(a_1^-)^\sim = a_2$ and $(a_1^\sim)^- = a_1$, it follows that $\mathcal{A}$ is not good. $\mathcal{A}$ can be extended to the good pseudo-BCK algebra (see [182]) $\mathcal{A}_1 = (A_1, \leq, \rightarrow, \rightsquigarrow, 0, 1)$, where $A_1 = \{0, a_1, a_2, b_2, s, a, b, n, c, d, m, 1\}$ with $0 < a_1 < a_2 < b_2 < s < a, b < n < c, d < m < 1$ (see Fig. 1.6).

The operations $\rightarrow$ and $\rightsquigarrow$ are defined as follows:

$\rightarrow$	0	a_1	a_2	b_2	s	a	b	n	c	d	m	1
0	1	1	1	1	1	1	1	1	1	1	1	1
a_1	0	1	1	1	1	1	1	1	1	1	1	1
a_2	0	a_2	1	1	1	1	1	1	1	1	1	1
b_2	0	a_2	a_2	1	1	1	1	1	1	1	1	1
s	0	a_1	a_2	b_2	1	1	1	1	1	1	1	1
a	0	a_1	a_2	b_2	m	1	m	1	1	1	1	1
b	0	a_1	a_2	b_2	m	m	1	1	1	1	1	1
n	0	a_1	a_2	b_2	m	m	m	1	1	1	1	1
c	0	a_1	a_2	b_2	m	m	m	m	1	m	1	1
d	0	a_1	a_2	b_2	m	m	m	m	m	1	1	1
m	0	a_1	a_2	b_2	m	m	m	m	m	m	1	1
1	0	a_1	a_2	b_2	s	a	b	n	c	d	m	1

$\rightsquigarrow$	0	a_1	a_2	b_2	s	a	b	n	c	d	m	1
0	1	1	1	1	1	1	1	1	1	1	1	1
a_1	0	1	1	1	1	1	1	1	1	1	1	1
a_2	0	b_2	1	1	1	1	1	1	1	1	1	1
b_2	0	a_1	a_2	1	1	1	1	1	1	1	1	1
s	0	a_1	a_2	b_2	1	1	1	1	1	1	1	1
a	0	a_1	a_2	b_2	m	1	m	1	1	1	1	1
b	0	a_1	a_2	b_2	m	m	1	1	1	1	1	1
n	0	a_1	a_2	b_2	m	m	m	1	1	1	1	1
c	0	a_1	a_2	b_2	m	m	m	m	1	m	1	1
d	0	a_1	a_2	b_2	m	m	m	m	m	1	1	1
m	0	a_1	a_2	b_2	m	m	m	m	m	m	1	1
1	0	a_1	a_2	b_2	s	a	b	n	c	d	m	1

One can easily check that $\mathcal{A}_1 = (A_1, \leq, \rightarrow, \rightsquigarrow, 0, 1)$ is a good pseudo-BCK(pP) lattice. The operation $\odot$ is given by the following table:

$\odot$	0	a_1	a_2	b_2	s	a	b	n	c	d	m	1
0	0	0	0	0	0	0	0	0	0	0	0	0
a_1	0	a_1	a_1	a_1	a_1	a_1	a_1	a_1	a_1	a_1	a_1	a_1
a_2	0	a_1	a_1	a_1	a_2	a_2	a_2	a_2	a_2	a_2	a_2	a_2
b_2	0	a_1	a_2	b_2	b_2	b_2	b_2	b_2	b_2	b_2	b_2	b_2
s	0	a_1	a_2	b_2	s	s	s	s	s	s	s	s
a	0	a_1	a_2	b_2	s	s	s	s	s	s	s	a
b	0	a_1	a_2	b_2	s	s	s	s	s	s	s	b
n	0	a_1	a_2	b_2	s	s	s	s	s	s	s	n
c	0	a_1	a_2	b_2	s	s	s	s	s	s	s	c
d	0	a_1	a_2	b_2	s	s	s	s	s	s	s	d
m	0	a_1	a_2	b_2	s	s	s	s	s	s	s	m
1	0	a_1	a_2	b_2	s	a	b	n	c	d	m	1

Proposition 1.23 *In any good pseudo-BCK(pP) algebra the following properties hold*:

(1) $(x^{\sim} \odot y^{\sim})^{-} = (x^{-} \odot y^{-})^{\sim}$;
(2) $x^{-\sim} \odot y^{-\sim} \leq (x \odot y)^{-\sim}$.

Proof

(1) Applying ($psbck$-c_{37}) and ($psbck$-c_{17}) we have:

$$\left(x^{\sim} \odot y^{\sim}\right)^{-} = x^{\sim} \rightarrow y^{\sim -} = x^{\sim} \rightarrow y^{-\sim} = y^{-} \rightsquigarrow x^{\sim -}$$
$$= y^{-} \rightsquigarrow x^{-\sim} = \left(x^{-} \odot y^{-}\right)^{\sim}.$$

(2) Because the pseudo-BCK(pP) algebra is good, by ($psbck$-c_{25}), we have:

$$(x \odot y)^{-\sim} = (x \odot y)^{\sim -} \geq x^{\sim -} \odot \left(x^{\sim -} \leadsto (x \odot y)^{\sim -}\right)$$

$$= x^{\sim -} \odot \left(x^{\sim -} \leadsto (x \odot y)^{-\sim}\right)$$

$$= x^{\sim -} \odot \left(x^{\sim -} \leadsto (x \rightarrow y^-)^{\sim}\right).$$

Applying (*psbck-c*$_{30}$) we get:

$$x^{\sim -} \leadsto \left(x \rightarrow y^-\right)^{\sim} = x^{\sim -} \leadsto \left((x \rightarrow y^-) \leadsto 0\right) = (x \rightarrow y^-) \odot x^{\sim -} \leadsto 0$$

$$= \left[(x \rightarrow y^-) \odot x^{\sim -}\right]^{\sim} = \left[(x^{\sim -} \rightarrow y^-) \odot x^{\sim -}\right]^{\sim}$$

(by (*psbck-c*$_{19}$) replacing y with y^- we have $x \rightarrow y^- = x^{-\sim} \rightarrow y^-$).

Applying (*psbck-c*$_{25}$) we have $(x^{\sim -} \rightarrow y^-) \odot x^{\sim -} \leq y^-$, hence $[(x^{\sim -} \rightarrow y^-) \odot x^{\sim -}]^{\sim} \geq y^{-\sim}$. Thus

$$(x \odot y)^{-\sim} \geq x^{\sim -} \odot \left(x^{\sim -} \leadsto (x \rightarrow y^-)^{\sim}\right) = x^{\sim -} \odot \left[(x^{\sim -} \rightarrow y^-) \odot x^{\sim -}\right]^{\sim}$$

$$\geq x^{\sim -} \odot y^{-\sim}. \qquad \square$$

Proposition 1.24 *Let* $(A, \leq, \rightarrow, \leadsto, 0, 1)$ *be a good pseudo-BCK algebra. We define a binary operation* $\oplus$ *on* A *by* $x \oplus y := y^{\sim} \rightarrow x^{\sim -}$. *Then for all* $x, y \in A$ *the following hold*:

(1) $x \oplus y = x^- \leadsto y^{\sim -}$;
(2) $x, y \leq x \oplus y$;
(3) $x \oplus 0 = 0 \oplus x = x^{\sim -}$;
(4) $x \oplus 1 = 1 \oplus x = 1$;
(5) $(x \oplus y)^{-\sim} = x \oplus y = x^{-\sim} \oplus y^{-\sim}$;
(6) $\oplus$ *is associative*.

Proof

(1) This follows by (*psbck-c*$_{19}$), second identity, replacing x with x^-.
(2) Since $x \leq x^{\sim -} \leq y^{\sim} \rightarrow x^{\sim -}$, it follows that $x \leq x \oplus y$.
   Similarly, $y \leq y^{\sim -} \leq x^- \leadsto y^{\sim -}$, so $y \leq x \oplus y$.
(3) $x \oplus 0 = 0^{\sim} \rightarrow x^{\sim -} = 1 \rightarrow x^{\sim -} = x^{\sim -}$.
   Similarly, $0 \oplus x = x^{\sim} \rightarrow 0^{\sim -} = x^{\sim} \rightarrow 0 = x^{\sim -}$.
(4) $1 \oplus x = x^{\sim} \rightarrow 1^{\sim -} = x^{\sim} \rightarrow 1 = 1$. Similarly, $x \oplus 1 = 1$.
(5) Applying (*psbck-c*$_{21}$), we get:

$$(x \oplus y)^{-\sim} = \left(y^{\sim} \rightarrow x^{\sim -}\right)^{-\sim} = y^{\sim} \rightarrow x^{\sim -} = x \oplus y.$$

We also have: $x^{-\sim} \oplus y^{-\sim} = (y^{-\sim})^{\sim} \rightarrow (x^{-\sim})^{-\sim} = y^{\sim} \rightarrow x^{-\sim} = x \oplus y$.
(6) Applying (*psbck-c*$_{21}$) and (*psbck-c*$_3$) we get:

$$(x \oplus y) \oplus z = \left(x^- \leadsto y^{\sim -}\right) \oplus z = z^{\sim} \rightarrow \left(x^- \leadsto y^{\sim -}\right)^{\sim -}$$

$$= z^{\sim} \rightarrow \left(x^- \leadsto y^{\sim -}\right) = x^- \leadsto \left(z^{\sim} \rightarrow y^{\sim -}\right) = x^- \leadsto (y \oplus z)$$

$$= x^- \leadsto (y \oplus z)^{\sim -} = x \oplus (y \oplus z). \qquad \square$$

Proposition 1.25 *If* $(A, \leq, \rightarrow, \rightsquigarrow, 0, 1)$ *is a good pseudo-BCK(pP) algebra, then*

$$x \oplus y = \left(y^- \odot x^-\right)^\sim = \left(y^\sim \odot x^\sim\right)^-.$$

Proof This follows by applying (*psbck-c37*).   □

As in [242] for the case of bounded $R\ell$-monoids, a good pseudo-BCK(pP) algebra A which satisfies the identity $(x \odot y)^{-\sim} = x^{-\sim} \odot y^{-\sim}$ for all $x, y \in A$ will be called a *normal* pseudo-BCK(pP) algebra.

If A is a pseudo-BCK(pP) algebra, then for any $n \in \mathbb{N}$, $x \in A$ we put $x^0 = 1$ and $x^{n+1} = x^n \odot x = x \odot x^n$.

If A is a good pseudo-BCK algebra, then for any $n \in \mathbb{N}$, $x \in A$ we put $0x = 0$ and $(n + 1)x = nx \oplus x = x \oplus nx$ (the latter equality is a consequence of Proposition 1.24(6)).

Proposition 1.26 *If* $(A, \leq, \rightarrow, \rightsquigarrow, 0, 1)$ *is a normal pseudo-BCK(pP) algebra, then the following hold for all $x, y \in A$ and $n \in \mathbb{N}$:*

(1) $(x \odot y)^- = y^- \oplus x^-$ *and* $(x \odot y)^\sim = y^\sim \oplus x^\sim$;
(2) $((x \odot y)^n)^- = n(y^- \oplus x^-)$ *and* $((x \odot y)^n)^\sim = n(y^\sim \oplus x^\sim)$;
(3) $(x^n)^- = nx^-$ *and* $(x^n)^\sim = nx^\sim$.

Proof

(1) Applying Proposition 1.25 we have:

$$(x \odot y)^- = (x \odot y)^{-\sim-} = \left(x^{-\sim} \odot y^{-\sim}\right)^- = y^- \oplus x^-;$$

$$(x \odot y)^\sim = (x \odot y)^{\sim-\sim} = \left(x^{\sim-} \odot y^{\sim-}\right)^\sim = y^\sim \oplus x^\sim.$$

(2) For $n = 2$, applying (1) we get:

$$\begin{aligned}
\left((x \odot y)^2\right)^- &= \left[(x \odot y) \odot (x \odot y)\right]^- = \left[(x \odot y) \odot (x \odot y)\right]^{-\sim-} \\
&= \left[(x \odot y)^{-\sim} \odot (x \odot y)^{-\sim}\right]^- = (x \odot y)^{-\sim-} \oplus (x \odot y)^{-\sim-} \\
&= (x \odot y)^- \oplus (x \odot y)^- = \left(y^- \oplus x^-\right) \oplus \left(y^- \oplus x^-\right) \\
&= 2\left(y^- \oplus x^-\right).
\end{aligned}$$

By induction we get $((x \odot y)^n)^- = n(y^- \oplus x^-)$.
Similarly, $((x \odot y)^n)^\sim = n(y^\sim \oplus x^\sim)$.
(3) This follows from (2) for $y = 1$.   □

Definition 1.11 Let A be a good pseudo-BCK algebra. If $x, y \in A$, we say that x is *orthogonal* to y, denoted $x \perp y$, iff $x^{-\sim} \leq y^\sim$. We can define a partial operation $+$ on A, namely if $x \perp y$, then $x + y := x \oplus y$.

Lemma 1.6 *Let A be a good pseudo-BCK algebra. Then the following properties hold for all $x, y \in A$:*

(1) $x \perp y$ *iff* $y^{-\sim} \leq x^{-}$;
(2) $x \perp x^{-}$ *and* $x + x^{-} = 1$;
(3) $x^{\sim} \perp x$ *and* $x^{\sim} + x = 1$;
(4) $x \perp 0$ *and* $x + 0 = x^{-\sim}$;
(5) $0 \perp x$ *and* $0 + x = x^{\sim -}$;
(6) *if* $x \leq y$, *then* $x \perp y^{-}$, $y^{\sim} \perp x$ *and* $x + y^{-} = y \rightarrow x^{-\sim}$, $y^{\sim} + x = y \leadsto x^{\sim -}$;
(7) *if* $x \perp y$, *then* $x^{-\sim} \perp y^{-\sim}$;
(8) $x^{-} \perp y^{-}$ *iff* $x^{\sim} \perp y^{\sim}$.

Proof

(1) $x \perp y$ iff $x^{-\sim} \leq y^{\sim}$ iff $y^{-\sim} \leq x^{-\sim -} = x^{-}$.
(2) Since $x^{-\sim} \leq x^{-\sim} = (x^{-})^{\sim}$, it follows that $x \perp x^{-}$ and

$$x + x^{-} = x^{-\sim} \rightarrow x^{-\sim} = 1.$$

(3) Similarly, from $x^{-\sim} \leq x^{-\sim} = (x^{\sim})^{-}$ we get that $x^{\sim} \perp x$ and

$$x^{\sim} + x = x^{\sim} \rightarrow x^{\sim -\sim} = x^{\sim} \rightarrow x^{\sim} = 1.$$

(4) Since $x^{-\sim} \leq 1 = 0^{\sim}$, it follows that $x \perp 0$ and

$$x + 0 = 0^{\sim} \rightarrow x^{\sim -} = 1 \rightarrow x^{\sim -} = x^{-\sim}.$$

(5) Since $x^{-\sim} \leq 1 = 0^{-}$, it follows that $0 \perp x$ and

$$0 + x = x^{\sim} \rightarrow 0^{\sim -} = x^{\sim} \rightarrow 0 = x^{\sim -}.$$

(6) Since $x \leq y$, we have $y^{-} \leq x^{-}$, that is, $(y^{-})^{-\sim} \leq x^{-}$, so $x \perp y^{-}$.
    Moreover, $x + y^{-} = y^{-\sim} \rightarrow x^{-\sim} = y \rightarrow x^{-\sim}$ (by *(psbck-c_{19})*).
    Similarly we have $y^{\sim} \leq x^{\sim}$, so $(y^{\sim})^{-\sim} \leq x^{\sim}$, that is, $y^{\sim} \perp x$ and $y^{\sim} + x = x^{\sim} \rightarrow y^{\sim -\sim} = x^{\sim} \rightarrow y^{\sim} = y \leadsto x^{\sim -}$ (by *(psbck-c_{17})*).
(7) $x \perp y$ implies $x^{-\sim} \leq y^{\sim}$, so $(x^{-\sim})^{-\sim} \leq (y^{-\sim})^{\sim}$, that is, $x^{-\sim} \perp y^{-\sim}$.
(8) $x^{-} \perp y^{-}$ iff $x^{--\sim} \leq y^{-\sim}$ iff $x^{-} \leq y^{-\sim}$ iff $(y^{\sim})^{-\sim} \leq (x^{\sim})^{-}$ iff $x^{\sim} \perp y^{\sim}$. $\square$

Proposition 1.27 *In a good pseudo-BCK(pP) algebra A the following are equivalent for all $x, y \in A$:*

(a) $x \perp y$;
(b) $y^{-\sim} \odot x^{-\sim} = 0$.

Proof

$$x \perp y \quad \text{iff} \quad y^{-\sim} \leq x^{-} = x^{-\sim -} \quad \text{iff} \quad y^{-\sim} \leq x^{-\sim} \rightarrow 0 \quad \text{iff}$$
$$y^{-\sim} \odot x^{-\sim} \leq 0 \quad \text{iff} \quad y^{-\sim} \odot x^{-\sim} = 0. \qquad \square$$

Proposition 1.28 *Let A be a good pseudo-BCK(pP) algebra and $x, y \in A$ such that $x \perp y$. Then the following hold*:

(1) $y \odot x = 0$;
(2) $x^n \perp y^m$ *for all* $n, m \in \mathbb{N}$.

Proof

(1) Since $x \perp y$, then $y^{-\sim} \odot x^{-\sim} = 0$. Taking into consideration that $x \leq x^{-\sim}$ and $y \leq y^{-\sim}$, we get $y \odot x \leq y^{-\sim} \odot x^{-\sim} = 0$, so $y \odot x = 0$.
(2) From $x^n \leq x$ and $y^m \leq y$, it follows that $(x^n)^{-\sim} \leq x^{-\sim}$ and $(y^m)^{-\sim} \leq y^{-\sim}$. Hence $(y^m)^{-\sim} \odot (x^n)^{-\sim} \leq y^{-\sim} \odot x^{-\sim} = 0$, so $(y^m)^{-\sim} \odot (x^n)^{-\sim} = 0$ and applying Proposition 1.27 we get that $x^n \perp y^m$. $\square$

Definition 1.12 Let A be a good pseudo-BCK algebra.

(1) We say that x is *N-orthogonal* to y, denoted $x \perp_{no} y$, if $x^- \leq y^{-\sim}$.
(2) A has the *strong orthogonality* property ((SO) for short), if $x \perp y$ implies $x \perp_{no} y$ for all $x, y \in A$ such that $x \neq 0$ and $y \neq 0$.

Remark 1.10 If A is a good pseudo-BCK algebra, then:

(1) $x \perp_{no} y$ iff $y^\sim \leq x^{-\sim}$;
(2) $x \perp_{no} y$ iff $x^- \perp y^-$ (and according to Proposition 1.6(8) we also have $x \perp_{no} y$ iff $x^\sim \perp y^\sim$);
(3) $x \perp_{no} 1$ and $1 \perp_{no} x$ for all $x \in A$;
(4) $x \perp_{no} 0$ iff $x^- = x^\sim = 0$;
(5) $0 \perp_{no} x$ iff $x^- = x^\sim = 0$.

Remark 1.11 Let A be a good pseudo-BCK algebra.
   Then A has the (SO) property iff $x^- = y^{-\sim}$ iff $y^\sim = x^{-\sim}$ for all $x \neq 0$ and $y \neq 0$.

1.5 Deductive Systems and Congruences

Definition 1.13 Let A be a pseudo-BCK algebra. The subset $D \subseteq A$ is called a *deductive system* of A if it satisfies the following conditions:

(DS_1) $1 \in D$;
(DS_2) for all $x, y \in A$, if $x, x \to y \in D$, then $y \in D$.

Lemma 1.7 *Let A be a pseudo-BCK algebra. Then $D \subseteq A$ with $1 \in D$ is a deductive system of A if and only if it satisfies the condition*:

(DS_2') *for all* $x, y \in A$, *if* $x, x \rightsquigarrow y \in D$, *then* $y \in D$.

Proof Assume that $D \subseteq A$ is a deductive system of A.

First of all we observe that, if $x \in D$, $y \in A$ such that $x \leq y$, then $x \to y = 1 \in D$, hence by (DS_2), $y \in D$.

If $x, x \rightsquigarrow y \in D$, then according to $(psBCK_2)$ we have $x \leq (x \rightsquigarrow y) \to y$.

Hence $(x \rightsquigarrow y) \to y \in D$ and by (DS_2) we get $y \in D$.

Thus (DS_2) implies (DS_2'). Similarly, (DS_2') implies (DS_2). $\square$

The deductive system D of a pseudo-BCK algebra A is called *proper* if $D \neq A$. Obviously, the deductive system D is proper iff $0 \notin D$ iff there is no $x \in A$ such that $x, x^- \in D$ iff there is no $x \in A$ such that $x, x^\sim \in D$.

Definition 1.14 A deductive system D of a pseudo-BCK algebra A is said to be *normal* if it satisfies the condition:

(DS_3) for all $x, y \in A$, $x \to y \in D$ iff $x \rightsquigarrow y \in D$.

Remark 1.12 In [208] a normal deductive system is called a *compatible deductive system*.

We will denote by $\mathcal{DS}(A)$ the set of all deductive systems and by $\mathcal{DS}_n(A)$ the set of all normal deductive systems of a pseudo-BCK algebra A.

Obviously, $\{1\}$, $A \in \mathcal{DS}(A)$, $\mathcal{DS}_n(A)$ and $\mathcal{DS}_n(A) \subseteq \mathcal{DS}(A)$.

Definition 1.15 Let A be a pseudo-BCK(pP) algebra. A subset $\emptyset \neq F \subseteq A$ is called a *filter* of A if it satisfies the following conditions:

(F_1) $x, y \in F$ implies $x \odot y \in F$;
(F_2) $x \in F$, $y \in A$, $x \leq y$ implies $y \in F$.

Proposition 1.29 *Let A be pseudo-BCK(pP) algebra and F a nonempty subset of A. Then the following are equivalent*:

(a) *F is a deductive system of A;*
(b) *F is a filter of A.*

Proof

(a) $\Rightarrow$ (b) Assume that F is a deductive system of A.

Consider $x, y \in F$. According to $(psbck\text{-}c_{30})$, and taking into consideration (DS_1), we have $x \to (y \to x \odot y) = x \odot y \to x \odot y = 1 \in F$.

Since $x, y \in F$, applying (DS_2) it follows that $y \to x \odot y \in F$ and finally $x \odot y \in F$.

Thus (F_1) is satisfied.

If $x \in F$, $y \in A$ such that $x \leq y$, then $x \to y = 1 \in F$. Hence, by (DS_1), we get $y \in F$, that is, (F_2). We conclude that F is a filter of A.

(b) $\Rightarrow$ (a) Assume that F is a filter of A.

Since F is a nonempty subset of A, there exists an $x \in F$ and obviously, $x \leq 1$. Hence, by (F_2), it follows that $1 \in F$. Thus (DS_1) is satisfied.

Consider $x, y \in A$ such that $x, x \to y \in F$. According to (F_1), $(x \to y) \odot x \in F$. Since by (psbck-c_{25}) $(x \to y) \odot x \leq y$, applying (F_2) we get $y \in F$, that is, (DS_2).

Thus F is a deductive system of A. $\square$

We will denote by $\mathcal{F}(A)$ the set of all filters and by $\mathcal{F}_n(A)$ the set of all normal filters of a pseudo-BCK(pP) algebra A.

It follows that in the case of a pseudo-BCK(pP) algebra A we have $\mathcal{DS}(A) = \mathcal{F}(A)$ and $\mathcal{DS}_n(A) = \mathcal{F}_n(A)$.

By Proposition 1.29, in the case of pseudo-BCK algebras we will use the notion of a deductive system, while for pseudo-BCK(pP) algebras and its particular cases (pseudo-hoops, FL_w-algebras, pseudo-MTL algebras, $R\ell$-monoids, pseudo-BL algebras) we can use both notions of deductive system and filter.

Definition 1.16 A deductive system is called *maximal* or an *ultrafilter* if it is proper and is not strictly contained in any other proper deductive system.

We make the following definitions:

$$\text{Max}(A) = \{F \mid F \text{ is a maximal deductive system of } A\} \quad \text{and}$$

$$\text{Max}_n(A) = \{F \mid F \text{ is a maximal normal deductive system of } A\}.$$

Clearly, $\text{Max}_n(A) \subseteq \text{Max}(A)$.

Example 1.18 Consider the pseudo-BCK algebra A from Example 1.3.

(1) The deductive systems of A are the following: $D_1 = \{a, c, 1\}$, $D_2 = \{b, c, 1\}$, $D_3 = \{c, 1\}$ and $D_4 = \{1\}$.
(2) D_1 and D_2 are maximal deductive systems.
(3) D_3 is a normal deductive system.
(4) D_1 and D_2 are not normal deductive systems ($b \to a = a \in D_1$, while $b \rightsquigarrow 0 = 0 \notin D_1$ and $a \rightsquigarrow 0 = b \in D_2$, while $a \to 0 = 0 \notin D_2$).

Example 1.19 In the pseudo-BCK algebra A_1 from Example 1.16, the set $D = \{a, b, c, d, 1\}$ is a maximal normal deductive system.

Proposition 1.30 *If A is a bounded pseudo-BCK(pP) algebra, then the sets*

$$A_0^- = \{x \in A \mid x^- = 0\} \quad \text{and} \quad A_0^\sim = \{x \in A \mid x^\sim = 0\}$$

are proper deductive systems of A.

Proof If $x, y \in A_0^-$, then $(x \odot y)^- = x \to y^- = x \to 0 = x^- = 0$, so $x \odot y \in A_0^-$.

If $x \in A_0^-$, $y \in A$ such that $x \leq y$, then $y^- \leq x^- = 0$, so $y^- = 0$, that is, $y \in A_0^-$. Because $0 \notin A_0^-$, we conclude that A_0^- is a proper deductive system of A.

Similarly for the case of $A_0^\sim$. $\square$

Proposition 1.31 *Let A be a bounded pseudo-BCK algebra and $H \in \mathcal{DS}_n(A)$. Then*:

(1) $x^- \in H$ *iff* $x^\sim \in H$;
(2) $x \in H$ *implies* $(x^-)^- \in H$ *and* $(x^\sim)^\sim \in H$.

Proof

(1) This follows by taking $y = 0$ in the definition of a normal deductive system.
(2) From $x \in H$ and $x \leq (x^-)^\sim$ we get $(x^-)^\sim \in H$, that is, $x^- \rightsquigarrow 0 \in H$. Hence $x^- \rightarrow 0 \in H$, so $(x^-)^- \in H$. Similarly, $(x^\sim)^\sim \in H$. $\square$

Proposition 1.32 *Any proper deductive system of a pseudo-BCK algebra A can be extended to a maximal deductive system of A.*

Proof This is an immediate consequence of Zorn's lemma. $\square$

Example 1.20

(1) Let A be the pseudo-BCK(pP) algebra from Example 1.9, $D_1 = \{s, a, b, n, c, d, m, 1\}$ and $D_2 = \{a_2, s, a, b, n, c, d, m, 1\}$. Then: $\mathcal{DS}(A) = \{\{1\}, D_1, D_2, A\}$, $\mathcal{DS}_n(A) = \{\{1\}, D_1, A\}$, $\mathrm{Max}(A) = \{D_2\}$ and $\mathrm{Max}_n(A) = \emptyset$.
(2) In the case of the pseudo-BCK(pP) algebra A_1 from Example 1.17, with $D_1 = \{a_1, a_2, b_2, s, a, b, n, c, d, m, 1\}$, $D_2 = \{b_2, s, a, b, n, c, d, m, 1\}$ and $D_3 = \{s, a, b, n, c, d, m, 1\}$, we have: $\mathcal{DS}(A_1) = \{\{1\}, D_1, D_2, D_3, A_1\}$, $\mathcal{DS}_n(A_1) = \{\{1\}, D_1, D_3, A_1\}$, $\mathrm{Max}(A_1) = \{D_1\}$ and $\mathrm{Max}_n(A_1) = \{D_1\}$.

Let A be a good pseudo-BCK(pP) algebra. The notion dual to that of a filter is that of an ideal.

Definition 1.17 An *ideal* of a good pseudo-BCK(pP) algebra A is a nonempty subset I of A satisfying the conditions:

(I_1) If $x, y \in I$, then $x \oplus y \in I$.
(I_2) If $x \in A$, $y \in I$, $x \leq y$, then $x \in I$.

Recall that, if A is a pseudo-BCK(pP) algebra, then for any $n \in \mathbb{N}$, $x \in A$ we put $x^0 = 1$ and $x^{n+1} = x^n \odot x = x \odot x^n$. If A is bounded, the *order* of $x \in A$, denoted $\mathrm{ord}(x)$, is the smallest $n \in \mathbb{N}$ such that $x^n = 0$. If there is no such n, then $\mathrm{ord}(x) = \infty$.

Definition 1.18 For every subset $X \subseteq A$, the smallest deductive system of A containing X (i.e. the intersection of all deductive systems $D \in \mathcal{DS}(A)$ such that $X \subseteq D$) is called the deductive system *generated by* X and it will be denoted by $[X)$. If $X = \{x\}$ we write $[x)$ instead of $[\{x\})$.

Lemma 1.8 *Let A be a bounded pseudo-BCK(pP) algebra and $x, y \in A$. Then*:

(1) $[x)$ *is proper iff* $\operatorname{ord}(x) = \infty$;
(2) *if* $x \leq y$ *and* $\operatorname{ord}(y) < \infty$, *then* $\operatorname{ord}(x) < \infty$;
(3) *if* $x \leq y$ *and* $\operatorname{ord}(x) = \infty$, *then* $\operatorname{ord}(y) = \infty$.

Proof

(1) $[x)$ is proper iff $0 \notin [x)$ iff $x^n \neq 0$ for all $n \in \mathbb{N}$ iff $\operatorname{ord}(x) = \infty$.
(2) and (3) follow from the fact that $x \leq y$ implies $x^n \leq y^n$ for all $n \in \mathbb{N}$.   □

Let A be a pseudo-BCK algebra.
Given an integer $n \geq 1$, we define inductively

$$x \to^0 y := y, \qquad x \to^n y := x \to \left(x \to^{n-1} y\right), \quad n \geq 1 \quad \text{and}$$

$$x \leadsto^0 y := y, \qquad x \leadsto^n y := x \leadsto \left(x \leadsto^{n-1} y\right), \quad n \geq 1.$$

Lemma 1.9 *Let* $(A, \to, \leadsto, 1)$ *be a pseudo-BCK algebra. Then* $[\emptyset) = 1$ *and for any* $\emptyset \neq X \subseteq A$ *we have:*

(1)

$$[X) = \left\{a \in A \mid x_1 \to \left(\cdots \to (x_n \to a)\cdots\right) = 1 \right.$$

$$\left. \text{for some } x_1, \ldots, x_n \in X \text{ and } n \geq 1\right\}$$

$$= \left\{a \in A \mid x_1 \leadsto \left(\cdots \leadsto (x_n \leadsto a)\cdots\right) = 1 \right.$$

$$\left. \text{for some } x_1, \ldots, x_n \in X \text{ and } n \geq 1\right\};$$

(2)

$$[x) = \left\{a \in A \mid x \to^n a = 1 \text{ for some } n \geq 1\right\}$$

$$= \left\{a \in A \mid x \leadsto^n a = 1 \text{ for some } n \geq 1\right\}.$$

Proof It is clear that $[\emptyset) = 1$.

(1) Let $Y = \{a \in A \mid x_1 \to (\cdots \to (x_n \to a)\cdots) = 1, x_1, \ldots, x_n \in X, n \geq 1\}$.
Obviously, $1 \in Y$. Consider $a, a \leadsto b \in Y$, i.e.,

$$x_1 \to \left(\cdots \to \left(x_m \to (a \leadsto b)\right)\cdots\right) = 1 \quad \text{and}$$

$$y_1 \to \left(\cdots \to (y_n \to a)\cdots\right) = 1$$

for some $x_1, \ldots, x_m, y_1, \ldots, y_n \in X, m, n \in \mathbb{N}$.
Applying $(psbck\text{-}c_3)$ we get inductively

$$a \leadsto \left(x_1 \to \left(\cdots \to (x_m \to b)\cdots\right)\right) = x_1 \to \left(\cdots \to \left(x_m \to (a \leadsto b)\right)\cdots\right) = 1.$$

Hence $a \leq x_1 \to (\cdots \to (x_m \to b)\cdots)$.

Applying $(psbck\text{-}c_{10})$ we get inductively

$$1 = y_1 \to \left(\cdots \to (y_n \to a) \cdots\right)$$

$$\leq y_1 \to \left(\cdots \to \left(y_n \to \left(x_1 \to \left(\cdots \to (x_m \to b) \cdots\right)\right)\right) \cdots\right).$$

It follows that $y_1 \to (\cdots \to (y_n \to (x_1 \to (\cdots \to (x_m \to b) \cdots)))\cdots) = 1$, so $b \in Y$. Thus $Y \in \mathcal{DS}(A)$.

Moreover, we can see that $X \subseteq Y$ and $Y \subseteq D$ whenever $D \in \mathcal{DS}(A)$, so $Y = [X)$, which proves that $Y = [X)$. Similarly,

$$[X) = \left\{a \in A \mid x_1 \rightsquigarrow \left(\cdots \rightsquigarrow (x_n \rightsquigarrow a) \cdots\right) = 1, x_1, \ldots, x_n \in X, n \geq 1\right\}.$$

(2) This follows from (1). $\qquad\qquad\qquad\qquad\qquad\qquad\qquad\qquad\qquad\qquad\qquad\square$

Proposition 1.33 *If A is a pseudo-BCK(pP) algebra and $X \subseteq A$, then*

$$[X) = \{a \in A \mid a \geq x_1 \odot x_2 \odot \cdots \odot x_n \text{ for some } n \geq 1 \text{ and } x_1, x_2, \ldots, x_n \in X\}.$$

Proof If we put

$$X' = \{a \in A \mid a \geq x_1 \odot x_2 \odot \cdots \odot x_n \text{ for some } n \geq 1 \text{ and } x_1, x_2, \ldots, x_n \in X\},$$

it is obvious that X' is a filter of A which contains X.

Taking into consideration Proposition 1.29, X' is a deductive system of A containing X, that is, $[X) \subseteq X'$.

Now let $Y \in \mathcal{DS}(A)$ such that $X \subseteq Y$. If $a \in X'$, then there are $x_1, x_2, \ldots, x_n \in X$ such that $x_1 \odot x_2 \odot \cdots \odot x_n \leq a$. Since $x_1, x_2, \ldots, x_n \in Y$, it follows that $x_1 \odot x_2 \odot \cdots \odot x_n \in Y$, so $a \in Y$. Hence $X' \subseteq Y$.

We conclude that $X' \subseteq \bigcap\{Y \mid Y \in \mathcal{DS}(A), X \subseteq Y\} = [X)$.

Thus $X' = [X)$, that is,

$$[X) = \{a \in A \mid a \geq x_1 \odot x_2 \odot \cdots \odot x_n \text{ for some } n \geq 1 \text{ and } x_1, x_2, \ldots, x_n \in X\}.\square$$

Remark 1.13 Let A be a pseudo-BCK(pP) algebra. Then:

(1) If X is a deductive system of A, then $[X) = X$.
(2) $[x) = \{y \in A \mid y \geq x^n \text{ for some } n \geq 1\}$ ($[x)$ is called a *principal* deductive system).
(3) If D is a deductive system of A and $x \in A \setminus D$, then

$$D(x) = \left[D \cup \{x\}\right)$$

$$= \left\{y \in A \mid y \geq \left(d_1 \odot x^{n_1}\right) \odot \left(d_2 \odot x^{n_2}\right) \odot \cdots \odot \left(d_m \odot x^{n_m}\right)\right.$$

$$\left.\text{for some } m \geq 1, n_1, n_2, \ldots, n_m \geq 0, d_1, d_2, \ldots, d_m \in D\right\}.$$

Lemma 1.10 *Let A be a pseudo-BCK(pP) algebra and D a proper deductive system of A. Then the following are equivalent:*

(a) *D is maximal*;
(b) *for all $x \in A$, if $x \notin D$ then $[D \cup \{x\}) = A$.*

Proof

(a) $\Rightarrow$ (b) Obviously, $D \subseteq [D \cup \{x\})$, $D \neq [D \cup \{x\})$. Since D is maximal, it follows
that $[D \cup \{x\}) = A$.

(b) $\Rightarrow$ (a) Suppose that there exists a proper deductive system E of A such that
$D \subset E$ and $D \neq E$. It follows that there is an $x \in E \setminus D$ and applying (b) we get
that $[D \cup \{x\}) = A$. But $[D \cup \{x\}) \subseteq E$, hence $E = A$. Thus E is not proper, a
contradiction. Therefore $D \in \mathrm{Max}(A)$. $\square$

Proposition 1.34 *If D_1, D_2 are nonempty subsets of a pseudo-BCK(pP) algebra A
such that $1 \in D_1 \cap D_2$, then*

$$[D_1 \cup D_2) = \left\{ x \in A \mid x \geq \left(d_1 \odot d_1'\right) \odot \left(d_2 \odot d_2'\right) \odot \cdots \odot \left(d_n \odot d_n'\right) \right.$$

$$\left. \text{for some } n \geq 1, d_1, d_2, \ldots, d_n \in D_1, d_1', d_2', \ldots, d_n' \in D_2 \right\}.$$

Proof Let $H = \{x \in A \mid x \geq (d_1 \odot d_1') \odot \cdots \odot (d_n \odot d_n')$ for some $d_1, d_2, \ldots, d_n \in$
$D_1, d_1', d_2', \ldots, d_n' \in D_2\}$. We prove that $H \in \mathcal{DS}(A)$. Let $x \in H$, $y \in A$, $x \leq y$.
Since $x \geq (d_1 \odot d_1') \odot \cdots \odot (d_n \odot d_n')$ for some $d_1, d_2, \ldots, d_n \in D_1, d_1', d_2', \ldots,$
$d_n' \in D_2$, we have $y \geq (d_1 \odot d_1') \odot \cdots \odot (d_n \odot d_n')$, that is, $y \in H$.

For $x, y \in H$ there exist $n, m \geq 1$, $d_1, d_2, \ldots, d_n, e_1, e_2, \ldots, e_m \in D_1, d_1', d_2',$
$\ldots, d_n', e_1', e_2', \ldots, e_m' \in D_2$ such that

$$x \geq \left(d_1 \odot d_1'\right) \odot \cdots \odot \left(d_n \odot d_n'\right), \qquad y \geq \left(e_1 \odot e_1'\right) \odot \cdots \odot \left(e_m \odot e_m'\right).$$

We get $x \odot y \geq (d_1 \odot d_1') \odot \cdots \odot (d_n \odot d_n') \odot (e_1 \odot e_1') \odot \cdots \odot (e_m \odot e_m')$, so
$x \odot y \in H$. Thus $H \in \mathcal{DS}(A)$.

Since $1 \in D_1 \cap D_2$, we deduce that $D_1, D_2 \subseteq H$ (if $x \in D_1$, then $x \geq x \odot 1$ with
$1 \in D_2$, so $x \in H$; if $x \in D_2$, then $x \geq 1 \odot x$ with $1 \in D_1$, so $x \in H$).

Hence $D_1 \cup D_2 \subseteq H$, so it follows that $[D_1 \cup D_2) \subseteq H$.

Now let $D \in \mathcal{DS}(A)$ such that $D_1 \cup D_2 \subseteq D$.

If $x \in H$, then there are $d_1, d_2, \ldots, d_n \in D_1, d_1', d_2', \ldots, d_n' \in D_2$ such that $x \geq$
$(d_1 \odot d_1') \odot \cdots \odot (d_n \odot d_n')$. Since $d_1, d_2, \ldots, d_n, d_1', d_2', \ldots, d_n' \in D$, it follows that
$(d_1 \odot d_1') \odot \cdots \odot (d_n \odot d_n') \in D$, hence $x \in D$.

Thus $H \subseteq D$, so $H \subseteq \bigcap \{D \in \mathcal{DS}(A) \mid D_1 \cup D_2 \subseteq D\} = [D_1 \cup D_2)$.

We conclude that $[D_1 \cup D_2) = H$. $\square$

Lemma 1.11 *Let A be a pseudo-BCK(pP) algebra and $H \in \mathcal{DS}_n(A)$. Then:*

(1) *For any $x \in A$ and $h \in H$, there is an $h' \in H$ such that $x \odot h \geq h' \odot x$;*
(2) *For any $x \in A$ and $h \in H$, there is an $h'' \in H$ such that $h \odot x \geq x \odot h''$.*

Proof

(1) Let $y = x \odot h$. Then $x \odot h = y \geq (x \rightarrow y) \odot x$.

But $h \leq x \rightsquigarrow x \odot h = x \rightsquigarrow y$. Since $h \in H$, it follows that $x \rightsquigarrow y \in H$.

Because H is a normal filter we have $h' = x \rightarrow y \in H$. Thus $x \odot h \geq h' \odot x$.

(2) Let $y = h \odot x$. Then $h \odot x = y \geq x \odot (x \rightsquigarrow y)$. But $h \leq x \rightarrow h \odot x = x \rightarrow y$. Since $h \in H$, it follows that $x \rightarrow y \in H$. Because H is a normal filter we have $h' = x \rightsquigarrow y \in H$. Thus $h \odot x \geq x \odot h'$. $\square$

Lemma 1.12 *Let A be a bounded pseudo-BCK(pP) algebra, H a normal proper deductive system of A and $x \in A$. Then the following are equivalent:*

(a) *there exists an $h \in H$ such that $x \leq h^-$;*
(b) *there exists an $h \in H$ such that $x \odot h = 0$;*
(c) *there exists an $h \in H$ such that $x \leq h^\sim$;*
(d) *there exists an $h \in H$ such that $h \odot x = 0$.*

Proof

(a) $\Leftrightarrow$ (b) and (c) $\Leftrightarrow$ (d) are obvious.
(b) $\Rightarrow$ (d) Assume that there exists an $h \in H$ such that $x \odot h = 0$. According to Lemma 1.11, there exists an $h' \in H$ such that $x \odot h \geq h' \odot x$, hence $h' \odot x = 0$.
(d) $\Rightarrow$ (b) Similar to (b) $\Rightarrow$ (d). $\square$

Proposition 1.35 *Let A be a pseudo-BCK(pP) algebra, $H \in \mathcal{DS}_n(A)$ and $x \in A$. Then*

$$H(x) = \big[H \cup \{x\}\big) = \big\{y \in A \mid y \geq h \odot x^n \text{ for some } n \in \mathbb{N}, h \in H\big\}$$
$$= \big\{y \in A \mid y \geq x^n \odot h \text{ for some } n \in \mathbb{N}, h \in H\big\}$$
$$= \big\{y \in A \mid x^n \rightarrow y \in H \text{ for some } n \geq 1\big\}$$
$$= \big\{y \in A \mid x^n \rightsquigarrow y \in H \text{ for some } n \geq 1\big\}.$$

Proof Let $y \in H(x)$. Then $y \geq (h_1 \odot x^{n_1}) \odot (h_2 \odot x^{n_2}) \odot \cdots \odot (h_m \odot x^{n_m})$ for some $m \geq 1, n_1, n_2, \ldots, n_m \geq 0, h_1, h_2, \ldots, h_m \in H$, by Remark 1.13(3).

If $m = 1$, then $y \geq h_1 \odot x^{n_1}$ and we take $h = h_1$ and $n = n_1$.

If $m = 2$, then $y \geq (h_1 \odot x^{n_1}) \odot (h_2 \odot x^{n_2}) = h_1 \odot (x^{n_1} \odot h_2) \odot x^{n_2}$.

According to Lemma 1.11, there is an $h'_2 \in H$ such that $x^{n_1} \odot h_2 \geq h'_2 \odot x^{n_1}$.

Hence $y \geq h_1 \odot (h'_2 \odot x^{n_1}) \odot x^{n_2} = (h_1 \odot h'_2) \odot x^{n_1 + n_2}$ and we take $h = h_1 \odot h'_2$ and $n = n_1 + n_2$. By induction we get $y \geq h \odot x^n$ for some $n \in \mathbb{N}, h \in H$.

Similarly, $y \geq x^n \odot h$ for some $n \in \mathbb{N}, h \in H$. Thus

$$H(x) = \big\{y \in A \mid y \geq h \odot x^n \text{ for some } n \in \mathbb{N}, h \in H\big\}$$
$$= \big\{y \in A \mid y \geq x^n \odot h \text{ for some } n \in \mathbb{N}, h \in H\big\}.$$

If $y \in H(x)$, then $h \odot x^n \leq y$ for some $n \geq 1, h \in H$. Thus $h \leq x^n \rightarrow y$, hence $x^n \rightarrow y \in H$.

Conversely, assume that $h = x^n \rightarrow y \in H$ for some $n \geq 1$.

We also have $(h \odot x^n) \to y = h \to (x^n \to y) = h \to h = 1$, hence $h \odot x^n \leq y$. Therefore, $y \in H(x)$ and we conclude that

$$H(x) = \{y \in A \mid x^n \to y \in H \text{ for some } n \geq 1\}.$$

Similarly, $H(x) = \{y \in A \mid x^n \rightsquigarrow y \in H \text{ for some } n \geq 1\}$.   □

Corollary 1.7 *Let A be a pseudo-BCK(pP) algebra and H a proper normal deductive system of A. Then the following are equivalent:*

(a) $H \in \mathrm{Max}_n(A)$;
(b) *for all $x \in A$, if $x \notin H$, then for any $y \in A$, $x^n \to y \in H$ for some $n \in \mathbb{N}, n \geq 1$;*
(c) *for all $x \in A$, if $x \notin H$, then for any $y \in A$, $x^n \rightsquigarrow y \in H$ for some $n \in \mathbb{N}, n \geq 1$.*

Proof

(a) $\Rightarrow$ (b) Since H is maximal, by Lemma 1.10, $[H \cup \{x\}) = A$ and applying Proposition 1.35 we get the assertion (b).
(b) $\Rightarrow$ (a) Let $x \in A \setminus H$. By (b), for all $y \in A$ we have $x^n \to y \in H$ for some $n \in \mathbb{N}, n \geq 1$. Since $(x^n \to y) \odot x^n \leq y$, by Proposition 1.35 it follows that $y \in [H \cup \{x\})$. Hence $[H \cup \{x\}) = A$. Applying Lemma 1.10 we get that $H \in \mathrm{Max}_n(A)$.
(a) $\Leftrightarrow$ (c) Similar to (a) $\Leftrightarrow$ (b).   □

Proposition 1.36 *If A is a pseudo-BCK(pP) algebra and $D_1, D_2 \in \mathcal{DS}_n(A)$, then*

$$[D_1 \cup D_2) = \{x \in A \mid x \geq u \odot v \text{ for some } u \in D_1, v \in D_2\}.$$

Proof By Proposition 1.34 we have:

$$[D_1 \cup D_2) = \{x \in A \mid x \geq (d_1 \odot d_1') \odot (d_2 \odot d_2') \odot \cdots \odot (d_n \odot d_n')$$
$$\text{for some } n \geq 1, d_1, d_2, \ldots, d_n \in D_1, d_1', d_2', \ldots, d_n' \in D_2\}.$$

Put $d = (d_1 \odot d_1') \odot (d_2 \odot d_2') \odot \cdots \odot (d_n \odot d_n') = d_1 \odot (d_1' \odot d_2) \odot \cdots \odot (d_{n-1}' \odot d_n) \odot d_n'$.

By Lemma 1.11, there is a $d_2'' \in D_2$ such that $d_1' \odot d_2 \geq d_2 \odot d_2''$. Hence

$$d \geq d_1 \odot d_2 \odot (d_2'' \odot d_3) \odot \cdots \odot (d_n \odot d_n').$$

Similarly, there is a $d_3'' \in D_2$ such that $d_2'' \odot d_3 \geq d_3 \odot d_3''$, so

$$d \geq d_1 \odot d_2 \odot d_3 \odot (d_3'' \odot d_4) \odot \cdots \odot (d_n \odot d_n').$$

Finally, $d \geq d_1 \odot d_2 \odot d_3 \odot \cdots \odot d_n \odot d_n''$ with $d_1, d_2, \ldots, d_n \in D_1, d_n'' \in D_2$. Taking $u = d_1 \odot d_2 \odot d_3 \odot \cdots \odot d_n$, $v = d_n''$, we get $x \geq d \geq u \odot v$ with $u \in D_1, v \in D_2$. □

Definition 1.19 A bounded pseudo-BCK(pP) algebra A is *locally finite* if for every $x \in A$, $x \neq 1$ we have $\mathrm{ord}(x) < \infty$.

Proposition 1.37 *Let A be a bounded pseudo-BCK(pP) algebra. The following are equivalent:*

(a) *A is locally finite;*
(b) *$\{1\}$ is the unique proper deductive system of A.*

Proof According to Lemma 1.8(1), A is locally finite iff for every $x \in A \setminus \{1\}$, $[x) = A$ iff $\{1\}$ is the unique proper deductive system of A.   □

Theorem 1.11 *If D is a proper deductive system of a bounded pseudo-BCK(pP) algebra A, then the following are equivalent:*

(1) $D \in \mathrm{Max}(A)$;
(2) *for any $x \notin D$ there is a $d \in D$, $n, m \in \mathbb{N}$, $n, m \geq 1$ such that $(d \odot x^n)^m = 0$.*

Proof

(a) $\Rightarrow$ (b) Since $D \in \mathrm{Max}(A)$ and $x \notin D$, we have $[D \cup \{x\}) = A$, so $0 \in [D \cup \{x\})$. By Remark 1.13 it follows that there exist $m \geq 1$, $n_1, n_2, \ldots, n_m \geq 0$, $d_1, d_2, \ldots, d_m \in D$ such that

$$\left(d_1 \odot x^{n_1}\right) \odot \left(d_2 \odot x^{n_2}\right) \odot \cdots \odot \left(d_m \odot x^{n_m}\right) \leq 0, \quad \text{so}$$

$$\left(d_1 \odot x^{n_1}\right) \odot \left(d_2 \odot x^{n_2}\right) \odot \cdots \odot \left(d_m \odot x^{n_m}\right) = 0.$$

Taking $n = \max\{n_1, n_2, \ldots, n_m\}$ and $d = d_1 \odot d_2 \odot \cdots \odot d_m \in D$ we get

$$\left(d \odot x^n\right)^m \leq \left(d_1 \odot x^{n_1}\right) \odot \left(d_2 \odot x^{n_2}\right) \odot \cdots \odot \left(d_m \odot x^{n_m}\right) = 0.$$

It follows that $(d \odot x^n)^m = 0$.

(b) $\Rightarrow$ (a) Assume that there is a proper deductive system E of A such that $D \subset E$, $D \neq E$. Then there exists an $x \in E$ such that $x \notin D$. By the hypothesis, there exist $d \in D$, $n, m \in \mathbb{N}$ such that $(d \odot x^n)^m = 0$. Since $x, d \in E$, it follows that $0 \in E$, hence $E = A$ which is a contradiction. Thus $D \in \mathrm{Max}(A)$.   □

Theorem 1.12 *If H is a proper normal deductive system of a bounded pseudo-BCK(pP) algebra A, then the following are equivalent:*

(a) $H \in \mathrm{Max}_n(A)$;
(b) *for any $x \in A$, $x \notin H$ iff $(x^n)^- \in H$ for some $n \in \mathbb{N}$;*
(c) *for any $x \in A$, $x \notin H$ iff $(x^n)^\sim \in H$ for some $n \in \mathbb{N}$.*

Proof

(a) $\Rightarrow$ (b) If $x \notin H$, then $[H \cup \{x\}) = A$, hence $0 \in [H \cup \{x\})$.
According to Proposition 1.35, there exists an $n \in \mathbb{N}$ and an $h \in H$ such that $h \odot x^n \leq 0$. Then $h \leq x^n \rightarrow 0 = (x^n)^-$, i.e. $(x^n)^- \in H$.
Conversely, consider $x \in A$ such that $(x^n)^- \in H$ for some $n \in \mathbb{N}$ and assume $x \in H$. Then $x^n \rightarrow 0 \in H$ and taking into consideration that $x^n \in H$, we get

$0 \in H$. This means that H is not a proper deductive system, which is a contradiction. Hence $x \notin H$.

(b) $\Rightarrow$ (a) Assume there is a proper deductive system F such that $H \subset F$ and $H \neq F$. Then there is an $x \in F$ such that $x \notin H$. Hence $(x^n)^- \in H \subset F$ for some $n \in N$. It follows that $x^n \in F$ and $x^n \odot (x^n)^- = 0 \in F$, which contradicts the fact that F is a proper deductive system. Thus $H \in \mathrm{Max}(A)$.

(a) $\Leftrightarrow$ (c) Similar to (a) $\Leftrightarrow$ (b). $\square$

Proposition 1.38 *If A is a bounded pseudo-BCK(pP) algebra and $D = A \setminus \{0\} \in \mathrm{Max}(A)$, then A is good.*

Proof Obviously $(0^-)^\sim = (0^\sim)^- = 0$. Assume $x > 0$, that is, $x \in D$. If $x^-, x^\sim \in D$ it follows that $x^- \odot x, x \odot x^\sim \in D$, that is, $0 \in D$, a contradiction.

Thus $x^- = x^\sim = 0$, hence $(x^-)^\sim = (x^\sim)^- = 1$. Therefore, $(x^-)^\sim = (x^\sim)^-$ for all $x \in A$, so A is a good pseudo-BCK(pP) algebra. $\square$

Proposition 1.39 *Let A be a bounded pseudo-BCK(pP) algebra, $D \in \mathrm{Max}(A)$ and $x, y \in A$. Then:*

(1) *$y \notin D$ and $y \odot x = x$ implies $x = 0$;*
(2) *$y \notin D$ and $x \odot y = x$ implies $x = 0$.*

Proof

(1) Consider $x \in A$, $y \in A \setminus D$ such that $y \odot x = x$. Assume $x > 0$ and consider $E = \{z \in A \mid z \odot x = x\}$. First we prove that E is a proper deductive system of A. Obviously, $1, y \in E$ and $0 \notin E$. Consider $z \in A$ such that $y \to z \in E$, so $(y \to z) \odot x = x$. Since $(y \to z) \odot y \odot x = (y \to z) \odot x = x$, it follows that $x = [(y \to z) \odot y] \odot x \leq z \odot x \leq x$. Thus $z \odot x = x$, hence $z \in E$. Therefore E is a proper deductive system. Since $y \in E$ and D is maximal, it follows that $y \in D$, a contradiction. Thus $x = 0$.

(2) Similar to (1). $\square$

A *congruence* on a pseudo-BCK algebra $(A, \to, \rightsquigarrow, 1)$ is an equivalence relation compatible with the operations $\to, \rightsquigarrow$. The set of all congruences of A is denoted by $\mathrm{Con}(A)$. The class of pseudo-BCK algebras is not closed under homomorphic images. In other words, there exist congruences $\theta \in \mathrm{Con}(A)$ such that the quotient algebra $(A/\theta, \to, \rightsquigarrow, 1/\theta)$ is not a pseudo-BCK algebra, as we can see in the following example (Example 2.2.3 in [208]).

Example 1.21 Let $(G, \cdot, ^{-1}, 1, \vee, \wedge)$ be an ℓ-group with an order-unit $u \in G^+$ and put $v = u^{-1}$. Then for every $g \in G$ there exists an $n \in \mathbb{N}$ with $v^n \leq g \leq v^{-n}$, hence for every $g \in G^-$ the maximum $\max\{n \in \mathbb{N} \mid g \leq v^n\}$ exists and we shall denote it by $m(g)$.

Let $X = \{x_i \mid i \in \mathbb{N}\}$ and $Y = \{y_i \mid i \in \mathbb{N}\}$ be two infinite sequences, and put $A = G^- \cup X \cup Y$. In order to turn A into a pseudo-BCK algebra we define the two binary operations $\to$ and $\rightsquigarrow$ as follows (where $g, h \in G^-$):

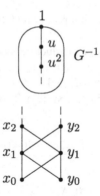

Fig. 1.7 Example of pseudo-BCK algebra whose homomorphic image is not pseudo-BCK algebra

(1) $g \to h := h \cdot (g \vee h)^{-1} = (h \cdot g^{-1}) \wedge 1$ and $g \rightsquigarrow h := (g \vee h)^{-1} \cdot h = (g^{-1} \cdot h) \wedge 1$;

(2) $x_i \to g = x_i \rightsquigarrow g = y_i \to g = y_i \rightsquigarrow g := 1$;

(3) $g \to x_i = g \rightsquigarrow x_i := x_{i+m(g)}$ and $g \to y_i = g \rightsquigarrow y_i := y_{i+m(g)}$;

(4)

$$x_i \to x_j = x_i \rightsquigarrow x_j = y_i \to y_j = y_i \rightsquigarrow y_j := \begin{cases} 1 & \text{if } i \leq j \\ v^{i-j} & \text{if } i > j; \end{cases}$$

(5) $x_i \to y_j = y_{i+1} \to y_j$, $x_i \rightsquigarrow y_j = y_{i+1} \rightsquigarrow y_j$, $y_i \to x_j = x_{i+1} \to x_j$ and $y_i \rightsquigarrow x_j = x_{i+1} \rightsquigarrow x_j$.

A straightforward verification yields that $(A, \to, \rightsquigarrow, 1)$ is a pseudo-BCK algebra (see Fig. 1.7).

It can easily be seen that the equivalence Φ with the partition $\{G^-, X, Y\}$ as well as Ψ with the partition $\{G^-, X \cup Y\}$ is a congruence on $(A, \to, \rightsquigarrow, 1)$ with kernel $G^- = [1]_\Phi = [1]_\Psi$.

At the same time, A gives an example of a pseudo-BCK algebra whose homomorphic image is not a pseudo-BCK algebra. Specifically, A/Φ does not satisfy the quasi-identity $(psBCK'_6)$, since $X \to Y = Y \to X = G^-$ and $X \neq Y$.

A congruence $\theta \in \mathrm{Con}(A)$ such that the quotient algebra $(A/\theta, \to, \rightsquigarrow, 1/\theta)$ is a pseudo-BCK algebra is called in [208] a *relative congruence*. With any $H \in \mathcal{DS}_n(A)$ we associate a binary relation $\equiv_H$ on A by defining $x \equiv_H y$ iff $x \to y, y \to x \in H$ iff $x \rightsquigarrow y, y \rightsquigarrow x \in H$.

The quotient of A by $\equiv_H$ will be denoted A/H. For any $x \in A$, let x/H be the congruence class $x/_{\equiv_H}$ of x, hence $A/H = \{x/H | x \in A\}$.

Denote by θ_H the congruence $\equiv_H$.

Proposition 1.40 *For a given $H \in \mathcal{DS}_n(A)$ the relation θ_H is a congruence on A.*

Proof Obviously, θ_H is reflexive and symmetric.

Assume $x \equiv_H y$ and $y \equiv_H z$. By $(psBCK_1)$ we have:

$$x \to y \le (y \to z) \rightsquigarrow (x \to z) \quad \text{and}$$

$$x \rightsquigarrow y \le (y \rightsquigarrow z) \to (x \rightsquigarrow z).$$

Since $x \to y \in H$, it follows that $(y \to z) \rightsquigarrow (x \to z) \in H$.

From $y \to z \in H$ and $(y \to z) \rightsquigarrow (x \to z) \in H$ we get $x \to z \in H$.

Similarly, $x \rightsquigarrow z \in H$. In the same manner, from

$$z \to y \le (y \to x) \rightsquigarrow (z \to x) \quad \text{and}$$

$$z \rightsquigarrow y \le (y \rightsquigarrow x) \to (z \rightsquigarrow x)$$

we get $z \to x \in H$ and $z \rightsquigarrow x \in H$.

Thus $x \equiv_H z$, that is, θ_H is transitive.

Hence θ_H is an equivalence relation on A.

We prove now that θ_H is compatible with $\to$ and $\rightsquigarrow$.

Consider $(x, y), (a, b) \in \theta_H$. Applying $(psBCK_1)$ we have:

$$x \to y \le (y \to a) \rightsquigarrow (x \to a) \quad \text{and}$$

$$y \to x \le (x \to a) \rightsquigarrow (y \to a).$$

Since $x \to y, y \to x \in H$, it follows that $(y \to a) \rightsquigarrow (x \to a) \in H$ and $(x \to a) \rightsquigarrow (y \to a) \in H$. Hence $(x \to a, y \to a) \in \theta_H$.

According to $(psbck\text{-}c_5)$ we have

$$a \to b \le (y \to a) \to (y \to b) \quad \text{and}$$

$$b \to a \le (y \to b) \to (y \to a).$$

Since $a \to b, b \to a \in H$, it follows that $(y \to a) \to (y \to b) \in H$ and $(y \to b) \to (y \to a) \in H$. Thus $(y \to a, y \to b) \in \theta_H$.

Due to the transitivity of θ_H we get $(x \to a, y \to b) \in \theta_H$, that is, θ_H is compatible with $\to$. We can prove similarly that θ_H is compatible with $\rightsquigarrow$. □

Proposition 1.41 *If $H \in \mathcal{DS}_n(A)$, then:*

(1) *θ_H is a relative congruence of $(A, \to, \rightsquigarrow, 1)$, that is, A/θ_H becomes a pseudo-BCK algebra with the natural operations induced from those of A;*

(2) *If A is a pseudo-BCK(pP) algebra, then θ_H is compatible with $\odot$.*

Proof

(1) If $[x]_{\theta_H} \to [y]_{\theta_H} = [1]_{\theta_H}$ and $[y]_{\theta_H} \to [x]_{\theta_H} = [1]_{\theta_H}$, then $[x \to y]_{\theta_H} = [1]_{\theta_H}$ and $[y \to x]_{\theta_H} = [1]_{\theta_H}$, so that recalling $H = [1]_{\theta_H}$ we have $x \to y, y \to x \in H$ and $(x, y) \in \theta_H$, that is, $[x]_{\theta_H} = [y]_{\theta_H}$.

Thus the quotient algebra A/θ_H is a pseudo-BCK algebra.

(2) If $x \equiv_{\theta_H} y$ and $a \equiv_{\theta_H} b$, we will prove that $x \odot a \equiv_{\theta_H} y \odot b$ and $a \odot x \equiv_{\theta_H} b \odot y$.

From $x \ge (x \to y) \odot x$ and $a \ge (b \to a) \odot b$, it follows that $x \odot a \ge (x \to y) \odot x \odot (b \to a) \odot b$. Since $b \to a \in H$, by Lemma 1.11 there exists an $h' \in H$

such that $x \odot (b \to a) \odot b \geq h' \odot x \odot b$. It follows that $x \odot a \geq (x \to y) \odot h' \odot x \odot b$, hence $(x \to y) \odot h' \leq x \odot b \to x \odot a$. Since $(x \to y) \odot h' \in H$, we get that $x \odot b \to x \odot a \in H$. Similarly, $x \odot a \to x \odot b \in H$, so $x \odot a \equiv_{\theta_H} x \odot b$.

One can analogously show that $x \odot b \equiv_{\theta_H} y \odot b$, hence $x \odot a \equiv_{\theta_H} y \odot b$. We prove in the same manner that $a \odot x \equiv_{\theta_H} b \odot y$.

We conclude that θ_H is compatible with $\odot$. $\square$

Corollary 1.8 *If A is a pseudo-BCK(pP) algebra and $H \in \mathcal{DS}_n(A)$, then A/θ_H is a pseudo-BCK(pP) algebra.*

Lemma 1.13 *If H is a normal deductive system of a bounded pseudo-BCK(pP) algebra A, then:*

(1) $x/H = 1/H$ *iff* $x \in H$;
(2) $x/H = 0/H$ *iff* $x^- \in H$ *iff* $x^\sim \in H$;
(3) $x/H \leq y/H$ *iff* $x \to y \in H$ *iff* $x \rightsquigarrow y \in H$.

Proof

(1) $x/H = 1/H$ iff $x \to 1, 1 \to x \in H$ iff $x \in H$.
(2) $x/H = 0/H$ iff $x \to 0, 0 \to x \in H$ iff $x^- \in H$ iff $x^\sim \in H$.
(3) $x/H \leq y/H$ iff $x/H \to y/H = 1/H$ iff $(x \to y)/H = 1/H$ iff $x \to y \in H$ iff $x \rightsquigarrow y \in H$. $\square$

Proposition 1.42 *If H is a proper normal deductive system of a bounded pseudo-BCK(pP) algebra A, then the following are equivalent:*

(a) $H \in \mathrm{Max}_n(A)$;
(b) A/H *is locally finite.*

Proof H is maximal iff the condition (b) from Theorem 1.12 is satisfied. This condition is equivalent to: for any $x \in A$, $x/H \neq 1/H$ iff $(x^n)^-/H = 1/H$ for some $n \in \mathbb{N}$ iff $(x/H)^n = 0/H$ for some $n \in \mathbb{N}$ iff A/H is locally finite. $\square$

Definition 1.20 Let A and B be two bounded pseudo-BCK(pP) algebras. A function $f : A \longrightarrow B$ is a *homomorphism* if it satisfies the following conditions, for all $x, y \in A$:

(H_1) $f(x \odot y) = f(x) \odot f(y)$;
(H_2) $f(x \to y) = f(x) \to f(y)$;
(H_3) $f(x \rightsquigarrow y) = f(x) \rightsquigarrow f(y)$;
(H_4) $f(0) = 0$.

Remark 1.14 If $f : A \longrightarrow B$ is a bounded pseudo-BCK(pP) algebra homomorphism, then one can easily prove that the following hold for all $x \in A$:

(H_5) $f(1) = 1$;
(H_6) $f(x^-) = (f(x))^-$;

(H_7) $f(x^\sim) = (f(x))^\sim$;
(H_8) if $x, y \in A$, $x \leq y$, then $f(x) \leq f(y)$.

The *kernel* of f is the set $\mathrm{Ker}(f) = f^{-1}(1) = \{x \in A \mid f(x) = 1\}$.

The function $\pi_H : A \longrightarrow A/H$ defined by $\pi_H(x) = x/H$ for any $x \in A$ is a surjective homomorphism which is called the *canonical projection* from A to A/H. One can easily prove that $\mathrm{Ker}(\pi_H) = H$.

Proposition 1.43 *If $f : A \longrightarrow B$ is a bounded pseudo-BCK(pP) algebra homomorphism, then the following hold:*

(1) $\mathrm{Ker}(f)$ *is a proper deductive system of A;*
(2) *f is injective iff $\mathrm{Ker}(f) = \{1\}$;*
(3) *If $G \in \mathcal{DS}(B)$, then $f^{-1}(G) \in \mathcal{DS}(A)$ and $\mathrm{Ker}(f) \subseteq f^{-1}(G)$.*
    If $G \in \mathcal{DS}_n(B)$, then $f^{-1}(G) \in \mathcal{DS}_n(A)$. In particular $\mathrm{Ker}(f) \in \mathcal{DS}_n(A)$;
(4) *If f is surjective and $D \in \mathcal{DS}(A)$ such that $\mathrm{Ker}(f) \subseteq D$, then $f(D) \in \mathcal{DS}(B)$.*

Proof

(1) We have $f(1) = 1$, so $1 \in \mathrm{Ker}(f)$. Let $x, y \in A$ such that $x, x \to y \in \mathrm{Ker}(f)$, that is, $f(x) = 1$ and $f(x \to y) = 1$. It follows that $1 = f(x) \to f(y) = f(y)$, so $y \in \mathrm{Ker}(f)$. Thus $\mathrm{Ker}(f)$ is a deductive system of A. Since $f(0) = 0 \neq 1$, it follows that $0 \notin \mathrm{Ker}(f)$. Hence $\mathrm{Ker}(f)$ is a proper deductive system.
(2) Suppose that f is injective and $x \in \mathrm{Ker}(f)$. It follows that $f(x) = 1 = f(1)$, so $x = 1$. Conversely, suppose that $\mathrm{Ker}(f) = \{1\}$ and $x, y \in A$ such that $f(x) = f(y)$. It follows that $f(x) \to f(y) = 1$, that is, $f(x \to y) = 1$. Thus $x \to y \in \mathrm{Ker}(f)$, so $x \to y = 1$, that is, $x \leq y$. Similarly we get $y \leq x$ and we conclude that $x = y$. Thus f is injective.
(3) If $G \in \mathcal{DS}(B)$, then $1 \in G$. Hence $f(x) = 1 \in G$ for all $x \in \mathrm{Ker}(f)$, so $\mathrm{Ker}(f) \subseteq f^{-1}(G)$. Let $x_1, x_2 \in A$ such that $x_1 \leq x_2$ and $x_1 \in f^{-1}(G)$. It follows that $f(x_1) \leq f(x_2)$ and $f(x_1) \in f(f^{-1}(G)) \subseteq G$, so $f(x_1) \in G$. Thus $f(x_2) \in G$ and $x_2 \in f^{-1}(G)$, so $f^{-1}(G) \in \mathcal{DS}(A)$.
    Suppose G is normal. Then $x \to y \in f^{-1}(G)$ iff $f(x \to y) \in G$ iff $f(x) \to f(y) \in G$ iff $f(x) \leadsto f(y) \in G$ iff $f(x \leadsto y) \in G$ iff $x \leadsto y \in f^{-1}(G)$, so $f^{-1}(G)$ is normal. Since $\{1\}$ is a normal deductive system of B, it follows that $f^{-1}(\{1\}) = \mathrm{Ker}(f)$ is normal.
(4) Since $1 \in D$, we have $1 = f(1) \in f(D)$. Let $y_1 \in f(D)$, $y_2 \in B$ such that $y_1 \to y_2 \in f(D)$. It follows that there is an $x_1 \in D$ such that $y_1 = f(x_1)$. Since f is surjective, there is an $x_2 \in A$ such that $y_2 = f(x_2)$. We have $y_1 \to y_2 \in f(D)$ iff $f(x_1) \to f(x_2) \in f(D)$ iff $f(x_1 \to x_2) \in f(D)$ iff $x_1 \to x_2 \in f^{-1}(D)$.
    Taking into consideration that $f^{-1}(D) \in \mathcal{DS}(A)$ and $x_1 \in f^{-1}(D)$, it follows that $x_2 \in f^{-1}(D)$, that is, $y_2 = f(x_2) \in f(D)$. Thus $f(D) \in \mathcal{DS}(B)$. $\quad\square$

Proposition 1.44 *If $f : A \longrightarrow B$ is a surjective bounded pseudo-BCK(pP) algebra homomorphism, then there is a bijective correspondence between $\{D \mid D \in \mathcal{DS}(A), \mathrm{Ker}(f) \subseteq D\}$ and $\mathcal{DS}(B)$.*

Proof By Proposition 1.43, for any $D \in \mathcal{DS}(A)$ such that $\mathrm{Ker}(f) \subseteq D$ and $G \in \mathcal{DS}(B)$ there is a correspondence $D \mapsto f(D)$ and $G \mapsto f^{-1}(G)$ between the two sets.

We have to prove that $f^{-1}(f(D)) = D$ and $f(f^{-1}(G)) = G$. Since f is surjective, it follows that $f(f^{-1}(G)) = G$. Obviously, $D \subseteq f^{-1}(f(D))$ always holds.

Suppose that $x \in f^{-1}(f(D))$, then $f(x) \in f(D)$, so there is a $x' \in D$ such that $f(x) = f(x')$. It follows that $f(x') \to f(x) = 1$, so $f(x' \to x) = 1$, that is, $x' \to x \in \mathrm{Ker}(f) \subseteq D$. From $x', x' \to x \in D$ we get $x \in D$. Thus $f^{-1}(f(D)) = D$. □

Corollary 1.9 *If $D \in \mathcal{DS}_n(A)$, then:*

(1) $\pi_D(E) \in \mathcal{DS}(A/D)$, *where* $E \in \mathcal{DS}(A)$ *such that* $D \subseteq E$;
(2) *The correspondence* $E \mapsto \pi_D(E)$ *is a bijection between* $\{F \mid F \in \mathcal{DS}(A), D \subseteq F\}$ *and* $\mathcal{DS}(A/D)$.

Proof

(1) This follows from Proposition 1.43(3).
(2) This follows from Proposition 1.44. □

Proposition 1.45 *If $D, H \in \mathcal{DS}_n(A)$ such that $H \subseteq D$, then $D \in \mathrm{Max}(A)$ iff $\pi_H(D) \in \mathrm{Max}(A/H)$.*

Proof We will apply Theorem 1.12. Suppose that $D \in \mathrm{Max}(A)$ and let $y \in A/H$, $y \notin \pi_H(D)$. It follows that there is an $x \in A$ such that $y = \pi_H(x) = x/H$. Obviously, $x \notin D$. Since $D \in \mathrm{Max}(A)$, it follows that:

$$\left(x^n\right)^- \in D \text{ for some } n \in \mathbb{N} \quad \text{iff} \quad \pi_H\left(\left(x^n\right)^-\right) \in \pi_H(D) \text{ for some } n \in \mathbb{N} \quad \text{iff}$$

$$\pi_H\left(\left((x/H)^n\right)^-\right) \in \pi_H(D) \text{ for some } n \in \mathbb{N} \quad \text{iff}$$

$$\left(y^n\right)^- \in \pi_H(D) \text{ for some } n \in \mathbb{N}.$$

Thus $\pi_H(D) \in \mathrm{Max}(A/H)$. The converse can be proved in a similar way. □

Corollary 1.10 *If H is a proper normal deductive system of a bounded pseudo-BCK(pP) algebra A, then there is a bijection between $\{D \mid D \in \mathrm{Max}(A), H \subseteq D\}$ and $\mathrm{Max}(A/H)$.*

Proposition 1.46 *If P is a proper normal deductive system of a bounded pseudo-BCK(pP) algebra A, then the following are equivalent:*

(a) *for all $x, y \in A$, $((x \odot y)^n)^- \in P$ for some $n \in \mathbb{N}$ implies $(x^m)^- \in P$ or $(y^m)^- \in P$ for some $m \in \mathbb{N}$;*
(b) *for all $x, y \in A$, $((x \odot y)^n)^\sim \in P$ for some $n \in \mathbb{N}$ implies $(x^m)^\sim \in P$ or $(y^m)^\sim \in P$ for some $m \in \mathbb{N}$.*

Proof This is obvious taking into consideration that, since P is a normal deductive system, $x^- \in P$ iff $x^\sim \in P$ for all $x \in A$. $\square$

Definition 1.21 A proper normal deductive system of a bounded pseudo-BCK(pP) algebra A is called *primary* if it satisfies one of the equivalent conditions from Proposition 1.46.

Remark 1.15 If the bounded pseudo-BCK(pP) algebra A is normal, then its primary deductive systems can be dually characterized by means of the operation $\oplus$. Indeed, if P is a proper normal deductive system of A, applying Proposition 1.26 we have:

$$\left((x \odot y)^n\right)^- = n\left(y^- \oplus x^-\right), \qquad \left(x^m\right)^- = mx^- \quad \text{and} \quad \left(y^m\right)^- = my^-$$

for all $n, m \in \mathbb{N}$.

Therefore, a proper normal deductive system P of the normal pseudo-BCK(pP) algebra A is primary if it satisfies the following condition for all $x, y \in A$:

$$\text{if } n\left(y^- \oplus x^-\right) \in P \text{ for some } n \in \mathbb{N}, \quad \text{then}$$

$$mx^- \in P \text{ or } my^- \in P \text{ for some } m \in \mathbb{N}.$$

Obviously, the above condition is equivalent to the following:

$$\text{if } n\left(y^\sim \oplus x^\sim\right) \in P \text{ for some } n \in \mathbb{N}, \quad \text{then}$$

$$mx^\sim \in P \text{ or } my^\sim \in P \text{ for some } m \in \mathbb{N}.$$

Proposition 1.47 *Let A be a bounded pseudo-BCK(pP) algebra and P a primary deductive system of A. Then the following are equivalent:*

(a) *for all $x \in A$, $(x^n)^- \in P$ for some $n \in \mathbb{N}$ implies $((x^-)^m)^- \notin P$ for all $m \in \mathbb{N}$;*
(b) *for all $x \in A$, $(x^n)^\sim \in P$ for some $n \in \mathbb{N}$ implies $((x^\sim)^m)^\sim \notin P$ for all $m \in \mathbb{N}$.*

Proof Since P is primary, it is a normal deductive system. Hence $x^- \in P$ iff $x^\sim \in P$ for all $x \in A$ and the assertion follows immediately. $\square$

Definition 1.22 A primary deductive system of a bounded pseudo-BCK(pP) algebra A is called *perfect* if it satisfies one of the equivalent conditions from Proposition 1.47.

Proposition 1.48 *If P is a perfect primary deductive system of a bounded pseudo-BCK(pP) algebra A, then:*

(1) *for all $x \in A$, $(x^n)^- \in P$ for some $n \in \mathbb{N}$ iff $((x^-)^m)^- \notin P$ for all $m \in \mathbb{N}$;*
(2) *for all $x \in A$, $(x^n)^\sim \in P$ for some $n \in \mathbb{N}$ iff $((x^\sim)^m)^\sim \notin P$ for all $m \in \mathbb{N}$.*

Proof

(1) The first implication follows immediately, since P is perfect.

Consider $x \in A$ such that $((x^-)^m)^- \notin P$ for all $m \in \mathbb{N}$. By $(psbck\text{-}c_{36})$, $x^- \odot x = 0$, so $((x^- \odot x)^m)^- = 0^- = 1 \in P$ for all $m \in \mathbb{N}$. Since P is primary, it follows that $((x^-)^n)^- \in P$ or $(x^n)^- \in P$ for some $n \in \mathbb{N}$. Taking into consideration that $((x^-)^n)^- \notin P$ for all $n \in \mathbb{N}$, we conclude that $(x^n)^- \in P$ for some $n \in \mathbb{N}$.

(2) Similar to (1).   □

Definition 1.23 Let A be a bounded pseudo-BCK(pP) algebra and $X \subseteq A \setminus \{0\}$. We say that an element $x \in A$ is:

(i) an X-*left zero divisor* if there is a $y \in X$ such that $x \odot y = 0$;
(ii) an X-*right zero divisor* if there is a $y \in X$ such that $y \odot x = 0$;
(iii) an X-*zero divisor* if there are $y_1, y_2 \in X$ such that $x \odot y_1 = y_2 \odot x = 0$.

If $X = A \setminus \{0\}$, then an X-zero divisor is called a *zero divisor* of A.

The element 0 is called the *trivial zero divisor*.

The set of all X-left zero divisors, X-right zero divisors and X-zero divisors of A will be denoted by $X_l\text{-Div}(A)$, $X_r\text{-Div}(A)$ and $X\text{-Div}(A)$, respectively.

By $\text{Div}(A)$ we will denote the set of all zero divisors of A.

Remark 1.16 Let A be a bounded pseudo-BCK(pP) algebra and $X \subseteq A \setminus \{0\}$. Then:

(1) $0 \in X_l\text{-Div}(A) \cap X_r\text{-Div}(A)$ and $1 \notin X_l\text{-Div}(A) \cup X_r\text{-Div}(A)$;
(2) if $x \in \text{Div}(A)$ and $y \in A$ such that $y \leq x$, then $y \in \text{Div}(A)$;
(3) if $x, y \in \text{Div}(A)$, then $x \odot y, y \odot x \in \text{Div}(A)$.

Proposition 1.49 *Let A be a bounded pseudo-BCK(pP) algebra satisfying the conditions*: $\text{Div}(A) = \{0\}$, $\text{ord}(x) = \infty$ *and* $x^- = x^{\sim} = 0$ *for all* $x \in A \setminus \{0\}$. *Then any proper normal deductive system of A is perfect.*

Proof We first prove that any proper normal deductive system P of A is primary.

Let $x, y \in A$ and consider the following cases:

(1) If $x, y > 0$, then $x \odot y > 0$, so $\text{ord}(x \odot y) = \infty$. It follows that $(x \odot y)^n \neq 0$ for all $n \in \mathbb{N}$. Hence $((x \odot y)^n)^- = 0 \notin P$.
(2) If $x = 0$, then $((x \odot y)^n)^- = 0^- = 1 \in P$ for all $n \in \mathbb{N}$. Moreover, $(x^m)^- = 0^- = 1 \in P$ for all $m \in \mathbb{N}$.
(3) If $y = 0$, then $((x \odot y)^n)^- = 0^- = 1 \in P$ for all $n \in \mathbb{N}$. Then we have $(y^m)^- = 0^- = 1 \in P$ for all $m \in \mathbb{N}$.

Thus P is a primary deductive system of A.

Since $x^n \neq 0$ for all $x \in A \setminus \{0\}$, it follows that $(x^n)^- = 0 \notin P$ for all $n \in \mathbb{N}$.

For $x = 0$ we have $(0^n)^- = 1 \in P$ for all $n \in \mathbb{N}$ and $((0^-)^m)^- = 0 \notin P$ for all $m \in \mathbb{N}$. Thus P is a perfect deductive system of A.   □

Example 1.22

(1) It is a simple routine to check that the normal deductive system $D = \{s, a, b, n, c, d, m, 1\}$ of the bounded pseudo-BCK(pP) algebra A from Example 1.9 is primary, but D is not perfect: $(a_1^2)^- = 0^- = 1 \in D$ and $((a_1^-)^2)^- = (a_1^2)^- = 0^- = 1 \in D$.

(2) According to Proposition 1.49, the normal deductive systems $D_1 = \{a_1, a_2, b_2, s, a, b, n, c, d, m, 1\}$ and $D_3 = \{s, a, b, n, c, d, m, 1\}$ of the pseudo-BCK(pP) algebra A_1 from Example 1.17 are perfect deductive systems.

Chapter 2
Pseudo-hoops

A generalization of pseudo-BL algebras was given in [148], where the *pseudo-hoops* were defined and studied. Pseudo-hoops were originally introduced by Bosbach in [15] and [16] under the name of *complementary semigroups*. It was proved that a pseudo-hoop has the pseudo-divisibility condition and it is a meet-semilattice, so a bounded $R\ell$-monoid can be viewed as a bounded pseudo-hoop together with the join-semilattice property. In other words, a bounded pseudo-hoop is a meet-semilattice ordered residuated, integral and divisible monoid.

In this chapter we present the main notions and results regarding pseudo-hoops from [148], we prove new properties of these structures, we introduce the notions of a join-center and cancellative-center of a pseudo-hoop and we define and study algebras on subintervals of pseudo-hoops.

2.1 Definitions and Properties

Definition 2.1 A *pseudo-hoop* is an algebra $(A, \odot, \rightarrow, \rightsquigarrow, 1)$ of the type $(2, 2, 2, 0)$ such that, for all $x, y, z \in A$:

$(psHOOP_1)$ $x \odot 1 = 1 \odot x = x$;
$(psHOOP_2)$ $x \rightarrow x = x \rightsquigarrow x = 1$;
$(psHOOP_3)$ $(x \odot y) \rightarrow z = x \rightarrow (y \rightarrow z)$;
$(psHOOP_4)$ $(x \odot y) \rightsquigarrow z = y \rightsquigarrow (x \rightsquigarrow z)$;
$(psHOOP_5)$ $(x \rightarrow y) \odot x = (y \rightarrow x) \odot y = x \odot (x \rightsquigarrow y) = y \odot (y \rightsquigarrow x)$.

In the sequel, we will agree that $\odot$ has higher priority than the operations $\rightarrow, \rightsquigarrow$, and those higher than $\wedge$ and $\vee$.

If the operation $\odot$ is commutative, or equivalently $\rightarrow = \rightsquigarrow$, then the pseudo-hoop is said to be *hoop*. Properties of hoops were studied in [1, 10, 15, 16].

On the pseudo-hoop A we define $x \leq y$ iff $x \rightarrow y = 1$ (equivalent to $x \rightsquigarrow y = 1$) and $\leq$ is a partial order on A.

In the sequel we will refer to the pseudo-hoop $(A, \odot, \rightarrow, \rightsquigarrow, 1)$ by its universe A.

L.C. Ciungu, *Non-commutative Multiple-Valued Logic Algebras*,
Springer Monographs in Mathematics, DOI 10.1007/978-3-319-01589-7_2,
© Springer International Publishing Switzerland 2014

Proposition 2.1 *In every pseudo-hoop* $(A, \odot, \rightarrow, \rightsquigarrow, 1)$ *the following hold*:

(*pshoop-c$_1$*) $x \odot y \leq z$ iff $x \leq y \rightarrow z$ iff $y \leq x \rightsquigarrow z$ (*pseudo-residuation*);
(*pshoop-c$_2$*) $(A, \odot, 1)$ *is a monoid and* $x \leq y$ *implies* $x \odot z \leq y \odot z, z \odot x \leq z \odot y$;
(*pshoop-c$_3$*) $(A, \leq)$ *is a meet-semilattice with*

$$x \wedge y = (x \rightarrow y) \odot x = (y \rightarrow x) \odot y = x \odot (x \rightsquigarrow y)$$

$$= y \odot (y \rightsquigarrow x) \quad (\textit{pseudo-divisibility});$$

(*pshoop-c$_4$*) *the element* 1 *is the greatest element of A and*

$$1 \rightarrow x = 1 \rightsquigarrow x = x, \qquad x \rightarrow 1 = x \rightsquigarrow 1 = 1;$$

(*pshoop-c$_5$*) $x \leq (x \rightarrow y) \rightsquigarrow y, x \leq (x \rightsquigarrow y) \rightarrow y$;
(*pshoop-c$_6$*) $x \rightarrow y \leq (y \rightarrow z) \rightsquigarrow (x \rightarrow z), x \rightsquigarrow y \leq (y \rightsquigarrow z) \rightarrow (x \rightsquigarrow z)$.

Proof

(*pshoop-c$_1$*) We have $x \odot y \leq z$ iff $x \odot y \rightarrow z = 1$ iff $x \rightarrow (y \rightarrow z) = 1$ iff $x \leq y \rightarrow z$. Similarly, $x \odot y \leq z$ iff $y \leq x \rightsquigarrow z$.

(*pshoop-c$_2$*) By (*psHOOP$_1$*), 1 is the neutral element of A.

Consider $u \in A$. We have:

$$(x \odot y) \odot z \leq u \quad \text{iff} \quad x \odot y \leq z \rightarrow u \quad \text{iff}$$

$$x \leq y \rightarrow (z \rightarrow u) = y \odot z \rightarrow u \quad \text{iff} \quad x \odot (y \odot z) \leq u.$$

It follows that $(x \odot y) \odot z = x \odot (y \odot z)$, so $\odot$ is associative.

Thus $(A, \odot, 1)$ is a monoid.

From $y \odot z \leq y \odot z$ we have $y \leq z \rightarrow y \odot z$. Since $x \leq y$, we get $x \leq z \rightarrow y \odot z$, hence $x \odot z \leq y \odot z$. Similarly, $z \odot x \leq z \odot y$.

(*pshoop-c$_3$*) From $x \rightarrow y \leq x \rightarrow y$ and $y \rightarrow x \leq y \rightarrow x$ we get $(x \rightarrow y) \odot x \leq y$ and $(y \rightarrow x) \odot y \leq x$, respectively. Hence, if we denote $x \wedge y = (x \rightarrow y) \odot x = (y \rightarrow x) \odot y$, it follows that $x \wedge y$ is a lower bound of $\{x, y\}$.

Let u be another lower bound of $\{x, y\}$, that is, $u \leq x$ and $u \leq y$.

From $u \rightarrow x = 1 = u \rightarrow u$ we get $(u \rightarrow x) \odot u \leq u \leq y$.

But $(u \rightarrow x) \odot u = (x \rightarrow u) \odot x$, hence $(x \rightarrow u) \odot x \leq y$, so $x \rightarrow u \leq x \rightarrow y$.

We have $(x \rightarrow u) \odot x = (u \rightarrow x) \odot u = 1 \odot u = u$.

It follows that $u = (x \rightarrow u) \odot x \leq (x \rightarrow y) \odot x = x \wedge y$. Thus $x \wedge y$ is the greatest lower bound of $\{x, y\}$ with respect to the order $\leq$.

We conclude that every pseudo-hoop is a meet-semilattice with respect to the order $\leq$, where the meet operation is given by

$$x \wedge y = (x \rightarrow y) \odot x = (y \rightarrow x) \odot y = x \odot (x \rightsquigarrow y) = y \odot (y \rightsquigarrow x).$$

(*pshoop-c$_4$*) Taking $y = 1$ in the equality $x \wedge y = (x \rightarrow y) \odot x = (y \rightarrow x) \odot y$, we get $x \wedge 1 = (x \rightarrow 1) \odot x = 1 \rightarrow x$. It follows that $1 \rightarrow x = x \wedge 1 \leq x$.

On the other hand, from $x \odot 1 = x$ we have $x \leq 1 \to x$.

Hence $1 \to x = x$ and similarly $1 \rightsquigarrow x = x$.

It follows that $x = 1 \to x = (x \to 1) \odot x = x \wedge 1 \leq 1$ for all $x \in A$.

We conclude that 1 is the greatest element of A.

From $1 \odot x = x \leq 1$ we get $1 \leq x \to 1$. But $x \to 1 \leq 1$, hence $x \to 1 = 1$.

Similarly, $x \rightsquigarrow 1 = 1$.

($pshoop$-c_5) From $(x \to y) \odot x = x \wedge y \leq y$, we get $x \leq (x \to y) \rightsquigarrow y$.

Similarly, from $x \odot (x \rightsquigarrow y) = x \wedge y \leq y$, we have $x \leq (x \rightsquigarrow y) \to y$.

($pshoop$-c_6) Applying ($pshoop$-c_3), we get:

$$(y \to z) \odot (x \to y) \odot x = (y \to z) \odot (x \wedge y) \leq (y \to z) \odot y = y \wedge z \leq z.$$

It follows that $(y \to z) \odot (x \to y) \leq x \to z$, so $x \to y \leq (y \to z) \rightsquigarrow (x \to z)$.

Similarly, $x \rightsquigarrow y \leq (y \rightsquigarrow z) \to (x \rightsquigarrow z)$.   □

Proposition 2.2 *Every pseudo-hoop is a pseudo-BCK(pP) algebra which is a meet-semilattice satisfying the pseudo-divisibility property.*

Proof Suppose that $(A, \odot, \to, \rightsquigarrow, 1)$ is a pseudo-hoop.

We will prove that $(A, \leq, \to, \rightsquigarrow, 1)$ is a pseudo-BCK(pP) algebra.

($psBCK_1$) follows from ($pshoop$-c_6);

($psBCK_2$) follows from ($pshoop$-c_5);

($psBCK_3$) follows from ($psHOOP_2$);

($psBCK_4$) follows from ($pshoop$-c_4);

($psBCK_5$) and ($psBCK_6$) follow by the definition of $\leq$ and from the fact that $\leq$ is a partial order on A.

The (pP) condition is a consequence of ($pshoop$-c_1) and Theorem 1.4.

Thus $(A, \leq, \to, \rightsquigarrow, 1)$ is a pseudo-BCK(pP) algebra.   □

It follows that all properties proved for a pseudo-BCK(pP) algebra are also valid in the case of pseudo-hoops.

Definition 2.2 A *bounded pseudo-hoop* is an algebra $(A, \odot, \to, \rightsquigarrow, 0, 1)$ such that $(A, \odot, \to, \rightsquigarrow, 1)$ is a pseudo-hoop and $0 \leq x$ for all $x \in A$.

Proposition 2.3 ([148]) *In any pseudo-hoop A, the following property holds for all $x, y \in A$:*

($pshoop$-c_7) $[((y \to x) \rightsquigarrow x) \to y] \to (y \to x) = y \to x$ *and* $[((y \rightsquigarrow x) \to x) \rightsquigarrow y] \rightsquigarrow (y \rightsquigarrow x) = y \rightsquigarrow x.$

Proof Let $z = (y \to x) \rightsquigarrow x$. By ($pshoop$-$c_5$) and ($psbck$-$c_{22}$) we have $y \leq z$ and $z \to x = y \to x$. It follows that

$$(z \to y) \to (y \to x) = (z \to y) \to (z \to x) = (z \to y) \odot z \to x = (z \land y) \to x$$

$$= y \to x.$$

Similarly, $[((y \rightsquigarrow x) \to x) \rightsquigarrow y] \rightsquigarrow (y \rightsquigarrow x) = y \rightsquigarrow x.$ □

Let $(A, \odot, \to, \rightsquigarrow, 0, 1)$ be a bounded pseudo-hoop.
We define two negations $^-$ and $^\sim$: for all $x \in A$, $x^- := x \to 0$, $x^\sim := x \rightsquigarrow 0$.
A bounded pseudo-hoop A is called *good* if $x^{-\sim} = x^{\sim-}$ for all $x \in A$.
If $x^{-\sim} = x^{\sim-} = x$ for all $x \in A$, then the bounded pseudo-hoop A is said to have
the *pseudo-double negation* condition, (pDN) for short.
Obviously, any bounded pseudo-hoop with (pDN) is good.

Proposition 2.4 *If A is a good pseudo-hoop, then for all $x, y \in A$:*

(*pshoop-c*$_8$) $(x^{-\sim} \to x)^\sim = (x^{-\sim} \rightsquigarrow x)^- = 0$;
(*pshoop-c*$_9$) $(x \to y)^{-\sim} = x^{-\sim} \to y^{-\sim}$ *and* $(x \rightsquigarrow y)^{-\sim} = x^{-\sim} \rightsquigarrow y^{-\sim}$;
(*pshoop-c*$_{10}$) $(x \land y)^{-\sim} = x^{-\sim} \land y^{-\sim}$;
(*pshoop-c*$_{11}$) $x \to y^- = x^{-\sim} \to y^-$ *and* $x \rightsquigarrow y^\sim = x^{-\sim} \rightsquigarrow y^\sim.$

Proof

(*pshoop-c*$_8$) From $0 \le x$ applying (*psbck-c*$_{10}$) we get $x^{-\sim} \to 0 \le x^{-\sim} \to x$, that
is, $x^- \le x^{-\sim} \to x$. Hence $(x^{-\sim} \to x)^\sim \le x^{-\sim}$.
It follows that:

$$\left(x^{-\sim} \to x\right)^\sim = \left(x^{-\sim} \to x\right)^\sim \land x^{-\sim} = x^{-\sim} \odot \left[\left(x^{-\sim} \rightsquigarrow \left(x^{-\sim} \to x\right)^\sim\right)\right]$$

$$= x^{-\sim} \odot \left[\left(x^{-\sim} \to x\right) \odot x^{-\sim}\right]^\sim = x^{-\sim} \odot \left(x^{-\sim} \land x\right)^\sim$$

$$= x^{-\sim} \odot x^\sim = 0$$

(we applied the axiom $v \odot (v \rightsquigarrow u) = u \land v$ for $v = x^{-\sim}$ and $u = (x^{-\sim} \to x)^\sim$).
Similarly, $x^{-\sim} \rightsquigarrow 0 \le x^{-\sim} \rightsquigarrow x$, so $x^\sim \le x^{-\sim} \rightsquigarrow x$. Hence $(x^{-\sim} \rightsquigarrow x)^- \le x^{-\sim}$.
Therefore

$$\left(x^{-\sim} \rightsquigarrow x\right)^- = \left(x^{-\sim} \rightsquigarrow x\right)^- \land x^{-\sim} = \left[x^{-\sim} \to \left(x^{-\sim} \rightsquigarrow x\right)^-\right] \odot x^{-\sim}$$

$$= \left[x^{-\sim} \odot \left(x^{-\sim} \rightsquigarrow x\right)\right]^- \odot x^{-\sim} = \left(x^{-\sim} \land x\right)^- \odot x^{-\sim}$$

$$= x^- \odot x^{-\sim} = 0.$$

Thus $(x^{-\sim} \to x)^\sim = (x^{-\sim} \rightsquigarrow x)^- = 0$.
(*pshoop-c*$_9$) Applying the axioms of a pseudo-hoop we have:

$$(x \land y) \odot (y \rightsquigarrow 0) \le y \odot (y \rightsquigarrow 0) = y \land 0 = 0.$$

But $x \land y = (x \to y) \odot x$, hence $(x \to y) \odot x \odot (y \rightsquigarrow 0) \le 0.$

It follows that $x \odot (y \rightsquigarrow 0) \leq (x \rightarrow y) \rightsquigarrow 0$. Applying $(psbck\text{-}c_1)$ we get

$$\left[(x \rightarrow y) \rightsquigarrow 0\right] \rightarrow 0 \leq x \odot (y \rightsquigarrow 0) \rightarrow 0 = x \rightarrow \left[(y \rightsquigarrow 0) \rightarrow 0\right].$$

Thus $(x \rightarrow y)^{-\sim} \leq x \rightarrow y^{-\sim}$. On the other hand we have:

$$\left(x \rightarrow y^{-\sim}\right)^{-\sim} = \left(x \odot y^{\sim}\right)^{--\sim} = \left(x \odot y^{\sim}\right)^{-} = x \rightarrow y^{-\sim}.$$

Replacing x with $x \rightarrow y^{-\sim}$ and y with $x \rightarrow y$ in the above identity we have:

$$\left(x \rightarrow y^{-\sim}\right) \rightarrow (x \rightarrow y)^{-\sim} = \left[\left(x \rightarrow y^{-\sim}\right) \rightarrow (x \rightarrow y)^{-\sim}\right]^{-\sim}.$$

Since $x \rightarrow y \leq (x \rightarrow y)^{-\sim}$, applying $(psbck\text{-}c_{10})$ we get:

$$\left(x \rightarrow y^{-\sim}\right) \rightarrow (x \rightarrow y)^{-\sim} \geq \left(x \rightarrow y^{-\sim}\right) \rightarrow (x \rightarrow y), \quad \text{so}$$

$$\left[\left(x \rightarrow y^{-\sim}\right) \rightarrow (x \rightarrow y)^{-\sim}\right]^{-\sim} \geq \left[\left(x \rightarrow y^{-\sim}\right) \rightarrow (x \rightarrow y)\right]^{-\sim}.$$

Taking into consideration the above identity and applying $(psHOOP_3)$ we get:

$$\left(x \rightarrow y^{-\sim}\right) \rightarrow (x \rightarrow y)^{-\sim} \geq \left[\left(x \rightarrow y^{-\sim}\right) \rightarrow (x \rightarrow y)\right]^{-\sim}$$

$$= \left[\left(x \rightarrow y^{-\sim}\right) \odot x \rightarrow y\right]^{-\sim} = \left(x \wedge y^{-\sim} \rightarrow y\right)^{-\sim}.$$

But $x \wedge y \leq y \leq y^{-\sim}$, hence by $(psbck\text{-}c_1)$ we have $y^{-\sim} \rightarrow y \leq x \wedge y^{-\sim} \rightarrow y$. Applying $(pshoop\text{-}c_8)$ we get:

$$\left(x \rightarrow y^{-\sim}\right) \rightarrow (x \rightarrow y)^{-\sim} \geq \left(y^{-\sim} \rightarrow y\right)^{-\sim} = 0^{-} = 1.$$

It follows that $(x \rightarrow y^{-\sim}) \rightarrow (x \rightarrow y)^{-\sim} = 1$, that is, $x \rightarrow y^{-\sim} \leq (x \rightarrow y)^{-\sim}$. We conclude that $(x \rightarrow y)^{-\sim} = x \rightarrow y^{-\sim}$.
From $(psbck\text{-}c_{19})$ we have $x \rightarrow y^{-\sim} = x^{-\sim} \rightarrow y^{-\sim}$, hence $(x \rightarrow y)^{-\sim} = x^{-\sim} \rightarrow y^{-\sim}$. Similarly, $(x \rightsquigarrow y)^{-\sim} = x^{-\sim} \rightsquigarrow y^{-\sim}$.
$(pshoop\text{-}c_{10})$ From $x \wedge y \leq x, y$ we get $(x \wedge y)^{-\sim} \leq x^{-\sim}, y^{-\sim}$.
Hence $(x \wedge y)^{-\sim} \leq x^{-\sim} \wedge y^{-\sim}$.
On the other hand, applying Proposition 1.23 and $(pshoop\text{-}c_9)$ we have:

$$(x \wedge y)^{-\sim} = \left((x \rightarrow y) \odot x\right)^{-\sim} \geq (x \rightarrow y)^{-\sim} \odot x^{-\sim}$$

$$= \left(x^{-\sim} \rightarrow y^{-\sim}\right) \odot x^{-\sim} = x^{-\sim} \wedge y^{-\sim}.$$

We conclude that $(x \wedge y)^{-\sim} = x^{-\sim} \wedge y^{-\sim}$.
$(pshoop\text{-}c_{11})$ Applying $(psbck\text{-}c_{19})$, $(pshoop\text{-}c_9)$ and $(psbck\text{-}c_{37})$ we have:

$$x^{-\sim} \rightarrow y^{-} = x^{-\sim} \rightarrow y^{--\sim} = \left(x \rightarrow y^{-}\right)^{-\sim}$$

$$= \left((x \odot y)^{-}\right)^{-\sim} = (x \odot y)^{-} = x \rightarrow y^{-}.$$

Similarly, $x \rightsquigarrow y^{\sim} = x^{-\sim} \rightsquigarrow y^{\sim}$. $\qquad\square$

Proposition 2.5 ([148]) *Let A be a pseudo-hoop and I an arbitrary set. Then*:

(1) $x \odot (\bigvee_{i \in I} y_i) = \bigvee_{i \in I} (x \odot y_i)$;
(2) $(\bigvee_{i \in I} y_i) \odot x = \bigvee_{i \in I} (y_i \odot x)$;
(3) $x \wedge (\bigvee_{i \in I} y_i) = \bigvee_{i \in I} (x \wedge y_i)$,

whenever the arbitrary unions exist.

Proof

(1) Since $y_i \le \bigvee_{i \in I} y_i$ for all $i \in I$, it follows that $x \odot y_i \le x \odot (\bigvee_{i \in I} y_i)$ for all $i \in I$. Thus $\bigvee_{i \in I} (x \odot y_i) \le x \odot (\bigvee_{i \in I} y_i)$.

Let $z \in A$ such that $\bigvee_{i \in I} (x \odot y_i) \le z$. It follows that $x \odot y_i \le z$ for all $i \in I$, so $y_i \le x \rightsquigarrow z$ for all $i \in I$. Thus $\bigvee_{i \in I} y_i \le x \rightsquigarrow z$, so $x \odot (\bigvee_{i \in I} y_i) \le z$.

So, for any $z \in A$, we have proved that $\bigvee_{i \in I} (x \odot y_i) \le z$ implies $x \odot (\bigvee_{i \in I} y_i) \le z$.

Taking $z = \bigvee_{i \in I} (x \odot y_i)$, we get $x \odot (\bigvee_{i \in I} y_i) \le \bigvee_{i \in I} (x \odot y_i)$.

We conclude that $x \odot (\bigvee_{i \in I} y_i) = \bigvee_{i \in I} (x \odot y_i)$.

(2) Similar to (1).

(3) From $y_i \le \bigvee_{i \in I} y_i$, we get $x \wedge y_i \le x \wedge \bigvee_{i \in I} y_i$ for all $i \in I$.

Hence $\bigvee_{i \in I} (x \wedge y_i) \le x \wedge (\bigvee_{i \in I} y_i)$.

Conversely, we have:

$$x \wedge \left(\bigvee_{i \in I} y_i \right) = \left(\bigvee_{i \in I} y_i \right) \wedge x = \left(\bigvee_{i \in I} y_i \right) \odot \left(\bigvee_{i \in I} y_i \rightsquigarrow x \right)$$

$$= \bigvee_{i \in I} \left(y_i \odot \left(\bigvee_{i \in I} y_i \rightsquigarrow x \right) \right).$$

From (*psbck-c*$_1$) we have $\bigvee_{i \in I} y_i \rightsquigarrow x \le y_i \rightsquigarrow x$ for all $i \in A$, so $y_i \odot (\bigvee_{i \in I} y_i \rightsquigarrow x) \le y_i \odot (y_i \rightsquigarrow x) = y_i \wedge x = x \wedge y_i$ for all $i \in I$.

It follows that

$$\bigvee_{i \in I} \left(y_i \odot \left(\bigvee_{i \in I} y_i \rightsquigarrow x \right) \right) \le \bigvee_{i \in I} (x \wedge y_i), \quad \text{so } x \wedge \left(\bigvee_{i \in I} y_i \right) \le \bigvee_{i \in I} (x \wedge y_i).$$

Thus $x \wedge (\bigvee_{i \in I} y_i) = \bigvee_{i \in I} (x \wedge y_i)$. $\square$

Proposition 2.6 ([148]) *Let A be a pseudo-hoop and $H \subseteq A$. The following are equivalent*:

(a) *H is a compatible deductive system*;
(b) *H is a normal filter (i.e. $x \odot H = H \odot x$ for all $x \in A$).*

Proof

(a) $\Rightarrow$ (b) Consider $y \in x \odot H$, $y = x \odot h$ with $x \in A$ and $h \in H$.

It follows that $x \odot h = y = x \wedge y = (x \rightarrow y) \odot x$.

From $h \leq x \rightsquigarrow x \odot h = x \rightsquigarrow y$ and $h \in H$ we have $x \rightsquigarrow y \in H$.
Since H is a compatible deductive system of A, we get $x \rightarrow y \in H$.
Consequently, if we let $h' = x \rightarrow y$, it follows that $y = h' \odot x \in H \odot x$.
Thus $x \odot H \subseteq H \odot x$.
Similarly, $H \odot x \subseteq x \odot H$, so $x \odot H = H \odot x$ for all $x \in A$.

(b) $\Rightarrow$ (a) Assume $x \rightsquigarrow y \in H$, so $x \wedge y = x \odot (x \rightsquigarrow y) \in x \odot H = H \odot x$, that is, $x \wedge y = h \odot x$ for some $h \in H$.
It follows that $x \rightarrow y = x \rightarrow x \wedge y = x \rightarrow h \odot x$. Since $h \leq x \rightarrow h \odot x$ and $h \in H$, we get $x \rightarrow y \in H$. Similarly, from $x \rightarrow y \in H$ we get $x \rightsquigarrow y \in H$.
Thus H is a compatible deductive system of A. $\square$

If $x, y \in A$, then we define the *pseudo-joins* of x and y by:

$$x \sqcup_1 y = \big((x \rightarrow y) \rightsquigarrow y\big) \wedge \big((y \rightarrow x) \rightsquigarrow x\big),$$
$$x \sqcup_2 y = \big((x \rightsquigarrow y) \rightarrow y\big) \wedge \big((y \rightsquigarrow x) \rightarrow x\big).$$

We will also use the notation:

$$x \vee_1 y = (x \rightarrow y) \rightsquigarrow y \quad \text{and} \quad x \vee_2 y = (x \rightsquigarrow y) \rightarrow y.$$

Obviously, $x \sqcup_1 y = (x \vee_1 y) \wedge (y \vee_1 x)$ and $x \sqcup_2 y = (x \vee_2 y) \wedge (y \vee_2 x)$.

Proposition 2.7 ([148]) *In any bounded pseudo-hoop A the following hold*:

(1) $x \sqcup_1 0 = x \sqcup_2 0 = 0 \sqcup_1 x = 0 \sqcup_2 x = x$;
(2) $x \sqcup_1 1 = x \sqcup_2 1 = 1 \sqcup_1 x = 1 \sqcup_2 x = 1$;
(3) $x \sqcup_1 x = x \sqcup_2 x = x$.

Proof

(1) $x \sqcup_1 0 = ((x \rightarrow 0) \rightsquigarrow 0) \wedge ((0 \rightarrow x) \rightsquigarrow x) = x^{-\sim} \wedge x = x$ (by $(psbck\text{-}c_{14})$).
      Similarly, $x \sqcup_2 0 = 0 \sqcup_1 x = 0 \sqcup_2 x = x$.
(2) $x \sqcup_1 1 = ((x \rightarrow 1) \rightsquigarrow 1) \wedge ((1 \rightarrow x) \rightsquigarrow x) = 1 \wedge 1 = 1$.
      Similarly, $x \sqcup_2 1 = 1 \sqcup_1 x = 1 \sqcup_2 x = 1$.
(3) This is obvious. $\square$

Proposition 2.8 ([148]) *In any bounded pseudo-hoop A the following hold*:

(1) $x \sqcup_1 y = y \sqcup_1 x$ *and* $x \sqcup_2 y = y \sqcup_2 x$;
(2) $x, y \leq x \sqcup_1 y$ *and* $x, y \leq x \sqcup_2 y$;
(3) $x \leq y$ *iff* $x \sqcup_1 y = y$;
(4) $x \leq y$ *iff* $x \sqcup_2 y = y$.

Proof

(1) This is obvious.
(2) By $(pshoop\text{-}c_5)$ and $(psbck\text{-}c_6)$ we have $x, y \leq (x \rightarrow y) \rightsquigarrow y$ and $x, y \leq (y \rightarrow x) \rightarrow x$. Hence $x, y \leq x \sqcup_1 y$. Similarly, $x, y \leq x \sqcup_2 y$.

(3) If $x \le y$, then $(x \to y) \rightsquigarrow y = 1 \rightsquigarrow y = y$.

   Hence $x \cup_1 y = y \land [(y \to x) \rightsquigarrow x] = y$, since by (pshoop-c5), $y \le (y \to x) \rightsquigarrow x$.

   Conversely, suppose that $x \cup_1 y = y$.

   It follows that $x \land y = x \land (x \cup_1 y) = x$, by (2). Thus $x \le y$.

(4) Similar to (3).   □

Proposition 2.9 ([148]) *Let A be a pseudo-hoop. The following are equivalent*:

(a) $\cup_1$ *is associative*;
(b) *for all* $x, y, z \in A$, $x \le y$ *implies* $x \cup_1 z \le y \cup_1 z$;
(c) *for all* $x, y, z \in A$, $x \cup_1 (y \land z) \le (x \cup_1 y) \land (x \cup_1 z)$;
(d) $\cup_1$ *is the join operation on A.*

Proof

(a) $\Rightarrow$ (d) We have $x, y \le x \cup_1 y$, so $x \cup_1 y$ is an upper bound of $\{x, y\}$.

   Let $z \in A$ such that $x, y \le z$. By (a), $(x \cup_1 y) \cup_1 z = x \cup_1 (y \cup_1 z)$.

   From $y \le z$ and (pshoop-c5) we have:

$$y \cup_1 z = \big((y \to z) \rightsquigarrow z\big) \land \big((z \to y) \rightsquigarrow y\big) = (1 \rightsquigarrow z) \land \big((z \to y) \rightsquigarrow y\big) = z.$$

   It follows that $x \cup_1 (y \cup_1 z) = x \cup_1 z = ((x \to z) \rightsquigarrow z) \land ((z \to x) \rightsquigarrow x) = z$.

   Thus $x \cup_1 y \le (x \cup_1 y) \cup_1 z = z$. Hence $x \cup_1 y$ is the l.u.b. of $\{x, y\}$, so $x \lor y$

   exists and $x \lor y = x \cup_1 y$.

(d) $\Rightarrow$ (a) Applying (d) we have $(x \cup_1 y) \cup_1 z = (x \lor y) \lor z = x \lor (y \lor z) = x \cup_1 (y \cup_1 z)$.

   Thus $\cup_1$ is associative.

(b) $\Rightarrow$ (d) We have $x, y \le x \cup_1 y$, so $x \cup_1 y$ is an upper bound of $\{x, y\}$.

   Let $z \in A$ such that $x, y \le z$. From $x \le z$, applying (b) we obtain:

$$x \cup_1 y \le z \cup_1 y = \big((z \to y) \rightsquigarrow y\big) \land \big((y \to z) \rightsquigarrow z\big) = z.$$

   We conclude that $x \cup_1 y$ is the l.u.b. of $\{x, y\}$, so $x \lor y$ exists and $x \lor y = x \cup_1 y$.

(d) $\Rightarrow$ (b) Let $x, y, z \in A$ such that $x \le y$.

   It follows that $x \cup_1 z = x \lor z \le y \lor z = y \cup_1 z$.

(c) $\Rightarrow$ (d) We have $x, y \le x \cup_1 y$, so $x \cup_1 y$ is an upper bound of $\{x, y\}$.

   Let $z \in A$ such that $x, y \le z$. We have $x \cup_1 (y \land z) \le (x \cup_1 y) \land (x \cup_1 z)$.

   Since $x \cup_1 z = ((x \to z) \rightsquigarrow z) \land ((z \to x) \rightsquigarrow x) = z$, we get $x \cup_1 y \le (x \cup_1 y) \land$

   $z \le z$. Thus $x \cup_1 y$ is the l.u.b. of $\{x, y\}$, so $x \lor y$ exists and $x \lor y = x \cup_1 y$.

(d) $\Rightarrow$ (c) For all $x, y, z \in A$, $x \cup_1 (y \land z) = x \lor (y \land z)$.

   Obviously, $y \land z \le y$ implies $x \lor (y \land z) \le x \lor y$.

   From $x \le x \lor z$ and $y \land z \le z \le x \lor z$ we get $x \lor (y \land z) \le x \lor z$.

   It follows that $x \lor (y \land z) \le (x \lor y) \land (x \lor z)$.

   Hence $x \cup_1 (y \land z) = x \lor (y \land z) \le (x \lor y) \land (x \lor z) = (x \cup_1 y) \land (x \cup_1 z)$. □

Proposition 2.10 ([148]) *Let A be a pseudo-hoop. The following are equivalent*:

(a) $\sqcup_2$ *is associative;*
(b) *for all* $x, y, z \in A$, $x \le y$ *implies* $x \sqcup_2 z \le y \sqcup_2 z$;
(c) *for all* $x, y, z \in A$, $x \sqcup_2 (y \wedge z) \le (x \sqcup_2 y) \wedge (x \sqcup_2 z)$;
(d) $\sqcup_2$ *is the join operation on* A.

Proof The proof is similar to that of Proposition 2.9. $\square$

Remark 2.1 Suppose that $\sqcup_1$ is associative. By Proposition 2.9 it follows that $\sqcup_1$ is the join operation on A, that is, $\vee$ exists and $\vee = \sqcup_1$.

Applying Proposition 2.5(3) we get that $(A, \wedge, \vee)$ is a distributive lattice.

The same result is obtained using $\sqcup_2$ and Proposition 2.10.

Theorem 2.1 *Every bounded locally finite pseudo-hoop has property* (pDN).

Proof Let A be a bounded locally finite pseudo-hoop and $x \in A$. If $x = 0$, then $0^{-\sim} = 0^{\sim-} = 0$. Suppose $x \ne 0$. We prove that $x^{-\sim} = x$. By (*psbck-c14*) we have $x \le x^{-\sim}$. Suppose that $x^{-\sim} \nleq x$, hence $x^{-\sim} \to x \ne 1$. Since A is locally finite, there is an $n \in \mathbb{N}$, $n \ge 1$, such that $(x^{-\sim} \to x)^n = 0$. We have:

$$\left(x^{-\sim} \to x\right) \to x^- = \left(x^{-\sim} \to x\right) \to x^{-\sim-} = \left(x^{-\sim} \to x\right) \to \left(x^{-\sim} \to 0\right)$$
$$= \left(x^{-\sim} \to x\right) \odot x^{-\sim} \to 0 = \left(x \wedge x^{-\sim}\right) \to 0$$
$$= x \to 0 = x^-,$$
$$\left(x^{-\sim} \to x\right)^2 \to x^- = \left(x^{-\sim} \to x\right) \to \left(\left(x^{-\sim} \to x\right) \to x^-\right)$$
$$= \left(x^{-\sim} \to x\right) \to x^- = x^-.$$

By induction we get $(x^{-\sim} \to x)^n \to x^- = x^-$. Thus $0 \to x^- = x^-$, so $x^- = 1$. Hence $x \le x^{-\sim} = 0$, that is, $x = 0$, a contradiction. Therefore $x^{-\sim} \le x$, so $x^{-\sim} = x$. Similarly, $x^{\sim-} = x$. Thus A satisfies (pDN). $\square$

We recall the notion of an ordinal sum of pseudo-hoops.

Let A_1 and A_2 be pseudo-hoops such that $A_1 \cap A_2 = \{1\}$. We set $A = A_1 \cup A_2$ and we define the operations $\odot$, $\to$, $\rightsquigarrow$ on A as follows:

$$x \odot y := \begin{cases} x \odot_i y & \text{if } x, y \in A_i, i = 1, 2 \\ x & \text{if } x \in A_1 \setminus \{1\}, y \in A_2 \\ y & \text{if } x \in A_2, y \in A_1 \setminus \{1\} \end{cases}$$

$$x \to y := \begin{cases} x \to_i y & \text{if } x, y \in A_i, i = 1, 2 \\ y & \text{if } x \in A_2, y \in A_1 \setminus \{1\} \\ 1 & \text{if } x \in A_1 \setminus \{1\}, y \in A_2 \end{cases}$$

$$x \rightsquigarrow y := \begin{cases} x \rightsquigarrow_i y & \text{if } x, y \in A_i, i = 1, 2 \\ y & \text{if } x \in A_2 \, y \in A_1 \setminus \{1\} \\ 1 & \text{if } x \in A_1 \setminus \{1\}, y \in A_2. \end{cases}$$

Then $(A, \odot, \rightarrow, \rightsquigarrow, 1)$ is a pseudo-hoop called the *ordinal sum* of A_1 and A_2 and we denote it by $A = A_1 \oplus A_2$. The construction can of course be extended to arbitrarily many systems of pseudo-hoops.

Definition 2.3 A pseudo-hoop A is called:

(1) *simple* if $\{1\}$ is the unique proper normal filter of A;
(2) *strongly simple* if $\{1\}$ is the unique proper filter of A.

Obviously, any strongly simple pseudo-hoop is simple.

When A is a hoop, since filters and normal filters coincide, the notions of simple and strongly simple hoop coincide.

Proposition 2.11 ([148]) *For any pseudo-hoop A the following are equivalent*:

(a) *A is strongly simple*;
(b) *for all $x \in A$, if $x \neq 1$ then $[x] = A$*;
(c) *for all $x, y \in A$, if $x \neq 1$ then there exists an $n \in \mathbb{N}$, $n > 0$, such that $y \geq x^n$*;
(d) *for all $x, y \in A$, if $x \neq 1$ then there exists an $n \in \mathbb{N}$, $n > 0$, such that $x \rightarrow^n y = 1$ for some $n \in \mathbb{N}$, $n \geq 1$*;
(e) *for all $x, y \in A$, if $x \neq 1$ then there exists an $n \in \mathbb{N}$, $n \geq 1$, such that $x \rightsquigarrow^n y = 1$ for some $n \in \mathbb{N}$, $n > 0$*.

Proof (a) $\Leftrightarrow$ (b) is obvious.

By Lemma 1.9 and Proposition 1.33 any one of the conditions (c), (d) and (e) is equivalent to condition (b).   □

Lemma 2.1 ([148]) *In any strongly simple pseudo-hoop A the following hold for all $x, y \in A$*:

(1) *$y \rightarrow x = x$ implies $x = 1$ or $y = 1$*;
(2) *$y \rightsquigarrow x = x$ implies $x = 1$ or $y = 1$*.

Proof

(1) Consider $x, y \in A$ such that $y \rightarrow x = x$. Applying $(psHOOP_3)$, it follows by induction that $y^n \rightarrow x = x$ for all $n \in \mathbb{N}$, $n > 0$. If $y \neq 1$, then according to Proposition 2.11(c), there exists an $n_0 \in \mathbb{N}$, $n_0 > 0$, such that $y^{n_0} \leq x$, that is, $y^{n_0} \rightarrow x = 1$. Hence $x = 1$.
(2) Similar to (1).   □

Definition 2.4 A pseudo-hoop $(A, \odot, \rightarrow, \rightsquigarrow, 1)$ is said to be *cancellative* if the monoid $(A, \odot, 1)$ is cancellative, that is, $x \odot a = y \odot a$ implies $x = y$ and $a \odot x = a \odot y$ implies $x = y$ for all $x, y, a \in A$.

Proposition 2.12 ([148]) *A pseudo-hoop* A *is cancellative iff the following identities hold*:

(C_1) $y \rightarrow x \odot y = x$;
(C_2) $y \rightsquigarrow y \odot x = x$,

for all $x, y \in A$.

Proof Suppose that A is cancellative.

It follows that $x \odot y = y \wedge (x \odot y) = (y \rightarrow x \odot y) \odot y$, hence $x = y \rightarrow x \odot y$.

Similarly, $y \odot x = y \wedge (y \odot x) = y \odot (y \rightsquigarrow y \odot x)$, hence $x = y \rightsquigarrow y \odot x$.

Conversely, suppose that A satisfies (C_1) and (C_2).

If $x \odot z = y \odot z$, then applying (C_1) twice we get $x = z \rightarrow x \odot z = z \rightarrow y \odot z = y$.

Similarly, from $z \odot x = z \odot y$ and (C_2) it follows that $x = z \rightsquigarrow z \odot x = z \rightsquigarrow z \odot y = y$. $\square$

Example 2.1 ([148]) Let $\mathbf{G} = (G, \vee, \wedge, +, -, 0)$ be an arbitrary ℓ-group and $\mathbf{G}^-$ be the negative cone of $\mathbf{G}$, that is, $\mathbf{G}^- = \{x \in G \mid x \leq 0\}$.

On G^- we define the following operations:

$$x \odot y := x + y,$$

$$x \rightarrow y := (y - x) \wedge 0,$$

$$x \rightsquigarrow y := (-x + y) \wedge 0.$$

Then $\mathbf{G}^- = (G^-, \odot, \rightarrow, \rightsquigarrow, 0)$ is a cancellative pseudo-hoop.

We shall verify the conditions $(psHOOP_1)$–$(psHOOP_5)$.

Consider $x, y, z \in G^-$.

$(psHOOP_1)$ $x \odot 0 = x + 0 = x = 0 + x = 0 \odot x$.
$(psHOOP_2)$ $x \rightarrow x = x \rightsquigarrow x = 0 \wedge 0 = 0$.
$(psHOOP_3)$ $x \odot y \rightarrow z = [z - (x + y)] \wedge 0 = (z - y - x) \wedge 0$ and

$$x \rightarrow (y \rightarrow z) = \left[(z - y) \wedge 0 - x\right] \wedge 0 = (z - y - x) \wedge (-x) \wedge 0$$
$$= (z - y - x) \wedge 0$$

(since $-x \geq 0$, we have $(-x) \wedge 0 = 0$). Thus $x \odot y \rightarrow z = x \rightarrow (y \rightarrow z)$.

$(psHOOP_4)$ Similar to $(psHOOP_3)$.
$(psHOOP_5)$ $(x \rightarrow y) \odot x = (y - x) \wedge 0 + x = y \wedge x$.

Similarly, $(y \rightarrow x) \odot y = x \wedge y$, $x \odot (x \rightsquigarrow y) = y \wedge x$, $y \odot (y \rightsquigarrow x) = x \wedge y$.

Thus $(x \rightarrow y) \odot x = (y \rightarrow x) \odot y = x \odot (x \rightsquigarrow y) = y \odot (y \rightsquigarrow x)$.

It follows that $\mathbf{G}^-$ is a pseudo-hoop.

We will verify conditions (C_1) and (C_2).

If $x, y \in G^-$, then $y \rightarrow x \odot y = (x + y - y) \wedge 0 = x \wedge 0 = x$ and $y \rightsquigarrow y \odot x = (-y + y + x) \wedge 0 = x \wedge 0 = x$.

Thus $\mathbf{G}^-$ is a cancellative pseudo-hoop.

Proposition 2.13 ([148]) *Let A be a cancellative pseudo-hoop. Then for all* $x, y, z \in A$ *the following hold*:

(1) $x \to y = x \odot z \to y \odot z$ *and* $x \rightsquigarrow y = z \odot x \rightsquigarrow z \odot y$;
(2) $x \odot z \le y \odot z$ *iff* $x \le y$ *and* $z \odot x \le z \odot y$ *iff* $x \le y$.

Proof

(1) By (C_1) and $(psHOOP_3)$ we get

$$x \to y = x \to (z \to y \odot z) = x \odot z \to y \odot z.$$

By (C_2) and $(psHOOP_4)$ we get

$$x \rightsquigarrow y = x \rightsquigarrow (z \rightsquigarrow z \odot y) = z \odot x \rightsquigarrow z \odot y.$$

(2) Applying (1), $x \odot z \le y \odot z$ iff $x \odot z \to y \odot z = 1$ iff $x \to y = 1$ iff $x \le y$.
    The second inequality can be proved in the same way.   □

Definition 2.5 A pseudo-hoop $(A, \odot, \to, \rightsquigarrow, 1)$ is said to be *Wajsberg* if it satisfies the following conditions:

(Wa_1) $(x \to y) \rightsquigarrow y = (y \to x) \rightsquigarrow x$;
(Wa_2) $(x \rightsquigarrow y) \to y = (y \rightsquigarrow x) \to x$,

i.e. $x \vee_1 y = y \vee_1 x$ and $x \vee_2 y = y \vee_2 x$.

Remark 2.2 Taking $y = 0$ in (Wa_1) and (Wa_2), it follows that a bounded Wajsberg pseudo-hoop satisfies (pDN).

Example 2.2 ([148]) Let $\mathbf{G} = (G, \vee, \wedge, +, -, 0)$ be an arbitrary ℓ-group. For an arbitrary element $u \in G$, $u \ge 0$ define on the set $G[u] = [0, u]$ the operations:

$$x \odot y := (x - u + y) \vee 0,$$

$$x \to y := (y - x + u) \wedge u,$$

$$x \rightsquigarrow y := (u - x + y) \wedge u.$$

Then $\mathbf{G[u]} = (G[u], \odot, \to, \rightsquigarrow, u)$ is a bounded Wajsberg pseudo-hoop.
Indeed, we check the conditions $(psHOOP_1)$–$(psHOOP_5)$.

$(psHOOP_1)$ $x \odot u = (x - u + u) \vee 0 = x \vee 0 = x$ and $u \odot x = (u - u + x) \vee 0 = x \vee 0 = x$, since $x \ge 0$.
$(psHOOP_2)$ $x \to x = x \rightsquigarrow x = u \wedge u = u$.
$(psHOOP_3)$ Applying the properties of an ℓ-group we have:

$$x \odot y \to z = \left[(x - u + y) \vee 0\right] \to z$$
$$= \left[z - (x - u + y) \vee 0 + u\right] \wedge u$$

$$= (z - y + u - x + u) \wedge (z + u) \wedge u$$

$$= (z - y + u - x + u) \wedge u \quad \text{and}$$

$$x \to (y \to z) = \big[(y \to z) - x + u\big] \wedge u = \big[(z - y + u) \wedge u - x + u\big] \wedge u$$

$$= (z - y + u - x + u) \wedge (u - x + u) \wedge u$$

$$= (z - y + u - x + u) \wedge u.$$

Thus $(x \odot y) \to z = x \to (y \to z)$.

($psHOOP_4$) Can be proved in a similar way as ($psHOOP_3$).

($psHOOP_5$) We have:

$$(x \to y) \odot x = \big[(x \to y) - u + x\big] \vee 0 = \big[(y - x + u) \wedge u - u + x\big] \vee 0$$

$$= \big[(y - x + u - u + x) \wedge (u - u + x)\big] \vee 0 = (y \wedge x) \vee 0$$

$$= y \wedge x.$$

Similarly, $(y \to x) \odot y = x \wedge y$, $x \odot (x \rightsquigarrow y) = y \wedge x$, $y \odot (y \rightsquigarrow x) = x \wedge y$.
Thus $(x \to y) \odot x = (y \to x) \odot y = x \odot (x \rightsquigarrow y) = y \odot (y \rightsquigarrow x)$.
It follows that $\mathbf{G[u]} = (G[u], \odot, \to, \rightsquigarrow, u)$ is a pseudo-hoop.
Obviously, it is bounded.
We will prove that $\mathbf{G[u]}$ satisfies conditions (Wa_1) and (Wa_2).
Let $x, y \in G[u]$. We have:

$$(x \to y) \rightsquigarrow y = \big[u - (x \to y) + y\big] \wedge u = \big[u - (y - x + u) \wedge u + y\big] \wedge u$$

$$= \big[(u - u + x - y + y) \vee (u - u + y)\big] \wedge u$$

$$= (x \vee y) \wedge u = x \vee y \quad \text{and}$$

$$(y \to x) \rightsquigarrow x = \big[u - (y \to x) + x\big] \wedge u = \big[u - (x - y + u) \wedge u + x\big] \wedge u$$

$$= \big[(u - u + y - x + x) \vee (u - u + x)\big] \wedge u$$

$$= (y \vee x) \wedge u = y \vee x = x \vee y.$$

Thus $(x \to y) \rightsquigarrow y = (y \to x) \rightsquigarrow x$, hence $\mathbf{G[u]}$ satisfies (Wa_1).
We can similarly prove that condition (Wa_2) is also satisfied.
Therefore $\mathbf{G[u]} = (G[u], \odot, \to, \rightsquigarrow, u)$ is a bounded Wajsberg pseudo-hoop.

Proposition 2.14 ([148]) *Let A be a Wajsberg pseudo-hoop. Then for all $x, y \in A$ the following hold*:

(cw_1) $x \cup_1 y = (x \to y) \rightsquigarrow y = (y \to x) \rightsquigarrow x$;
(cw_2) $x \cup_2 y = (x \rightsquigarrow y) \to y = (y \rightsquigarrow x) \to x$;
(cw_3) $\cup_1$ *and* $\cup_2$ *are associative*;
(cw_4) $x \vee y = x \cup_1 y = x \cup_2 y$.

Proof

(cw_1) This follows from the definition of $\cup_1$ and (Wa_1).

(cw_2) This follows from the definition of $\cup_2$ and (Wa_2).

(cw_3) If $x \leq y$ and $z \in A$, then applying ($psbck$-c_1) twice we get $y \to z \leq x \to z$
and $(x \to z) \rightsquigarrow z \leq (y \to z) \rightsquigarrow z$, that is, $x \cup_1 z \leq y \cup_1 z$.
By Proposition 2.9, $\cup_1$ is associative. Similarly, $\cup_2$ is associative.

(cw_4) This follows by Remark 2.1. $\square$

Corollary 2.1 *If A is a Wajsberg pseudo-hoop, then*

$$x \vee y \to y = x \to y \quad and \quad x \vee y \rightsquigarrow y = x \rightsquigarrow y$$

for all $x, y \in A$.

Definition 2.6 A pseudo-hoop A is called *basic* if it satisfies the following conditions:

(Ba_1) $(x \to y) \to z \leq ((y \to x) \to z) \to z$;

(Ba_2) $(x \rightsquigarrow y) \rightsquigarrow z \leq ((y \rightsquigarrow x) \rightsquigarrow z) \rightsquigarrow z$.

We say that a pseudo-hoop is *representable* if it can be represented as a subdirect product of linearly ordered pseudo-hoops (see Chap. 3).

It is straightforward to verify that any linearly ordered pseudo-hoop and hence any representable pseudo-hoop is basic.

Proposition 2.15 ([148]) *Let A be a basic pseudo-hoop. For any $x, y, z \in A$, the following hold:*

(1) $(x \to y) \cup_1 (y \to x) = 1$ *and* $(x \rightsquigarrow y) \cup_2 (y \rightsquigarrow x) = 1$;

(2) $x \to y = (x \cup_1 y) \to y$ *and* $x \rightsquigarrow y = (x \cup_2 y) \rightsquigarrow y$;

(3) $(x \cup_1 y) \to z = (x \to z) \wedge (y \to z)$ *and* $(x \cup_2 y) \rightsquigarrow z = (x \rightsquigarrow z) \wedge (y \rightsquigarrow z)$.

Proof

(1) Let $u = (x \to y) \cup_1 (y \to x)$. According to ($Ba_1$) we have $(x \to y) \to u \leq ((y \to x) \to u) \to u$. Applying Proposition 2.8(2) we have $x \to y, y \to x \leq u$, hence $(x \to y) \to u = (y \to x) \to u = 1$. It follows that $1 \leq 1 \to u = u$, so $u = 1$, that is, $(x \to y) \cup_1 (y \to x) = 1$. Similarly, $(x \rightsquigarrow y) \cup_2 (y \rightsquigarrow x) = 1$.

(2) Since $x \leq x \cup_1 y$, applying ($psbck$-c_1) we get $(x \cup_1 y) \to y \leq x \to y$.

From ($pshoop$-c_5) and ($psbck$-c_1) it follows that

$$x \to y \leq ((x \to y) \rightsquigarrow y) \to y \leq (x \cup_1 y) \to y,$$

since $x \cup_1 y = ((x \to y) \rightsquigarrow y) \wedge ((y \to x) \rightsquigarrow x) \leq (x \to y) \rightsquigarrow y$.

Hence $x \to y = (x \cup_1 y) \to y$.

We can prove in the same manner that $x \rightsquigarrow y = (x \cup_2 y) \rightsquigarrow y$.

(3) Since $x, y \leq x \cup_1 y$, applying $(psbck\text{-}c_1)$ we have $(x \cup_1 y) \to z \leq x \to z$ and $(x \cup_1 y) \to z \leq y \to z$. Hence $(x \cup_1 y) \to z \leq (x \to z) \wedge (y \to z)$.

Let $u = [(x \to z) \wedge (y \to z)] \rightsquigarrow [(x \cup_1 z) \to z]$. We will prove that $u = 1$. We have:

$$\left[(x \to z) \wedge (y \to z)\right] \odot \left[(x \cup_1 y) \to y\right] \odot (x \cup_1 y)$$
$$= \left[(x \to z) \wedge (y \to z)\right] \odot \left[(x \cup_1 y) \wedge y\right]$$
$$= \left[(x \to z) \wedge (y \to z)\right] \odot y \leq (y \to z) \odot y = y \wedge z \leq z, \quad \text{so}$$
$$\left[(x \to z) \wedge (y \to z)\right] \odot \left[(x \cup_1 y) \to y\right] \leq (x \cup_1 y) \to z.$$

It follows that $(x \cup_1 y) \to y \leq [(x \to z) \wedge (y \to z)] \rightsquigarrow [(x \cup_1 y) \to z] = u$.

Applying (2) it follows that $x \to y = (x \cup_1 y) \to y \leq u$, that is, $(x \to y) \to u = 1$.

Similarly, $(y \to x) \to u = 1$.

By (Ba_1) we get $1 = (x \to y) \to u \leq ((y \to x) \to u) \to u = 1 \to u = u$, so $u = 1$.

Hence $(x \to z) \wedge (y \to z) \leq (x \cup_1 z) \to z$.

We conclude that $(x \cup_1 y) \to z = (x \to z) \wedge (y \to z)$.

Similarly, $(x \cup_2 y) \rightsquigarrow z = (x \rightsquigarrow z) \wedge (y \rightsquigarrow z)$. $\qquad \square$

Proposition 2.16 ([148]) *Let A be a basic pseudo-hoop. Then, for any $x, y \in A$, $x \vee y$ exists and $x \vee y = x \cup_1 y = x \cup_2 y$. The lattice $(A, \wedge, \vee)$ is distributive.*

Proof We have $x, y \leq x \cup_1 y$ and $x, y \leq x \cup_2 y$. Let $z \in A$ such that $x, y \leq z$, that is, $x \to z = y \to z = 1$.

According to Proposition 2.15(3) we have $(x \cup_1 y) \to z = (x \to z) \wedge (y \to z) = 1 \wedge 1 = 1$, so $x \cup_1 y \leq z$. Similarly, $x \cup_2 y \leq z$. Thus $x \vee y = x \cup_1 y = x \cup_2 y$.

Finally, applying Proposition 2.5(3) we conclude that $(A, \wedge, \vee)$ is a distributive lattice. $\qquad \square$

Proposition 2.17 ([148]) *Let A be a pseudo-hoop. The following are equivalent:*

(a) *A is a basic pseudo-hoop;*
(b) *$\cup_1$ and $\cup_2$ are associative and $(x \to y) \cup_1 (y \to x) = 1$ for all $x, y \in A$;*
(c) *$\cup_1$ and $\cup_2$ are associative and $(x \rightsquigarrow y) \cup_2 (y \rightsquigarrow x) = 1$ for all $x, y \in A$.*

Proof

(a) $\Rightarrow$ (b) Applying Proposition 2.16 it follows that $\vee = \cup_1 = \cup_2$ is the join operation on A. Taking into consideration Propositions 2.9 and 2.10 we get that $\cup_1$ and $\cup_2$ are associative. The second assertion follows by Proposition 2.15(1).

(b) $\Rightarrow$ (a) By Remark 2.1 we have $\vee = \cup_1 = \cup_2$. Applying $(psbck\text{-}c_{24})$ and $(psbck\text{-}c_{12})$ we get

$$\big((x \to y) \to z\big) \odot \big((y \to x) \to z\big) \le \big((x \to y) \to z\big) \wedge \big((y \to x) \to z\big)$$

$$= \big((x \to y) \vee (y \to x)\big) \to z = 1 \to z = z.$$

Hence $((x \to y) \to z) \le ((y \to x) \to z) \to z$, that is, (Ba_1).
(Ba_2) can be proved similarly.
(a) $\Leftrightarrow$ (c) follows in the same manner as (a) $\Leftrightarrow$ (b).   □

Proposition 2.18 ([148]) *In any basic pseudo-hoop A the following hold*:

(1) $x \odot (y \wedge z) = (x \odot y) \wedge (x \odot z)$ *and* $(y \wedge z) \odot x = (y \odot x) \wedge (z \odot x)$;
(2) $(x \to y) \to (y \to x) = y \to x$ *and* $(x \rightsquigarrow y) \rightsquigarrow (y \rightsquigarrow x) = y \rightsquigarrow x$.

Proof

(1) According to Proposition 2.16, $\vee$ exists, $\vee = \cup_1 = \cup_2$ and $(A, \wedge, \vee)$ is distributive. Applying Propositions 2.15, 2.5 we get:

$$(x \odot y) \wedge (x \odot z) = \big[(x \odot y) \wedge (x \odot z)\big] \odot 1$$

$$= \big[(x \odot y) \wedge (x \odot z)\big] \odot \big[(y \rightsquigarrow z) \vee (z \rightsquigarrow y)\big]$$

$$= \big[((x \odot y) \wedge (x \odot z)) \odot (y \rightsquigarrow z)\big]$$

$$\vee \big[((x \odot y) \wedge (x \odot z)) \odot (z \rightsquigarrow y)\big]$$

$$\le \big[x \odot y \odot (y \rightsquigarrow z)\big] \vee \big[x \odot z \odot (z \rightsquigarrow y)\big]$$

$$= \big[x \odot (y \wedge z)\big] \vee \big[x \odot (z \wedge y)\big] = x \odot (y \wedge z).$$

On the other hand, from $x \odot (y \wedge z) \le x \odot y$ and $x \odot (y \wedge z) \le x \odot z$ we get $x \odot (y \wedge z) \le (x \odot y) \wedge (x \odot z)$.

Thus $x \odot (y \wedge z) = (x \odot y) \wedge (x \odot z)$ and similarly $(y \wedge z) \odot x = (y \odot x) \wedge (z \odot x)$.

(2) According to $(psbck\text{-}c_6)$ we have $y \to x \le (x \to y) \to (y \to x)$.

Applying Proposition 2.15, we have

$$1 = (y \to x) \cup_1 (x \to y)$$

$$= \big[((y \to x) \to (x \to y)) \rightsquigarrow (x \to y)\big]$$

$$\wedge \big[((x \to y) \to (y \to x)) \rightsquigarrow (y \to x)\big].$$

Hence $(x \to y) \to (y \to x) \le y \to x$.
Thus $(x \to y) \to (y \to x) = y \to x$.
Similarly, $(x \rightsquigarrow y) \rightsquigarrow (y \rightsquigarrow x) = y \rightsquigarrow x$.   □

Proposition 2.19 ([148]) *Any Wajsberg pseudo-hoop is a basic pseudo-hoop.*

Proof Let $x, y \in A$. By $(pshoop\text{-}c_7)$, (Wa_1) and $(psbck\text{-}c_{22})$ we have:

$$y \rightarrow x = \left[\left((y \rightarrow x) \rightsquigarrow x \right) \rightarrow y \right] \rightarrow (y \rightarrow x)$$
$$= \left[\left((x \rightarrow y) \rightsquigarrow y \right) \rightarrow y \right] \rightarrow (y \rightarrow x) = (x \rightarrow y) \rightarrow (y \rightarrow x).$$

Applying (cw_1) we get

$$(x \rightarrow y) \cup_1 (y \rightarrow x) = \left((x \rightarrow y) \rightarrow (y \rightarrow x) \right) \rightsquigarrow (y \rightarrow x)$$
$$= (y \rightarrow x) \rightsquigarrow (y \rightarrow x) = 1.$$

By (cw_3) it follows that $\cup_1$ and $\cup_2$ are associative and applying Proposition 2.17 we get that A is a basic pseudo-hoop. $\qquad\square$

Proposition 2.20 ([148]) *Let A be a basic pseudo-hoop satisfying the conditions*:

$$y \rightarrow x = x \quad implies \quad x = 1 \ or \ y = 1 \quad and$$

$$y \rightsquigarrow x = x \quad implies \quad x = 1 \ or \ y = 1$$

for all $x, y \in A$. Then A is a linearly ordered Wajsberg pseudo-hoop.

Proof Consider $x, y \in A$. Applying Proposition 2.18(2) we have $(x \rightarrow y) \rightarrow (y \rightarrow x) = y \rightarrow x$. Taking into consideration the hypothesis, we get $x \rightarrow y = 1$ or $y \rightarrow x = 1$, that is, $x \leq y$ or $y \leq x$. It follows that A is a linearly ordered pseudo-hoop.

We will now prove that A is a Wajsberg pseudo-hoop.

Let $x, y \in A$. If $x = y$, then (Wa_1) is obvious. Assume $x \neq y$. Since A is linear, we can suppose that $x < y$. It follows that $(x \rightarrow y) \rightsquigarrow y = 1 \rightsquigarrow y = y$.

By $(pshoop\text{-}c_7)$ we have $\left[\left((y \rightarrow x) \rightsquigarrow x \right) \rightarrow y \right] \rightarrow (y \rightarrow x) = y \rightarrow x$, so by hypothesis and the fact that $y \rightarrow x \neq 1$, we get $\left((y \rightarrow x) \rightsquigarrow x \right) \rightarrow y = 1$, hence $(y \rightarrow x) \rightsquigarrow x \leq y$. But, from $(pshoop\text{-}c_5)$ we have $y \leq (y \rightarrow x) \rightsquigarrow x$, so $(y \rightarrow x) \rightsquigarrow x = y$. Hence $(x \rightarrow y) \rightsquigarrow y = (y \rightarrow x) \rightsquigarrow x$.

Thus A satisfies (Wa_1). Similarly, A satisfies (Wa_2).

We conclude that A is a Wajsberg pseudo-hoop. $\qquad\square$

Corollary 2.2 *Every strongly simple basic pseudo-hoop is a linearly ordered Wajsberg pseudo-hoop.*

Proof This follows from Lemma 2.1 and Proposition 2.20. $\qquad\square$

Example 2.3 ([148]) The pseudo-hoop $\mathbf{G}^-$ from Example 2.1 is a basic pseudo-hoop. Indeed, consider $x, y, z \in G^-$. We have

$$(x \rightarrow y) \rightarrow z = \left(z - (x \rightarrow y) \right) \wedge 0 = \left(z - (y - x) \wedge 0 \right) \wedge 0 = \left[(z - x + y) \vee z \right] \wedge 0.$$

Similarly, $(y \rightarrow x) \rightarrow z = \left[(z - y + x) \vee z \right] \wedge 0$. It follows that:

$$\left((y \rightarrow x) \rightarrow z \right) \rightarrow z = \left[z - \left(((z - y + x) \vee z) \wedge 0 \right) \right] \wedge 0$$
$$= \left[(z - (z - y + x) \vee z) \vee z \right] \wedge 0$$

$$= \left[\left((z - x + y - z) \wedge 0\right) \vee z\right] \wedge 0$$
$$= \left[(z - x + y - z) \vee z\right] \wedge (0 \vee z) \wedge 0$$
$$= \left[(z - x + y - z) \vee z\right] \wedge 0.$$

Since $z \leq 0$, we have $0 \leq -z$, hence $z - x + y \leq z - x + y - z$. Thus

$$(x \to y) \to z = \left[(z - x + y) \vee z\right] \wedge 0 \leq \left[(z - x + y - z) \vee z\right] \wedge 0$$
$$= \left((y \to x) \to z\right) \to z.$$

It follows that $\mathbf{G}^-$ satisfies (Ba_1) and similarly $\mathbf{G}^-$ satisfies (Ba_2). We conclude that $\mathbf{G}^-$ is a basic pseudo-hoop.

Example 2.4 The pseudo-hoop $\mathbf{G[u]}$ from Example 2.2 is a basic pseudo-hoop. Indeed, we have proved that $\mathbf{G[u]}$ is a Wajsberg pseudo-hoop and applying Proposition 2.19 it follows that $\mathbf{G[u]}$ is a basic pseudo-hoop.

Definition 2.7 An element a of a pseudo-hoop A is said to be an *idempotent* if $a^2 = a$. The set of all idempotents of A is denoted by $\mathrm{Id}(A)$.

A pseudo-hoop A is called an *idempotent pseudo-hoop* if $\mathrm{Id}(A) = A$, that is, all elements of A are idempotent.

On the other hand, an idempotent pseudo-hoop A is a Gödel pseudo-hoop, that is, a pseudo-hoop with condition (Gödel) ($a \odot a = a$ for all $a \in A$).

Lemma 2.2 (Proposition 3.1 in [106]) *If $a \in \mathrm{Id}(A)$, then for all $x \in A$ we have:*

(1) $a \odot x = a \wedge x = x \odot a$;
(2) $a \to x = a \rightsquigarrow x$.

Proof

(1) We have:

$$a \odot x \leq a \wedge x = a \odot (a \rightsquigarrow x) = a \odot a \odot (a \rightsquigarrow x) = a \odot (a \wedge x) \leq a \odot x.$$

Thus $a \odot x = a \wedge x$ and similarly $x \odot a = a \wedge x$.
(2) For an arbitrary $z \in A$ we have:

$$z \leq a \to x \quad \text{iff} \quad z \odot a \leq x \quad \text{iff} \quad a \odot z \leq x \quad \text{iff} \quad z \leq a \rightsquigarrow x,$$

that is, $a \to x = a \rightsquigarrow x$.   □

Remark 2.3

(1) Representable Brouwerian algebras are idempotent basic hoops and generalized Boolean algebras are idempotent Wajsberg hoops ([198]).
(2) Any bounded idempotent pseudo-hoop A is good.
    Indeed, applying the identity $a \to x = a \rightsquigarrow x$ for $x = 0$, we get $a^- = a^\sim$, so $a^{-\sim} = a^{--} = a^{\sim-}$ for all $a \in A$.

2.2 Join-Center and Cancellative-Center of Pseudo-hoops

We introduce the notions of join-center and cancellative-center of a pseudo-hoop and we prove some of their properties.

Definition 2.8 If A is a pseudo-hoop, then the set $JC(A) = \{a \in A \mid a \vee x$ exists for all $x \in A\}$ is called the *join-center* of A.

Obviously, $1 \in JC(A)$ and, if A is bounded, then $0 \in JC(A)$.
If A is a Wajsberg pseudo-hoop, then $JC(A) = A$.

Definition 2.9 If A is a pseudo-hoop, then the set $CC(A) = \{a \in A \mid x \odot a = y \odot a$ implies $x = y$ and $a \odot x = a \odot y$ implies $x = y$ for all $x, y \in A\}$ is called the *cancellative-center* of A.

Obviously, $1 \in CC(A)$ and $0 \notin CC(A)$.
If A is a cancellative pseudo-hoop, then $CC(A) = A$.

Proposition 2.21 *If A is a pseudo-hoop and $a \in CC(A)$, then the following hold for all $x \in A$:*

(1) $a \rightarrow (x \odot a) = x$;
(2) $a \rightsquigarrow (a \odot x) = x$.

Proof

(1) We have $x \odot a = a \wedge (x \odot a) = (a \rightarrow (x \odot a)) \odot a$.
   Taking into consideration that $a \in CC(A)$, we get $a \rightarrow (x \odot a) = x$.
(2) Similarly, from $a \odot x = a \wedge (a \odot x) = a \odot (a \rightsquigarrow (a \odot x))$ we have $a \rightsquigarrow (a \odot x) = x$. □

Corollary 2.3 *If A is a pseudo-hoop and $a \in CC(A)$, then*

$$a^n \rightarrow a^{n+1} = a^n \rightsquigarrow a^{n+1} = a \quad \text{for all } n \in \mathbb{N}.$$

Proof Applying Proposition 2.21(1) and ($psHOOP_3$) for $x = a, a^2, a^3, \ldots$ we get:

$$a = a \rightarrow a^2 = a \rightarrow (a \rightarrow a^3) = a^2 \rightarrow a^3 = a^2 \rightarrow (a \rightarrow a^4)$$

$$= a^3 \rightarrow a^4 = \cdots = a^n \rightarrow a^{n+1}.$$

Similarly, by Proposition 2.21(2) and ($psHOOP_4$) we get $a = a^n \rightsquigarrow a^{n+1}$. □

Proposition 2.22 *If A is a pseudo-hoop and $a \in CC(A)$, then the following hold for all $x, y \in A$:*

(1) $x \rightarrow y = (x \odot a) \rightarrow (y \odot a)$;

(2) $x \rightsquigarrow y = (a \odot x) \rightsquigarrow (a \odot y)$;
(3) $x \leq y$ iff $x \odot a \leq y \odot a$ iff $a \odot x \leq a \odot y$.

Proof

(1) Applying Proposition 2.21(1) and $(psHOOP_3)$ we get:

$$x \rightarrow y = x \rightarrow (a \rightarrow (y \odot a)) = (x \odot a) \rightarrow (y \odot a).$$

(2) Similarly, $x \rightsquigarrow y = x \rightsquigarrow (a \rightsquigarrow (a \odot y)) = (a \odot x) \rightsquigarrow (a \odot y)$.
(3) Applying (1) we have $x \odot a \leq y \odot a$ iff $(x \odot a) \rightarrow (y \odot a) = 1$ iff $x \rightarrow y = 1$
iff $x \leq y$. Similarly, applying (2) we get $a \odot x \leq a \odot y$ iff $x \leq y$. $\square$

2.3 Algebras on Subintervals of Pseudo-hoops

The problem of introducing an MV-algebra structure and a pseudo-MV algebra
structure on subintervals of algebras was solved in [37] and respectively in [195]
and [196]. It was proved in [105] that for a bounded Rℓ-monoid or a pseudo-BL
algebra A, for any $a, b \in A$, $a \leq b$, the subinterval $[a, b]$ can be endowed with a
structure of the same kind as that on A. For the case of an FL_w-algebra A it was
proved in [55] that, if a, b with $a \leq b$ belonging to the Boolean center of A, then the
subinterval $[a, b]$ of A can be endowed with a structure of an FL_w-algebra. In this
section we will establish some conditions on $a, b \in A$ for the subinterval $[a, b]$ of A
to be endowed with a structure of a pseudo-hoop.

Theorem 2.2 *Let $(A, \odot, \rightarrow, \rightsquigarrow, 0, 1)$ be a bounded pseudo-hoop and $a \in JC(A)$.
Then the algebra $A_a^1 = ([a, 1], \odot_a^1, \rightarrow_a^1, \rightsquigarrow_a^1, a, 1)$ is a bounded pseudo-hoop, where
$x \odot_a^1 y := (x \odot y) \vee a, x \rightarrow_a^1 y := x \rightarrow y$ and $x \rightsquigarrow_a^1 y := x \rightsquigarrow y$.*

Proof First, we observe that $a \leq y \leq x \rightarrow y, x \rightsquigarrow y$ implies $x \rightarrow y, x \rightsquigarrow y \in [a, 1]$
for all $x, y \in [a, 1]$. We will check conditions $(psHOOP_1)$–$(psHOOP_5)$ from the
definition of a pseudo-hoop:

$(psHOOP_1)$ For all $x \in [a, 1]$ we have:

$$x \odot_a^1 1 = (x \odot 1) \vee a = x \vee a = x \quad \text{and}$$

$$1 \odot_a^1 x = (1 \odot x) \vee a = x \vee a = x.$$

$(psHOOP_2)$ $x \rightarrow_a^1 x = x \rightarrow x = 1$ and $x \rightsquigarrow_a^1 x = x \rightsquigarrow x = 1$;
$(psHOOP_3)$

$$x \odot_a^1 y \rightarrow_a^1 z = (x \odot y) \vee a \rightarrow z = (x \odot y \rightarrow z) \wedge (a \rightarrow z)$$

$$= (x \odot y) \rightarrow z = x \rightarrow (y \rightarrow z) = x \rightarrow_a^1 (y \rightarrow_a^1 z)$$

(since $a \leq z$, it follows that $a \rightarrow z = 1$).

$(psHOOP_4)$ can be proved in a similar way as $(psHOOP_3)$.

$(psHOOP_5)$ Since $(x \to y) \odot x = (y \to x) \odot y = x \odot (x \leadsto y) = y \odot (y \leadsto x)$, we get

$$\begin{aligned}
\left[(x \to y) \odot x \right] \vee a &= \left[(y \to x) \odot y \right] \vee a \\
&= \left[x \odot (x \leadsto y) \right] \vee a = \left[y \odot (y \leadsto x) \right] \vee a,
\end{aligned}$$

that is,

$$\left(x \to_a^1 y \right) \odot_a^1 x = \left(y \to_a^1 x \right) \odot_a^1 y = x \odot_a^1 \left(x \leadsto_a^1 y \right) = y \odot_a^1 \left(y \leadsto_a^1 x \right).$$

Thus $A_a^1 = ([a, 1], \odot_a^1, \to_a^1, \leadsto_a^1, a, 1)$ is a bounded pseudo-hoop. $\qquad\square$

Obviously, $A = A_0^1$ and $\{1\} = A_1^1$.

Theorem 2.3 *Let* $(A, \odot, \to, \leadsto, 0, 1)$ *be a bounded pseudo-hoop,* $a \in CC(A)$ *and* $A_0^a = ([0, a], \odot_0^a, \to_0^a, \leadsto_0^a, 0, a)$, *where:* $x \odot_0^a y := x \odot (a \leadsto y)$, $x \to_0^a y := (x \to y) \odot a$ *and* $x \leadsto_0^a y := a \odot (x \leadsto y)$. *Then* A_0^a *is a bounded pseudo-hoop.*

Proof We will verify the axioms of a pseudo-hoop:

$(psHOOP_1)$ For all $x \in [0, a]$ we have:

$$x \odot_0^a a = x \odot (a \leadsto a) = x \odot 1 = x \quad \text{and}$$
$$a \odot_0^a x = a \odot (a \leadsto x) = a \wedge x = x.$$

$(psHOOP_2)$ For all $x \in [0, a]$ we have:

$$x \to_0^a x = (x \to x) \odot a = 1 \odot a = a \quad \text{and}$$
$$x \leadsto_0^a x = a \odot (x \leadsto x) = a \odot 1 = a.$$

$(psHOOP_3)$ First of all we note that $x \odot_0^a y = (a \to x) \odot y$.
    Indeed, from $(a \to x) \odot a \odot (a \leadsto y) = (a \to x) \odot a \odot (a \leadsto y)$ we get $(a \wedge x) \odot (a \leadsto y) = (a \to x) \odot (a \wedge y)$. Hence $x \odot (a \leadsto y) = (a \to x) \odot y$, that is, $x \odot_0^a y = (a \to x) \odot y$.
    Applying the rules of calculus in pseudo-hoops, we get:

$$x \odot_0^a y \to_0^a z = x \to_0^a \left(y \to_0^a z \right) \quad \text{iff}$$
$$(a \to x) \odot y \to_0^a z = x \to_0^a (y \to z) \odot a \quad \text{iff}$$
$$\left[(a \to x) \odot y \to z \right] \odot a = \left[x \to (y \to z) \odot a \right] \odot a \quad \text{iff}$$
$$(a \to x) \odot y \to z = x \to (y \to z) \odot a \quad \text{iff}$$
$$(a \to x) \to (y \to z) = x \to (y \to z) \odot a.$$

For any $u \in A$ we have:

$$u \leq (a \to x) \to (y \to z) \quad \Rightarrow \quad u \odot (a \to x) \leq y \to z$$
$$\Rightarrow \quad u \odot (a \to x) \odot a \leq (y \to z) \odot a$$
$$\Rightarrow \quad u \odot (a \wedge x) \leq (y \to z) \odot a$$
$$\Rightarrow \quad u \odot x \leq (y \to z) \odot a$$
$$\Rightarrow \quad u \leq x \to (y \to z) \odot a.$$

Conversely,

$$u \leq x \to (y \to z) \odot a \quad \Rightarrow \quad u \odot x \leq (y \to z) \odot a$$
$$\Rightarrow \quad u \odot (a \wedge x) \leq (y \to z) \odot a$$
$$\Rightarrow \quad u \odot (a \to x) \odot a \leq (y \to z) \odot a$$
$$\Rightarrow \quad u \odot (a \to x) \leq y \to z$$
$$\Rightarrow \quad u \leq (a \to x) \to (y \to z).$$

Since u is arbitrary, it follows that $(a \to x) \to (y \to z) = x \to (y \to z) \odot a$.
Thus $x \odot_0^a y \to_0^a z = x \to_0^a (y \to_0^a z)$.
($psHOOP_4$) This can be proved in a similar way as ($psHOOP_3$).
($psHOOP_5$) For all $x, y \in [0, a]$ we have:

$$\left(x \to_0^a y\right) \odot_0^a x = \left(y \to_0^a x\right) \odot_0^a y \quad \text{iff}$$
$$\left[(x \to y) \odot a\right] \odot_0^a x = \left[(y \to x) \odot a\right] \odot_0^a y \quad \text{iff}$$
$$(x \to y) \odot a \odot (a \rightsquigarrow x) = (y \to x) \odot a \odot (a \rightsquigarrow y) \quad \text{iff}$$
$$(x \to y) \odot (a \wedge x) = (y \to x) \odot (a \wedge y) \quad \text{iff}$$
$$(x \to y) \odot x = (y \to x) \odot y.$$

The last identity is true, since $(A, \odot, \to, \rightsquigarrow, 0, 1)$ is a pseudo-hoop. The remaining identities in ($psHOOP_5$) can be proved in a similar manner as the above. We conclude that A_0^a is a bounded pseudo-hoop. $\qquad \square$

Theorem 2.4 *Let $(A, \odot, \to, \rightsquigarrow, 0, 1)$ be a bounded pseudo-hoop, $a, b \in CC(A) \cap JC(A)$, $a \leq b$ and $A_a^b = ([a, b], \odot_a^b, \to_a^b, \rightsquigarrow_a^b, a, b)$, where: $x \odot_a^b y := (x \odot (b \rightsquigarrow y)) \vee a$, $x \to_a^b y := (x \to y) \odot b$ and $x \rightsquigarrow_a^b y := b \odot (x \rightsquigarrow y)$. Then A_a^b is a bounded pseudo-hoop.*

Proof According to Theorem 2.2, the algebra $([a, 1], \odot_a^1, \wedge, \vee, \to_a^1, \rightsquigarrow_a^1, a, 1)$ with the operations $x \odot_a^1 y = (x \odot y) \vee a$, $x \to_a^1 y = x \to y$ and $x \rightsquigarrow_a^1 y = x \rightsquigarrow y$ is a bounded pseudo-hoop. Let $x, y \in [a, b]$. Since $x \leq b$, by $(psbck\text{-}c_1)$ we have $b \to y \leq x \to y$, hence

$$(x \to y) \odot b \geq (b \to y) \odot b = b \wedge y \geq a.$$

Similarly, $b \rightsquigarrow y \leq x \rightsquigarrow y$, so:

$$b \odot (x \rightsquigarrow y) \geq b \odot (b \rightsquigarrow y) = b \wedge y \geq a.$$

By Theorem 2.3 it follows that the algebra $([a, b], \odot_a^b, \to_a^b, \rightsquigarrow_a^b, a, b)$ is a bounded pseudo-hoop with the operations:

$$x \odot_a^b y = x \odot_a^1 \left(b \rightsquigarrow_a^1 y \right) = x \odot_a^1 (b \rightsquigarrow y) = \left(x \odot (b \rightsquigarrow y) \right) \vee a,$$

$$x \to_a^b y = \left(x \to_a^1 y \right) \odot_a^1 b = (x \to y) \odot_a^1 b = \left((x \to y) \odot b \right) \vee a = (x \to y) \odot b,$$

$$x \rightsquigarrow_a^b y = b \odot_a^1 \left(x \rightsquigarrow_a^1 y \right) = b \odot_a^1 (x \rightsquigarrow y)$$

$$= \left(b \odot (x \rightsquigarrow y) \right) \vee a = b \odot (x \rightsquigarrow y). \qquad \square$$

Chapter 3
Residuated Lattices

Residuation is a fundamental concept of ordered structures and the *residuated lattices*, obtained by adding a residuated monoid operation to lattices, have been applied in several branches of mathematics, including ℓ-groups, ideal lattices of rings and multiple-valued logics. The study of commutative residuated lattices was initiated in the late 1930s by Krull, Dilworth and Ward ([93, 204, 261, 262]) and recently they have been investigated in [35, 121, 145, 152, 165].

Non-commutative residuated lattices, sometimes called pseudo-residuated lattices, biresiduated lattices or generalized residuated lattices, are the algebraic counterparts of substructural logics, i.e. logics which lack at least one of the three structural rules, namely contraction, weakening and exchange.

In this book residuated lattice means non-commutative residuated lattice.

Complete studies on residuated lattices were developed by Ono, Jipsen, Galatos, Tsinakis and Kowalski ([3, 11, 124–126, 129, 200, 235–237]). Particular cases of residuated lattices are the full Lambek algebras (FL-algebras), integral residuated lattices and bounded integral residuated lattices (FL_w-algebras).

In the present chapter we investigate the properties of a residuated lattice and the lattice of filters of a residuated lattice, we study the Boolean center of an FL_w-algebra and we define and study the directly indecomposable FL_w-algebras. We prove that any linearly ordered FL_w-algebra is directly indecomposable and any subdirectly irreducible FL_w-algebra is directly indecomposable. Finally, the FL_w-algebras of fractions relative to a meet-closed system is introduced and investigated.

3.1 Definitions and Properties

We recall some basic notions and results regarding residuated lattices and FL_w-algebras and we give examples of proper FL_w-algebras.

Definition 3.1 A *residuated lattice* is an algebra $\mathcal{A} = (A, \wedge, \vee, \odot, \rightarrow, \rightsquigarrow, e)$ of type $(2, 2, 2, 2, 2, 0)$ satisfying the following axioms:

(A_1) $(A, \wedge, \vee)$ is a lattice;

L.C. Ciungu, *Non-commutative Multiple-Valued Logic Algebras*,
Springer Monographs in Mathematics, DOI 10.1007/978-3-319-01589-7_3,
© Springer International Publishing Switzerland 2014

(A_2) $(A, \odot, e)$ is a monoid (i.e. $\odot$ is associative, with identity element e);
(A_3) $x \odot y \leq z$ iff $x \leq y \to z$ iff $y \leq x \rightsquigarrow z$ for any $x, y, z \in A$ (*pseudo-residuation*).

(Our notation differs slightly from that of [129]; we write $x \to y$ and $x \rightsquigarrow y$ instead of y/x and $x \backslash y$, respectively.)

Note that generally, e is not the top element of A.

We will agree that, in the absence of parentheses, the operation $\odot$ is performed first, followed by $\to$, $\rightsquigarrow$ and $\wedge$, $\vee$.

In the next proposition we prove some properties of residuated lattices (see [11, 129, 200]).

Proposition 3.1 *In any residuated lattice* $(A, \wedge, \vee, \odot, \to, \rightsquigarrow, e)$ *the following hold*:

$(rl\text{-}c_1)$ *if* $x \leq y$, *then* $x \odot z \leq y \odot z$ *and* $z \odot x \leq z \odot y$;
$(rl\text{-}c_2)$ $x \odot (\bigvee_{i \in I} y_i) = \bigvee_{i \in I}(x \odot y_i)$ *and* $(\bigvee_{i \in I} y_i) \odot x = \bigvee_{i \in I}(y_i \odot x)$;
$(rl\text{-}c_3)$ $y \to (\bigwedge_{i \in I} x_i) = \bigwedge_{i \in I}(y \to x_i)$ *and* $y \rightsquigarrow (\bigwedge_{i \in I} x_i) = \bigwedge_{i \in I}(y \rightsquigarrow x_i)$;
$(rl\text{-}c_4)$ $(\bigvee_{i \in I} x_i) \to y = \bigwedge_{i \in I}(x_i \to y)$ *and* $(\bigvee_{i \in I} x_i) \rightsquigarrow y = \bigwedge_{i \in I}(x_i \rightsquigarrow y)$,

whenever the arbitrary meets and unions exist.

Proof

$(rl\text{-}c_1)$ From $y \odot z \leq y \odot z$ we have $y \leq z \to y \odot z$ and taking into consideration that $x \leq y$ we get $x \leq z \to y \odot z$.

Hence $x \odot z \leq y \odot z$ and similarly $z \odot x \leq z \odot y$.

$(rl\text{-}c_2)$, $(rl\text{-}c_3)$, $(rl\text{-}c_4)$ follow similarly as in Proposition 1.14. $\square$

Proposition 3.2 *In any residuated lattice* $(A, \wedge, \vee, \odot, \to, \rightsquigarrow, e)$ *the following hold*:

$(rl\text{-}c_5)$ $(x \to y) \odot x \leq y$ *and* $x \odot (x \rightsquigarrow y) \leq y$;
$(rl\text{-}c_6)$ $x \odot (y \to z) \leq y \to (x \odot z)$ *and* $(y \rightsquigarrow z) \odot x \leq y \rightsquigarrow (z \odot x)$;
$(rl\text{-}c_7)$ $(y \to z) \odot (x \to y) \leq x \to z$ *and* $(x \rightsquigarrow y) \odot (y \rightsquigarrow z) \leq x \rightsquigarrow z$;
$(rl\text{-}c_8)$ $x \to y \leq (y \to z) \rightsquigarrow (x \to z)$ *and* $x \rightsquigarrow y \leq (y \rightsquigarrow z) \to (x \rightsquigarrow z)$;
$(rl\text{-}c_9)$ $x \to y \leq (z \to x) \to (z \to y)$ *and* $x \rightsquigarrow y \leq (z \rightsquigarrow x) \rightsquigarrow (z \rightsquigarrow y)$;
$(rl\text{-}c_{10})$ $x \to y \leq (x \odot z) \to (y \odot z)$ *and* $x \rightsquigarrow y \leq (z \odot x) \rightsquigarrow (z \odot y)$;
$(rl\text{-}c_{11})$ $x \to (y \to z) = (x \odot y) \to z$ *and* $x \rightsquigarrow (y \rightsquigarrow z) = (y \odot x) \rightsquigarrow z$;
$(rl\text{-}c_{12})$ $x \leq (x \to y) \rightsquigarrow y$ *and* $x \leq (x \rightsquigarrow y) \to y$;
$(rl\text{-}c_{13})$ $e \to x = e \rightsquigarrow x = x$;
$(rl\text{-}c_{14})$ $x \to x \geq e$ *and* $x \rightsquigarrow x \geq e$;
$(rl\text{-}c_{15})$ $(x \to y) \odot (z \to e) \leq (z \odot x) \to y$ *and* $(z \rightsquigarrow e) \odot (x \rightsquigarrow y) \leq (x \odot z) \rightsquigarrow y$;
$(rl\text{-}c_{16})$ $(x \to x) \odot x = x$ *and* $x \odot (x \rightsquigarrow x) = x$;
$(rl\text{-}c_{17})$ $(x \to x)^2 = x \to x$ *and* $(x \rightsquigarrow x)^2 = x \rightsquigarrow x$;
$(rl\text{-}c_{18})$ $z \odot (x \wedge y) \leq (z \odot x) \wedge (z \odot y)$ *and* $(x \wedge y) \odot z \leq (x \odot z) \wedge (y \odot z)$;
$(rl\text{-}c_{19})$ $x \to (y \rightsquigarrow z) = y \rightsquigarrow (x \to z)$ *and* $x \rightsquigarrow (y \to z) = y \to (x \rightsquigarrow z)$;
$(rl\text{-}c_{20})$ $y \odot (x \to e) \leq x \to y$ *and* $(x \rightsquigarrow e) \odot y \leq x \rightsquigarrow y$.

Proof

(*rl-c*$_5$) From $x \to y \le x \to y$ we get $(x \to y) \odot x \le y$ and from $x \rightsquigarrow y \le x \rightsquigarrow y$ we have $x \odot (x \rightsquigarrow y) \le y$.

(*rl-c*$_6$) Applying (*rl-c*$_5$) we have $(y \to z) \odot y \le z$, and by (*rl-c*$_1$) it follows that $x \odot (y \to z) \odot y \le x \odot z$. Thus $x \odot (y \to z) \le y \to (x \odot z)$.

Similarly, $(y \rightsquigarrow z) \odot x \le y \rightsquigarrow (z \odot x)$.

(*rl-c*$_7$) Applying (*rl-c*$_5$) we have:

$$(y \to z) \odot (x \to y) \odot x \le (y \to z) \odot y \le z, \quad \text{hence}$$

$$(y \to z) \odot (x \to y) \le x \to z.$$

Similarly, $x \odot (x \rightsquigarrow y) \odot (y \rightsquigarrow z) \le y \odot (y \rightsquigarrow z) \le z$, hence $(x \rightsquigarrow y) \odot (y \rightsquigarrow z) \le x \rightsquigarrow z$.

(*rl-c*$_8$) This follows from (*rl-c*$_7$).

(*rl-c*$_9$) Applying (*rl-c*$_5$) we have $(x \to y) \odot (z \to x) \odot z \le (x \to y) \odot x \le y$, so $(x \to y) \odot (z \to x) \le z \to y$ and finally $x \to y \le (z \to x) \to (z \to y)$.

Similarly, $x \rightsquigarrow y \le (z \rightsquigarrow x) \rightsquigarrow (z \rightsquigarrow y)$.

(*rl-c*$_{10}$) Applying (*rl-c*$_5$) we get $(x \to y) \odot x \odot z \le y \odot z$, hence $x \to y \le (x \odot z) \to (y \odot z)$. Similarly, $x \rightsquigarrow y \le (z \odot x) \rightsquigarrow (z \odot y)$.

(*rl-c*$_{11}$) For any $u \in A$ we have:

$$u \le x \to (y \to z) \quad \text{iff} \quad u \odot x \le y \to z \quad \text{iff} \quad (u \odot x) \odot y \le z \quad \text{iff}$$

$$u \odot (x \odot y) \le z \quad \text{iff} \quad u \le (x \odot y) \to z,$$

so $x \to (y \to z) = (x \odot y) \to z$.

Similarly, $x \rightsquigarrow (y \rightsquigarrow z) = (y \odot x) \rightsquigarrow z$.

(*rl-c*$_{12}$) The inequalities follow applying (*rl-c*$_5$).

(*rl-c*$_{13}$) For any $u \in A$ we have $u \le e \to x$ iff $u \odot e \le x$ iff $u \le x$.

Thus $e \to x = x$ and similarly $e \rightsquigarrow x = x$.

(*rl-c*$_{14}$) From $x \le x$ we get $e \odot x \le x$, hence $e \le x \to x$.

Similarly, $x \odot e \le x$ implies $e \le x \rightsquigarrow x$.

(*rl-c*$_{15}$) Applying (*rl-c*$_6$) and (*rl-c*$_{11}$) we have:

$$(x \to y) \odot (z \to e) \le z \to (x \to y) \odot e = z \to (x \to y) = z \odot x \to y.$$

Similarly, $(z \rightsquigarrow e) \odot (x \rightsquigarrow y) \le (x \odot z) \rightsquigarrow y$.

(*rl-c*$_{16}$) From $x \to x \le x \to x$ we get $(x \to x) \odot x \le x$.

On the other hand, from $e \le x \to x$ we have $x \le (x \to x) \odot x$.

Thus $(x \to x) \odot x = x$. Similarly, $x \odot (x \rightsquigarrow x) = x$.

(*rl-c*$_{17}$) Applying (*rl-c*$_7$) we have $(x \to x)^2 = (x \to x) \odot (x \to x) \le x \to x$.

On the other hand, from $e \le x \to x$ we get $x \to x \le (x \to x)^2$, hence $(x \to x)^2 = x \to x$. Similarly, $(x \rightsquigarrow x)^2 = x \rightsquigarrow x$.

(*rl-c*$_{18}$) From $x \wedge y \le x$ and $x \wedge y \le y$, it follows that $z \odot (x \wedge y) \le z \odot x$ and $z \odot (x \wedge y) \le z \odot y$. Thus $z \odot (x \wedge y) \le (z \odot x) \wedge (z \odot y)$.

Similarly, $(x \wedge y) \odot z \le (x \odot z) \wedge (y \odot z)$.

(rl-c_{19}) For any $u \in A$ we have:

$$u \le x \to (y \leadsto z) \quad \text{iff} \quad u \odot x \le y \leadsto z \quad \text{iff} \quad y \odot u \odot x \le z \quad \text{iff}$$

$$y \odot u \le x \to z \quad \text{iff} \quad u \le y \leadsto (x \to z).$$

Thus $x \to (y \leadsto z) = y \leadsto (x \to z)$. Similarly, $x \leadsto (y \to z) = y \to (x \leadsto z)$.
(rl-c_{20}) By (rl-c_5) we have $(x \to e) \odot x \le e$ and applying (rl-c_1) we get $y \odot (x \to e) \odot x \le y \odot e = y$. Hence $y \odot (x \to e) \le x \to y$.
Similarly, from $x \odot (x \leadsto e) \le e$ we get $x \odot (x \leadsto e) \odot y \le e \odot y = y$, so $(x \leadsto e) \odot y \le x \leadsto y$.   □

Example 3.1 ([124]) Let $(G, \vee, \wedge, \odot, ^{-1}, e)$ be an ℓ-group.
   Define: $x \to y := y \odot x^{-1}$ and $x \leadsto y := x^{-1} \odot y$.
   Then $(G, \wedge, \vee, \odot, \to, \leadsto, e)$ is a residuated lattice.

Example 3.2 ([200]) Let $(A, \wedge, \vee, \odot, \to, \leadsto, e)$ be a residuated lattice and $A^- = \{x \in A \mid x \le e\}$. Then the *negative cone* of A is defined as $(A^-, \wedge, \vee, \odot, \to_{A^-}, \leadsto_{A^-}, e)$, where $x \to_{A^-} y := (x \to y) \wedge e$ and $x \leadsto_{A^-} y := (x \leadsto y) \wedge e$. It is easy to check that $(A^-, \wedge, \vee, \odot, \to_{A^-}, \leadsto_{A^-}, e)$ is again a residuated lattice.

Proposition 3.3 *If a residuated lattice $(A, \wedge, \vee, \odot, \to, \leadsto, e)$ has a bottom element* 0, *then it also has a top element* 1 *and for every $x \in A$ the following hold:*

(1) $x \odot 0 = 0 \odot x = 0$;
(2) $0 \to x = 0 \leadsto x = 1$;
(3) $x \to 1 = x \leadsto 1 = 1$.

Proof

(1) Since 0 is a bottom element, we have $0 \le x \leadsto 0$, hence $x \odot 0 \le 0$, so $x \odot 0 = 0$ for all $x \in A$. Similarly, from $0 \le x \to 0$ we get $0 \odot x \le 0$, that is, $0 \odot x = 0$. Thus $x \odot 0 = 0 \odot x = 0$.
(2) First of all we show that $0 \to 0 = 0 \leadsto 0$. Indeed, for any $u \in A$ we have:

$$u \le 0 \to 0 \quad \text{iff} \quad u \odot 0 \le 0 \quad \text{iff} \quad u \odot 0 = 0 \quad \text{iff}$$

$$0 \odot u = 0 \quad \text{iff} \quad u \le 0 \leadsto 0.$$

It follows that $0 \to 0 = 0 \leadsto 0$.
   Take $1 = 0 \to 0 = 0 \leadsto 0$. From $x \odot 0 = 0$ we get $x \le 0 \to 0 = 1$ for all $x \in A$. Hence 1 is the top element of A.
   Now, from $1 \odot 0 = 0 \le x$ we get $1 \le 0 \to x$, so $0 \to x = 1$.
   Similarly, from $0 \odot 1 = 0 \le x$, we have $1 \le 0 \leadsto x$, hence $0 \leadsto x = 1$.
(3) Since $1 \odot x \le 1$, we have $1 \le x \to 1$, hence $x \to 1 = 1$.
   Similarly, from $x \odot 1 \le 1$ we get $1 \le x \leadsto 1$, that is, $x \leadsto 1 = 1$.   □

A residuated lattice $\mathcal{A}$ with a constant 0 (which can denote any element) is called a *pointed residuated lattice* or *full Lambek algebra* (*FL-algebra*, for short). If $x \leq e$ for all $x \in A$, then $\mathcal{A}$ is called an *integral residuated lattice*.

Clearly, an integral residuated lattice is a porim.

Remark 3.1 If $(A, \wedge, \vee, \odot, \rightarrow, \rightsquigarrow, e)$ is a residuated lattice, then $x \leq y$ iff $x \rightarrow y \geq e$ iff $x \rightsquigarrow y \geq e$.

Indeed, $x \leq y$ iff $e \odot x \leq y$ iff $e \leq x \rightarrow y$. Similarly, $x \leq y$ iff $e \leq x \rightsquigarrow y$.

If $(A, \wedge, \vee, \odot, \rightarrow, \rightsquigarrow, e)$ is an integral residuated lattice, then $x \leq y$ iff $x \rightarrow y = x \rightsquigarrow y = e$.

Remark 3.2 Every porim is a pseudo-BCK(pP) algebra.

Indeed, $(psBCK_1)$ and $(psBCK_2)$ can be proved in the same way as $(rl\text{-}c_8)$ and $(rl\text{-}c_{12})$, respectively. The remaining axioms of a pseudo-BCK algebra are consequences of Remark 3.1.

An FL-algebra $\mathcal{A}$ which satisfies the condition $0 \leq x \leq e$ for all $x \in A$ is called an *FL$_w$-algebra* or *bounded integral residuated lattice*. According to Proposition 3.3, in this case we have $e = 1$, so that an FL$_w$-algebra will be denoted $(A, \wedge, \vee, \odot, \rightarrow, \rightsquigarrow, 0, 1)$. Clearly, if $\mathcal{A}$ is an FL$_w$-algebra, then $(A, \wedge, \vee, 0, 1)$ is a bounded lattice.

A totally ordered FL$_w$-algebra is called a *chain* or *linearly ordered* FL$_w$-algebra.

An FL$_w$-algebra is *commutative* if the operation $\odot$ is commutative (iff $\rightarrow = \rightsquigarrow$) and we shall call such algebras FL$_{ew}$-algebras.

In the sequel we will refer to the FL$_w$-algebra $(A, \wedge, \vee, \odot, \rightarrow, \rightsquigarrow, 0, 1)$ by its universe A.

Example 3.3 Consider $A = \{0, a, b, c, 1\}$ with $0 < a < b < c < 1$ and the operations $\odot, \rightarrow, \rightsquigarrow$ given by the following tables:

$\odot$	0	a	b	c	1
0	0	0	0	0	0
a	0	0	0	a	a
b	0	0	0	b	b
c	0	a	a	c	c
1	0	a	b	c	1

$\rightarrow$	0	a	b	c	1
0	1	1	1	1	1
a	b	1	1	1	1
b	b	c	1	1	1
c	0	a	b	1	1
1	0	a	b	c	1

$\rightsquigarrow$	0	a	b	c	1
0	1	1	1	1	1
a	b	1	1	1	1
b	b	b	1	1	1
c	0	b	b	1	1
1	0	a	b	c	1

Then $(A, \wedge, \vee, \odot, \rightarrow, \rightsquigarrow, 0, 1)$ is an FL$_w$-algebra.

Example 3.4 The bounded pseudo-BCK(pP) lattice $(A, \wedge, \vee, \odot, \rightarrow, \rightsquigarrow, 0, 1)$ from Example 1.9 is a proper FL$_w$-algebra.

Example 3.5 The good pseudo-BCK(pP) lattice $(A_1, \wedge, \vee, \odot, \rightarrow, \rightsquigarrow, 0, 1)$ from Example 1.17 is a proper good FL$_w$-algebra.

Remark 3.3

(1) According to Theorem 1.4, any bounded pseudo-BCK(pP) lattice is an FL_w-algebra.
(2) Taking into consideration $(rl\text{-}c_8)$, $(rl\text{-}c_{12})$, Proposition 3.3 and Remark 3.1, it follows that any FL_w-algebra is a bounded pseudo-BCK(pP) lattice.

Theorem 3.1 *Bounded pseudo-BCK(pP) lattices are categorically isomorphic with FL_w-algebras.*

Proof This is a consequence of Remark 3.3. $\square$

Recall that $x^- = x \to 0$ and $x^\sim = x \rightsquigarrow 0$.

Proposition 3.4 *In any FL_w-algebra the following hold*:

$(rl\text{-}c_{21})$ $x \to y = x \to (x \wedge y)$ *and* $x \rightsquigarrow y = x \rightsquigarrow (x \wedge y)$;
$(rl\text{-}c_{22})$ *if* $x \vee y = 1$, *then for each* $n \in \mathbb{N}, n \geq 1$, $x^n \vee y^n = 1$;
$(rl\text{-}c_{23})$ $x \vee (y \odot z) \geq (x \vee y) \odot (x \vee z)$;
$(rl\text{-}c_{24})$ $x \vee y \leq [(x \to y) \rightsquigarrow y] \wedge [(y \to x) \rightsquigarrow x]$;
$(rl\text{-}c_{25})$ $x \vee y \leq [(x \rightsquigarrow y) \to y] \wedge [(y \rightsquigarrow x) \to x]$;
$(rl\text{-}c_{26})$ $x^m \vee y^n \geq (x \vee y)^{mn}$ *for all* $m, n \in \mathbb{N}, m, n \geq 1$;
$(rl\text{-}c_{27})$ $(x^{-\sim} \vee y^{-\sim})^{-\sim} = (x^{-\sim} \vee y)^{-\sim} = (x \vee y)^{-\sim}$ *and* $(x^{\sim-} \vee y^{\sim-})^{\sim-} = (x^{\sim-} \vee y)^{\sim-} = (x \vee y)^{\sim-}$.

Proof

$(rl\text{-}c_{21})$ By $(psbck\text{-}c_{25})$ we have $(x \to y) \odot x \leq x \wedge y$, so $x \to y \leq x \to (x \wedge y)$.
   Applying $(psbck\text{-}c_{10})$, $x \wedge y \leq y$ implies $x \to (x \wedge y) \leq x \to y$.
   Thus $x \to y = x \to (x \wedge y)$. Similarly, $x \rightsquigarrow y = x \rightsquigarrow (x \wedge y)$.
$(rl\text{-}c_{22})$ If $x \vee y = 1$, then $x = x \odot 1 = x \odot (x \vee y) = x \odot x \vee x \odot y \leq x^2 \vee y$.
   Hence $(x^2 \vee y) \vee y \geq x \vee y = 1$, so $x^2 \vee y = 1$.
   It follows that $y = 1 \odot y = (x^2 \vee y) \odot y = x^2 \odot y \vee y \odot y \leq x^2 \vee y^2$.
   Thus $x^2 \vee (x^2 \vee y^2) \geq x^2 \vee y = 1$, so $x^2 \vee y^2 = 1$.
   Now we have $1 = x \vee y = x^2 \vee y^2 = (x^2)^2 \vee (y^2)^2 = ((x^2)^2)^2 \vee ((y^2)^2)^2 = \cdots$.
   We conclude that $x^{2^n} \vee y^{2^n} = 1$ for all $n \in \mathbb{N}, n \geq 1$.
   Taking into consideration that $n \leq 2^n$ we get $x^n \vee y^n \geq x^{2^n} \vee y^{2^n} = 1$.
   Thus $x^n \vee y^n = 1$ for all $n \in \mathbb{N}, n \geq 1$.
$(rl\text{-}c_{23})$ Applying $(rl\text{-}c_2)$ we have:

$$(x \vee y) \odot (x \vee z) = \big((x \vee y) \odot x\big) \vee \big((x \vee y) \odot z\big)$$
$$= \big((x \odot x) \vee (y \odot x)\big) \vee \big((x \odot z) \vee (y \odot z)\big)$$
$$\leq (x \vee x) \vee \big(x \vee (y \odot z)\big) = x \vee (y \odot z).$$

$(rl\text{-}c_{24})$ From $(x \to y) \odot x \leq y$ we have $x \leq (x \to y) \rightsquigarrow y$.

Taking into consideration that $y \leq (x \to y) \rightsquigarrow y$, we get $x \vee y \leq (x \to y) \rightsquigarrow y$.
From $(y \to x) \odot y \leq x$ we get $y \leq (y \to x) \rightsquigarrow x$.
Since $x \leq (y \to x) \rightsquigarrow x$, we get $x \vee y \leq (y \to x) \rightsquigarrow x$.
Thus $x \vee y \leq [(x \to y) \rightsquigarrow y] \wedge [(y \to x) \rightsquigarrow x]$.

(rl-c_{25}) Similar to (rl-c_{24}).

(rl-c_{26}) Applying (rl-c_{23}), we show that $x \vee y^n \geq (x \vee y)^n$ for all $n \in \mathbb{N}$, $n \geq 1$.
For $n = 1$, it is obvious that $x \vee y \geq x \vee y$.
For $n = 2$ we have: $x \vee y^2 \geq (x \vee y) \odot (x \vee y) = (x \vee y)^2$.
Suppose that $x \vee y^n \geq (x \vee y)^n$ and we have:

$$x \vee y^{n+1} = x \vee (y^n \odot y) \geq (x \vee y^n) \vee (x \vee y) \geq (x \vee y)^n \vee (x \vee y) = (x \vee y)^{n+1}.$$

We conclude that $x \vee y^n \geq (x \vee y)^n$ for all $n \in \mathbb{N}$, $n \geq 1$.
It follows that:

$$x^m \vee y^n \geq (x^m \vee y)^n = (y \vee x^m)^n \geq ((y \vee x)^m)^n = (y \vee x)^{mn} = (x \vee y)^{mn}.$$

(rl-c_{27}) Applying ($psbck$-c_{41}) we have:

$$\begin{aligned}
(x^{-\sim} \vee y^{-\sim})^{-\sim} &= (x^{-\sim-} \wedge y^{-\sim-})^\sim \\
&= (x^- \wedge y^-)^\sim = ((x \vee y)^-)^\sim = (x \vee y)^{-\sim} \quad \text{and}
\end{aligned}$$

$$\begin{aligned}
(x^{-\sim} \vee y)^{-\sim} &= (x^{-\sim-} \wedge y^-)^\sim \\
&= (x^- \wedge y^-)^\sim = ((x \vee y)^-)^\sim = (x \vee y)^{-\sim}.
\end{aligned}$$

Similarly, $(x^{\sim-} \vee y^{\sim-})^{\sim-} = (x^{\sim-} \vee y)^{\sim-} = (x \vee y)^{\sim-}$. $\square$

Theorem 3.2 *Let* $(A, \wedge, \vee, \odot, \to, \rightsquigarrow, 0, 1)$ *be an* FL_w*-algebra. Then the algebra* $A'_a = ([a, 1], \wedge, \vee, \to, \rightsquigarrow, \odot^1_a, a, 1)$ *is an* FL_w*-algebra, where* $x \odot^1_a y := (x \odot y) \vee a$ *for all* $x, y \in [a, 1]$.

Proof We will check the conditions (A_1)–(A_3) from the definition of an FL_w-algebra:

(A_1) Obviously, $([a, 1], \wedge, \vee, a, 1)$ is a bounded lattice with smallest element a and greatest element 1.

(A_2) Since $x \odot^1_a 1 = (x \odot 1) \vee a = x \vee a = x$ and $1 \odot^1_a x = (1 \odot x) \vee a = x \vee a = x$, it follows that 1 is the unit element in $([a, 1], \odot^1_a, 1)$.
For any $x, y, z \in [a, 1]$ we have:

$$\begin{aligned}
(x \odot^1_a y) \odot^1_a z &= (((x \odot y) \vee a) \odot z) \vee a \\
&= ((x \odot y \odot z) \vee (a \odot z)) \vee a \\
&= (x \odot y \odot z) \vee ((a \odot z) \vee a) \\
&= (x \odot y \odot z) \vee a,
\end{aligned}$$

$$x \odot_a^1 \left(y \odot_a^1 z\right) = \left(x \odot \left((y \odot z) \vee a\right)\right) \vee a$$
$$= \left((x \odot y \odot z) \vee (x \odot a)\right) \vee a$$
$$= (x \odot y \odot z) \vee \left((x \odot a) \vee a\right)$$
$$= (x \odot y \odot z) \vee a.$$

Thus $\odot_a^1$ is associative.

It follows that $([a, 1], \odot_a^1, 1)$ is a monoid.

(A_3) $x \odot_a^1 y \leq z \Rightarrow (x \odot y) \vee a \leq z \Rightarrow x \odot y \leq z \Rightarrow x \leq y \to z$ and $y \leq x \leadsto z$.
Conversely, $x \leq y \to z \Rightarrow x \odot y \leq z$ and considering that $a \leq z$ we get $(x \odot y) \vee a \leq z \vee a = z \Rightarrow x \odot_a^1 y \leq z$.
Similarly, from $y \leq x \leadsto z$ we get $x \odot_a^1 y \leq z$.
We conclude that A_a^1 is an FL_w-algebra.
It is trivial to see that $A = A_0^1$ and $\{1\} = A_1^1$. $\square$

The next result is proved in a similar manner as in [110] for the case of bounded $R\ell$-monoids.

Proposition 3.5 *In any locally finite FL_w-algebra the following hold*:

(1) $0 < x < 1$ *iff* $0 < x^- < 1$;
(1') $0 < x < 1$ *iff* $0 < x^\sim < 1$;
(2) $x^- = 0$ *iff* $x = 1$;
(2') $x^\sim = 0$ *iff* $x = 1$;
(3) $x^- = 1$ *iff* $x = 0$;
(3') $x^\sim = 1$ *iff* $x = 0$.

Proof

(1) Consider $0 < x < 1$ and let $n > 0$ be the least integer such that $x^n = 0$.
   It follows that $x^{n-1} \neq 0$ and $x^{n-1} \odot x = x^n = 0$, hence $0 < x^{n-1} \leq x \to 0 = x^-$.
   If $x^- = 1$, then $0 < x \leq x^{-\sim} = 0$, a contradiction. Thus $0 < x^- < 1$.
   Conversely, consider $0 < x^- < 1$, so $0 < x^{-\sim} < 1$ and $0 \leq x \leq x^{-\sim}$.
   If $x = 0$, then $x^- = 1$, which is a contradiction. Thus $0 < x < 1$.
(2) Assume $x^- = 0$. If $x < 1$, then similarly as in (1) we get $0 < x^-$, a contradiction. Hence $x = 1$. Conversely, if $x = 1$, then $x^- = 0$.
(3) Assume $x^- = 1$, so $x^{-\sim} = 0$. Taking into consideration that $x \leq x^{-\sim}$, we get $x = 0$. Conversely, if $x = 0$, then $x^- = 1$.
(1'), (2'), (3') can be proved in a similar way to (1), (2) and (3), respectively. $\square$

For any FL_w-algebra A, we make the following definitions:

$$\mathrm{Reg}(A) := \left\{x \in A \mid x = x^{-\sim} = x^{\sim-}\right\},$$
$$\mathrm{Id}(A) := \{x \in A \mid x \odot x = x\}.$$

Proposition 3.6 *If $x \in \mathrm{Id}(A)$ and $y \in A$, then:*

$$(x \rightarrow y) \odot x = y \odot x \quad \text{and} \quad x \odot (x \rightsquigarrow y) = x \odot y.$$

Proof Taking $z = x$ in $(psbck\text{-}c_{27})$ we get $x \rightarrow y \leq x \odot x \rightarrow y \odot x$. Since $x \odot x = x$, it follows that $x \rightarrow y \leq x \rightarrow y \odot x$, so $(x \rightarrow y) \odot x \leq y \odot x$. On the other hand, from $y \leq x \rightarrow y$ we get $y \odot x \leq (x \rightarrow y) \odot x$. Thus $(x \rightarrow y) \odot x = y \odot x$.

The second identity follows similarly from the second inequality $(psbck\text{-}c_{27})$. $\qquad\square$

Proposition 3.7 *If for all $x, y \in A$:*

$$(x \rightarrow y) \odot x = y \odot x \quad \text{or} \quad x \odot (x \rightsquigarrow y) = x \odot y,$$

then $A = \mathrm{Id}(A)$.

Proof Taking $y = x$ in any of the two identities and applying $(rl\text{-}c_{16})$, we get $x \odot x = x$ for all $x \in A$. Hence $A = \mathrm{Id}(A)$. $\qquad\square$

Example 3.6 If A is the FL_w-algebra from Example 3.4, then $\mathrm{Id}(A) = \{0, a_2, s, 1\}$ and $\mathrm{Reg}(A) = \{0, 1\}$.

Definition 3.2 The *distance functions* on the FL_w-algebra A are the functions $d_1, d_2 : A \times A \longrightarrow A$ defined by:

$$d_1(x, y) := (x \rightarrow y) \wedge (y \rightarrow x) = x \vee y \rightarrow x \wedge y;$$

$$d_2(x, y) := (x \rightsquigarrow y) \wedge (y \rightsquigarrow x) = x \vee y \rightsquigarrow x \wedge y.$$

Proposition 3.8 *The two distance functions satisfy the following properties:*

(1) $d_1(x, y) = d_1(y, x)$ and $d_2(x, y) = d_2(y, x)$;
(2) $d_1(x, y) = 1$ iff $x = y$ iff $d_2(x, y) = 1$;
(3) $d_1(x, 0) = x^-$ and $d_2(x, 0) = x^\sim$;
(4) $d_1(x, 1) = x = d_2(x, 1)$;
(5) $d_1(x, y) \leq d_2(x^-, y^-)$;
(6) $d_2(x, y) \leq d_1(x^\sim, y^\sim)$;
(7) $d_1(x, y) \leq d_1(x^{\sim-}, y^{\sim-})$;
(8) $d_2(x, y) \leq d_2(x^{-\sim}, y^{-\sim})$;
(9) $d_2(x^-, y^-) = d_1(x^{-\sim}, y^{-\sim})$;
(10) $d_1(x^\sim, y^\sim) = d_2(x^{\sim-}, y^{\sim-})$.

Proof

(1) This is obvious.
(2) $d_1(x, y) = 1 \Leftrightarrow x \rightarrow y = 1$ and $y \rightarrow x = 1 \Leftrightarrow x \leq y$ and $y \leq x \Leftrightarrow x = y$.
Similarly, $d_2(x, y) = 1 \Leftrightarrow x = y$.

(3) By the definition of the distance functions we have:

$$d_1(x, 0) = (x \to 0) \wedge (0 \to x) = x^- \wedge 1 = x^-;$$
$$d_2(x, 0) = (x \rightsquigarrow 0) \wedge (0 \rightsquigarrow x) = x^\sim \wedge 1 = x^\sim.$$

(4) Similarly we get:

$$d_1(x, 1) = (x \to 1) \wedge (1 \to x) = 1 \wedge x = x;$$
$$d_2(x, 1) = (x \rightsquigarrow 1) \wedge (1 \rightsquigarrow x) = 1 \wedge x = x.$$

(5) By (*psbck-c$_{15}$*) we have:

$$d_1(x, y) = (x \to y) \wedge (y \to x) \le \left(y^- \rightsquigarrow x^- \right) \wedge \left(x^- \rightsquigarrow y^- \right) = d_2\left(x^-, y^- \right).$$

(6) By (*psbck-c$_{15}$*) we have:

$$d_2(x, y) = (x \rightsquigarrow y) \wedge (y \rightsquigarrow x) \le \left(y^\sim \to x^\sim \right) \wedge \left(x^\sim \to y^\sim \right) = d_1\left(x^\sim, y^\sim \right).$$

(7) By (5) and (6) we get $d_1(x, y) \le d_2(x^-, y^-) \le d_1(x^{-\sim}, y^{-\sim})$.
(8) Similarly, $d_2(x, y) \le d_1(x^\sim, y^\sim) \le d_2(x^{\sim-}, y^{\sim-})$.
(9) By the above properties we get:

$$d_2\left(x^-, y^- \right) \le d_1\left(x^{-\sim}, y^{-\sim} \right) \le d_2\left(x^{-\sim-}, y^{-\sim-} \right) = d_2\left(x^-, y^- \right),$$

hence $d_2(x^-, y^-) = d_1(x^{-\sim}, y^{-\sim})$.
(10) Similarly, $d_1(x^\sim, y^\sim) \le d_2(x^{\sim-}, y^{\sim-}) \le d_1(x^{\sim-\sim}, y^{\sim-\sim}) = d_1(x^\sim, y^\sim)$,
hence $d_1(x^\sim, y^\sim) = d_2(x^{\sim-}, y^{\sim-})$.   □

3.2   The Lattice of Filters of an FL$_w$-Algebra

Recall that a nonempty subset F of a lattice L is a *filter* of L if it satisfies the conditions: (i) $x, y \in F$ implies $x \wedge y \in F$ and (ii) $x \in F$, $y \in L$, $x \le y$ implies $y \in F$.

Let $(A, \wedge, \vee, \odot, \to, \rightsquigarrow, 0, 1)$ be an FL$_w$-algebra. We recall that (see Definition 1.15) a nonempty set F of A is called a *filter* of A if the following conditions hold:

(F_1) If $x, y \in F$, then $x \odot y \in F$;
(F_2) If $x \in F$, $y \in A$, $x \le y$, then $y \in F$.

Remark 3.4 If F is a filter of A, then:

(F_3) $1 \in F$.
(F_4) If $x \in F$, $y \in A$, then $y \to x \in F$, $y \rightsquigarrow x \in F$.
(F_5) If $x, y \in F$, then $x \wedge y \in F$.

Remark 3.5

(1) Any filter of A is a filter for the lattice $(A, \wedge, \vee)$, but the converse is not true. Indeed, let F be a filter of an FL_w-algebra A and $x, y \in A$. Since $x \odot y \in F$ and $x \odot y \leq x \wedge y$, we get $x \wedge y \in F$, so F is a filter of the lattice $(A, \wedge, \vee)$.

Consider now $A = \{0, a, b, c, 1\}$ where $0 < a < b < c < 1$ and the operations $\odot, \rightarrow, \rightsquigarrow$ are given by the following tables:

$\odot$	0	a	b	c	1
0	0	0	0	0	0
a	0	0	0	0	a
b	0	0	0	0	b
c	0	0	a	a	c
1	0	a	b	c	1

$\rightarrow$	0	a	b	c	1
0	1	1	1	1	1
a	c	1	1	1	1
b	b	c	1	1	1
c	b	c	c	1	1
1	0	a	b	c	1

$\rightsquigarrow$	0	a	b	c	1
0	1	1	1	1	1
a	c	1	1	1	1
b	c	c	1	1	1
c	a	c	c	1	1
1	0	a	b	c	1

Then $(A, \wedge, \vee, \odot, \rightarrow, \rightsquigarrow, 0, 1)$ is an FL_w-algebra (we will see later that it is a pseudo-MTL chain).

One can easily prove that $F = \{c, 1\}$ is a filter of the lattice $(A, \vee, \wedge)$, but F is not a filter of the FL_w-algebra A ($c \in F$, but $c \odot c = a \notin F$).

(2) In FL_w algebras filters coincide with deductive systems, being pseudo-BCK(pP) algebras.

Proposition 3.9 *Let A be an FL_w-algebra. Then $[x \vee y) = [x) \cap [y)$ for all $x, y \in A$.*

Proof Consider $z \in [x \vee y)$, so $z \geq (x \vee y)^n$ for some $n \geq 1$. It follows that $z \geq x^n$ and $z \geq y^n$ for some $n \geq 1$, that is, $z \in [x)$ and $z \in [y)$.

Thus $z \in [x) \cap [y)$, so $[x \vee y) \subseteq [x) \cap [y)$.

Conversely, if $z \in [x) \cap [y)$, then $z \in [x)$ and $z \in [y)$, so $z \geq x^n$ and $z \geq y^m$ for some $n, m \geq 1$. Applying (rl-c26) we get $z \geq x^n \vee y^m \geq (x \vee y)^{mn}$, that is, $z \in [x \vee y)$. Hence $[x) \cap [y) \subseteq [x \vee y)$. Thus $[x \vee y) = [x) \cap [y)$. $\square$

If F_1 and F_2 are filters of A, we define $F_1 \wedge F_2 = F_1 \cap F_2$ and $F_1 \vee F_2 = [F_1 \cup F_2)$.

Recall that, if $F \in \mathcal{F}(A)$ and $x \in A \setminus F$ we denote $[F \cup \{x\})$ by $F(x)$. Then according to Propositions 1.35 and 1.36 we have $F(x) = F \vee [x)$.

Definition 3.3 Let $\mathcal{L} = (L, \wedge, \vee)$ be a lattice.

(1) For every $y, z \in L$, the *relative pseudocomplement of y with respect to z*, provided it exists, is the greatest element x such that $x \wedge y \leq z$. It is denoted by $y \Rightarrow z$ (i.e. $y \Rightarrow z = \max\{x \mid x \wedge y \leq z\}$).

(2) $\mathcal{L}$ is said to be *relatively pseudocomplemented* provided the relative pseudo-complement $y \Rightarrow z$ exists for every $y, z \in L$.

(3) A *Heyting algebra* is a relatively pseudocomplemented lattice with 0, i.e. it is bounded.

If $\mathcal{L}$ is a relatively pseudocomplemented lattice, then $\Rightarrow$ can be viewed as a binary operation on L and there exists the greatest element, 1, of the lattice: $1 = x \Rightarrow x$ for all $x \in L$. Consequently, we have the following equivalent definition, with $\odot = \wedge$:

Definition 3.4

(1) A *relatively pseudocomplemented lattice* is an algebra $\mathcal{L} = (L, \wedge, \vee, \Rightarrow, 1)$, where $(L, \wedge, \vee, 1)$ is a lattice with greatest element and the binary operation $\Rightarrow$ on L satisfies: for all $x, y, z \in L$, $x \leq y \Rightarrow z$ if and only if $x \wedge y \leq z$.

(2) A *Heyting algebra* is an alternative name for a bounded relatively pseudocomplemented lattice. For any $x \in L$, the element $x^* = x \Rightarrow 0$ is called the pseudo-complement of x.

Remark 3.6

(1) A *Brouwer algebra* is the dual of a Heyting algebra ($\vee$ instead of $\wedge$, $\geq$ instead of $\leq$, $y \rightarrow z = \min\{x \mid z \leq x \vee y\}$ instead of $y \Rightarrow z$).

(2) Recall that *Gödel algebras* are Heyting algebras verifying the condition $(x \Rightarrow y) \vee (y \Rightarrow x) = 1$ and that the *Gödel t-norm* and its associated residuum (implication) on $[0, 1]$ are:

$$x \odot_G y := \min\{x, y\} = x \wedge y,$$

$$x \rightarrow_G y := \begin{cases} 1 & \text{if } x \leq y \\ y & \text{if } x > y. \end{cases}$$

(Gödel implication).

Note also that a proper Heyting algebra (i.e. which is not a Gödel algebra) is not linearly ordered.

(3) An FL_w-algebra A satisfying $A = \mathrm{Id}(A)$ is a Heyting algebra.

According to [9], a complete lattice is a Heyting algebra if and only if it satisfies the identity

$$a \wedge \left(\bigvee_{i \in I} b_i \right) = \bigvee_{i \in I} (a \wedge b_i).$$

Let A be an FL_w-algebra.

The proof of the next result is similar to that in [30] for the case of pseudo-BL algebras.

Theorem 3.3 $(\mathcal{F}(A), \wedge, \vee, \Rightarrow, \{1\}, A)$ *is a complete Heyting algebra, that is,*

$$F \wedge \left(\bigvee_{i \in I} G_i \right) = \bigvee_{i \in I} (F \wedge G_i)$$

for any filter F and for any family of filters $\{G_i\}_{i \in I}$ of A.

Proof We have $F \wedge (\bigvee_{i \in I} G_i) = \bigvee_{i \in I}(F \wedge G_i)$ iff $F \cap [\bigcup_{i \in I} G_i) = [\bigcup_{i \in I}(F \cap G_i))$ for any filter F and for any family of filters $\{G_i\}_{i \in I}$ of A.

   Clearly, $[\bigcup_{i \in I}(F \cap G_i)) \subseteq F \cap [\bigcup_{i \in I} G_i)$.

   Conversely, if $x \in F \cap [\bigcup_{i \in I} G_i)$, then $x \in F$ and there exist $i_1, i_2, \ldots, i_m \in I$, $x_{i_j} \in G_{i_j}$ $(1 \le j \le m)$ such that $x \ge x_{i_1} \odot \cdots \odot x_{i_m}$.

   By $(rl\text{-}c_{23})$ we have $x = x \vee (x_{i_1} \odot \cdots \odot x_{i_m}) \ge (x \vee x_{i_1}) \odot \cdots \odot (x \vee x_{i_m})$.

   Since $x \vee x_{i_j} \in F \cap G_{i_j}$ for every $1 \le j \le m$, it follows that $x \in [\bigcup_{i \in I}(F \cap G_i))$, hence $F \cap [\bigcup_{i \in I} G_i) \subseteq [\bigcup_{i \in I}(F \cap G_i))$. Thus $F \cap [\bigcup_{i \in I} G_i) = [\bigcup_{i \in I}(F \cap G_i))$, that is, $F \wedge (\bigvee_{i \in I} G_i) = \bigvee_{i \in I}(F \wedge G_i)$. $\square$

Proposition 3.10 *If $F_1, F_2 \in \mathcal{F}_n(A)$ then:*

(1) $F_1 \wedge F_2 \in \mathcal{F}_n(A)$;
(2) $F_1 \vee F_2 \in \mathcal{F}_n(A)$.

Proof

(1) We have $F_1 \wedge F_2 = F_1 \cap F_2$. Consider $x, y \in A$ such that $x \to y \in F_1 \cap F_2$, that is, $x \to y \in F_1$ and $x \to y \in F_2$. It follows that $x \rightsquigarrow y \in F_1$ and $x \rightsquigarrow y \in F_2$, hence $x \rightsquigarrow y \in F_1 \cap F_2$.
   Similarly, $x \rightsquigarrow y \in F_1 \cap F_2$ implies $x \to y \in F_1 \cap F_2 = F_1 \wedge F_2$.
   Hence $F_1 \wedge F_2 \in \mathcal{F}_n(A)$.
(2) Let $x, y \in A$ such that $x \to y \in F_1 \vee F_2 = [F_1 \cup F_2)$. By Proposition 1.36, there are $u \in F_1$, $v \in F_2$ such that $u \odot v \le x \to y$.
   Hence $(u \odot v) \odot x \le y$, so $u \odot (v \odot x) \le y$.
   Applying Lemma 1.11, there is a $v' \in F_2$ such that $v \odot x \ge x \odot v'$.
   It follows that $y \ge (u \odot x) \odot v'$.
   Similarly, there is a $u' \in F_1$ such that $u \odot x \ge x \odot u'$, so $y \ge x \odot (u' \odot v')$.
   We get $u' \odot v' \le x \rightsquigarrow y$, hence $x \rightsquigarrow y \in F_1 \vee F_2$.
   Similarly, $x \rightsquigarrow y \in F_1 \vee F_2$ implies $x \to y \in F_1 \vee F_2$.
   Thus $F_1 \vee F_2 \in \mathcal{F}_n(A)$. $\square$

Proposition 3.11 *If $(F_i)_{i \in I}$ is a family of normal filters of A, then:*

(1) $\bigwedge_{i \in I} F_i \in \mathcal{F}_n(A)$;
(2) $\bigvee_{i \in I} F_i \in \mathcal{F}_n(A)$.

Proof Similar to the above argument. $\square$

As a consequence of the above result we get:

Theorem 3.4 $\mathcal{F}_n(A)$ *is a complete sublattice of* $(\mathcal{F}(A), \subseteq)$.

Proposition 3.12 *For a given* $H \in \mathcal{F}_n(A)$ *the relation* $\theta_H = \equiv_H$ *is a congruence on* A.

Proof According to Propositions 1.40 and 1.41, θ_H is an equivalence relation compatible with $\rightarrow$, $\rightsquigarrow$ and $\odot$.

Consider $(x, y), (a, b) \in \theta_H$.

We have $x \rightarrow y, y \rightarrow x, x \rightsquigarrow y, y \rightsquigarrow x \in H$ and $a \rightarrow b, b \rightarrow a, a \rightsquigarrow b, b \rightsquigarrow a \in H$.

From $y \leq y \vee b$, applying (*psbck-c$_{10}$*), it follows that $x \rightarrow y \leq x \rightarrow y \vee b$, so $x \rightarrow y \vee b \in H$. Similarly, from $b \leq y \vee b$ we get $a \rightarrow b \leq a \rightarrow y \vee b$, hence $a \rightarrow y \vee b \in H$.

According to (*rl-c$_4$*) we have $x \vee a \rightarrow y \vee b = (x \rightarrow y \vee b) \wedge (a \rightarrow y \vee b) \in H$.

We can prove similarly that $y \vee b \rightarrow x \vee a, x \vee a \rightsquigarrow y \vee b, y \vee b \rightsquigarrow x \vee a \in H$.

Thus $x \vee a \equiv_{\theta_H} y \vee b$. We conclude that θ_H is compatible with $\vee$.

From $x \wedge a \leq x$, applying (*psbck-c$_1$*) we get $x \rightarrow y \leq x \wedge a \rightarrow y$.

Hence $x \wedge a \rightarrow y \in H$.

Similarly, from $x \wedge a \leq a$ we have $a \rightarrow b \leq x \wedge a \rightarrow b$, so $x \wedge a \rightarrow b \in H$.

Since by (*rl-c$_3$*) we have $x \wedge a \rightarrow y \wedge b = (x \wedge a \rightarrow y) \wedge (x \wedge a \rightarrow b)$, it follows that $x \wedge a \rightarrow y \wedge b \in H$. Similarly, $y \wedge b \rightarrow x \wedge a, x \wedge a \rightsquigarrow y \wedge b, y \wedge b \rightsquigarrow x \wedge a \in H$.

Thus $x \wedge a \equiv_{\theta_H} y \wedge b$. Hence θ_H is compatible with $\wedge$.

We conclude that θ_H is a congruence on A. $\square$

3.3 Boolean Center of an FL$_w$-Algebra

Let $(A, \wedge, \vee, 0, 1)$ be a bounded lattice. Recall that (see [4, 156]) an element $a \in A$ is said to be *complemented* if there is an element $b \in A$ such that $a \vee b = 1$ and $a \wedge b = 0$; if such an element b exists, it is called a *complement* of a. We will denote the set of all complemented elements in A by $B(A)$. Complements are generally not unique unless the lattice is distributive. In FL$_w$-algebras however, although the underlying lattices need not be distributive, the complements are unique.

If a has a unique complement, we shall denote this complement by a'.

Proposition 3.13 *Let* $(A, \wedge, \vee, \odot, \rightarrow, \rightsquigarrow, 0, 1)$ *be an FL$_w$-algebra and suppose that* $a \in A$ *has the complement* $b \in A$. *Then the following hold:*

(1) *if* c *is a complement of* a *in* A, *then* $c = b$;
(2) $a^- = a^\sim = b$ *and* $b^- = b^\sim = a$;
(3) $a^2 = a$.

Proof

(1) Since b and c are complements of a, it follows that $a \vee b = 1 = a \vee c$ and $a \wedge b = 0 = a \wedge c$. Hence $c = c \odot 1 = c \odot (a \vee b) = (c \odot a) \vee (c \odot b) = c \odot b$ (since $c \odot a \le c \wedge a = 0$, we have $c \odot a = 0$).

Similarly, $b = 1 \odot b = (a \vee c) \odot b = (a \odot b) \vee (c \odot b) = c \odot b$. Thus $b = c$.

(2) We have:

$$a \odot b \le a \wedge b = 0, \quad \text{so } a \le b \to 0 = b^- \text{ and } b \le a \rightsquigarrow 0 = a^\sim;$$

$$b \odot a \le b \wedge a = 0, \quad \text{so } b \le a \to 0 = a^- \text{ and } a \le b \rightsquigarrow 0 = b^\sim.$$

But, $a^- = a^- \odot 1 = a^- \odot (a \vee b) = (a^- \odot a) \vee (a^- \odot b) = a^- \odot b$.

Hence $a^- \odot b = a^- \ge a^- \wedge b$ and using the fact that $a^- \odot b \le a^- \wedge b$, we get $a^- = a^- \odot b = a^- \wedge b$, so $b \ge a^- \wedge b = a^-$.

Thus $a^- = b$ and similarly $a^\sim = b$.

(3) Applying $(rl\text{-}c_2)$ we get:

$$a = a \odot (a \vee b) = a^2 \vee (a \odot b) = a^2$$

(since $a \odot b \le a \wedge b = 0$).   □

Let $B(A)$ be the set of all complemented elements of the lattice $L(A) = (A, \wedge, \vee, 0, 1)$. The set $B(A)$ is called the *Boolean center* of A.

Lemma 3.1 *Let A be an FL$_w$-algebra. Then the following are equivalent:*

(a) $x \in B(A)$;
(b) $x \vee x^- = 1$ *and* $x \wedge x^- = 0$;
(c) $x \vee x^\sim = 1$ *and* $x \wedge x^\sim = 0$.

Proof

(a) $\Rightarrow$ (b) Since $x \in B(A)$, there exists a $y \in A$ such that $x \vee y = 1$ and $x \wedge y = 0$. Hence $x^- = x^- \odot 1 = x^- \odot (x \vee y) = (x^- \odot x) \vee (x^- \odot y) = x^- \odot y$, so $y \ge x^- \odot y = x^-$. On the other hand, because $y \odot x \le x \wedge y = 0$, it follows that $y \odot x = 0$, so $y \le x^-$. Thus $x^- = y$, that is, $x \vee x^- = 1$ and $x \wedge x^- = 0$.
(b) $\Rightarrow$ (a) Obviously.
(a) $\Rightarrow$ (c) Similar to (a) $\Leftrightarrow$ (b).   □

Corollary 3.1 *If $x \in B(A)$, then:*

(1) $(x^-)^2 = x^-$ *and* $(x^\sim)^2 = x^\sim$;
(2) $x \to x^- = x \rightsquigarrow x^- = x^-$ *and* $x \to x^\sim = x \rightsquigarrow x^\sim = x^\sim$;
(3) $x^- \to x = x^- \rightsquigarrow x = x^\sim \to x = x^\sim \rightsquigarrow x = x$.

Proof We have:

(1) $x^- = x^- \odot 1 = x^- \odot (x^- \vee x) = (x^-)^2 \vee (x^- \odot x) = (x^-)^2; x^\sim = 1 \odot x^\sim = (x \vee x^\sim) \odot x^\sim = (x \odot x^\sim) \vee (x^\sim)^2 = (x^\sim)^2.$
(2) and (3) follow from $(rl\text{-}c_{24})$ and $(rl\text{-}c_{25})$, taking into consideration that $x \vee x^- = x \vee x^\sim = 1$ and applying $(psbck\text{-}c_6).$ □

Corollary 3.2 *If $x \in B(A)$, then $x^{-\sim} = x^{\sim-} = x.$*

Proof Since x and $x^{-\sim}$ are complements of x^-, then by Proposition 3.13 it follows that $x^{-\sim} = x$. Similarly, $x^{\sim-} = x.$ □

Proposition 3.14 *If $a \in B(A)$ and $x \in A$, then the following hold:*

(1) $a \odot x = x \odot a = a \odot x \odot a;$
(2) $a^- \odot x = x \odot a^- = a^- \odot x \odot a^- = a^\sim \odot x \odot a^\sim = a^\sim \odot x = x \odot a^\sim.$

Proof

(1) Applying the properties of an FL_w-algebra we get:

$$a \odot x = a \odot x \odot 1 = a \odot x \odot (a \vee a^\sim)$$
$$= (a \odot x \odot a) \vee (a \odot x \odot a^\sim) = a \odot x \odot a$$

(we applied the fact that $a \odot x \odot a^\sim \leq a \odot a^\sim = 0$, so $a \odot x \odot a^\sim = 0$).
   Similarly, $x \odot a = a \odot x \odot a.$
(2) We have:

$$a^- \odot x = a^- \odot x \odot 1 = a^- \odot x \odot (a^- \vee a)$$
$$= (a^- \odot x \odot a^-) \vee (a^- \odot x \odot a) = a^- \odot x \odot a^-$$

(since $a^- \odot x \odot a \leq a^- \odot a = 0$, we have $a^- \odot x \odot a = 0$).

$$x \odot a^\sim = (a^\sim \vee a) \odot (x \odot a^\sim)$$
$$= (a^\sim \odot x \odot a^\sim) \vee (a \odot x \odot a^\sim) = a^\sim \odot x \odot a^\sim$$

(since $a \odot x \odot a^\sim \leq a \odot a^\sim = 0$, it follows that $a \odot x \odot a^\sim = 0$).
   The assertion follows taking into consideration that $a^- = a^\sim.$ □

Proposition 3.15 *If $a \in B(A)$, then the filters $[a), [a^-)$ and $[a^\sim)$ are normal.*

Proof Since $a^2 = a$, we have:

$$[a) = \{x \in A \mid x^n \leq a \text{ for some } n \in \mathbb{N}, n \geq 1\} = \{x \in A \mid a \leq x\}.$$

It follows that:

$$x \rightarrow y \in [a) \quad \text{iff} \quad a \leq x \rightarrow y \quad \text{iff} \quad a \odot x \leq y \quad \text{iff}$$

$$x \odot a \leq y \quad \text{iff} \quad a \leq x \rightsquigarrow y \quad \text{iff} \quad x \rightsquigarrow y \in [a).$$

Thus $[a)$ is a normal filter of A. Using the identities $(a^-)^2 = a^-$ and $(a^\sim)^2 = a^\sim$, it follows that $[a^-) = \{x \in A \mid a^- \leq x\}$ and $[a^\sim) = \{x \in A \mid a^\sim \leq x\}$.

We prove in a similar manner that $[a^-)$ and $[a^\sim)$ are normal filters of A.   □

Proposition 3.16 *Let A be an FL_w-algebra, $x \in B(A)$ and $n \in \mathbb{N}$, $n \geq 1$. Then the following are equivalent:*

(a) $x^n \in B(A)$;
(b) $x \vee (x^n)^- = 1$ *and* $x \vee (x^n)^\sim = 1$.

Proof

(a) $\Rightarrow$ (b)   Let $x^n \in B(A)$. By Lemma 3.1 we have $x^n \vee (x^n)^- = 1$. Since $x^n \leq x$, we get $1 = x^n \vee (x^n)^- \leq x \vee (x^n)^-$, so $x \vee (x^n)^- = 1$.
Similarly, $x \vee (x^n)^\sim = 1$.

(b) $\Rightarrow$ (a)   Since $x \vee (x^n)^- = 1$, by (rl-c22) we have $x^n \vee ((x^n)^-)^n = 1$.
Because $((x^n)^-)^n \leq (x^n)^-$, we get $1 = x^n \vee ((x^n)^-)^n \leq x^n \vee (x^n)^-$, so $x^n \vee (x^n)^- = 1$.
Similarly $x^n \vee (x^n)^\sim = 1$, so by (psbck-c41) we get $(x^n)^- \wedge (x^n)^{\sim-} = 0$.
Because $x^n \leq (x^n)^{\sim-}$ we get $(x^n)^- \wedge x^n \leq (x^n)^- \wedge (x^n)^{\sim-} = 0$, so $(x^n)^- \wedge x^n = 0$.
From $x^n \vee (x^n)^- = 1$ and $x^n \wedge (x^n)^- = 0$ it follows that $x^n \in B(A)$.   □

Proposition 3.17 *If $x \in A$, $n \in \mathbb{N}$, $n \geq 1$ such that $x^n \in B(A)$ and $x^n \geq x^- \vee x^\sim$, then $x = 1$.*

Proof From $x^n \geq x^- \vee x^\sim$ we get $x^n \geq x^-$ and $x^n \geq x^\sim$.
Hence $(x^n)^\sim \leq x^{-\sim}$ and $(x^n)^- \leq x^{\sim-}$, respectively.
Since $x^n \in B(A)$, by Proposition 3.16 we have:

$$1 = x \vee (x^n)^- \leq x \vee x^{\sim-} = x^{\sim-} \quad \text{and}$$

$$1 = x \vee (x^n)^\sim \leq x \vee x^{-\sim} = x^{-\sim},$$

so $x^{\sim-} = x^{-\sim} = 1$, that is, $x^- = x^\sim = 0$.
Applying (psbck-c30) we have:

$$\left(x^2\right)^- = (x \odot x)^- = x \odot x \rightarrow 0 = x \rightarrow (x \rightarrow 0) = x \rightarrow x^- = x \rightarrow 0 = x^-$$

and similarly $(x^2)^\sim = x^\sim$, so $(x^2)^- = (x^2)^\sim = 0$.
Recursively we get $(x^n)^- = (x^n)^\sim = 0$. Hence $1 = x \vee (x^n)^- = x \vee 0 = x$.   □

3.4 Directly Indecomposable FL_w-Algebras

Recall that (see [19]), if A is an algebra and $\theta_1, \theta_2 \in \mathrm{Con}(A)$, then $\theta_1 \circ \theta_2$ is the binary relation on A defined by $(x, y) \in \theta_1 \circ \theta_2$ iff there exists a $z \in A$ such that $(x, z) \in \theta_1$ and $(z, y) \in \theta_2$. If $\theta_1, \theta_2 \in \mathrm{Con}(A)$ such that $\theta_1 \circ \theta_2 = \theta_2 \circ \theta_1$ we say that θ_1 and θ_2 are *permutable*. An algebra A is called *congruence permutable* if $\theta_1 \circ \theta_2 = \theta_2 \circ \theta_1$ for all $\theta_1, \theta_2 \in \mathrm{Con}(A)$.

According to [200] every FL_w-algebra is congruence permutable.

For any algebra A, two permutable congruences θ_1, θ_2 of A are *complementary factor congruences* if $\theta_1 \vee \theta_2 = \mathbf{1}$ and $\theta_1 \wedge \theta_2 = \mathbf{0}$ ($\mathbf{1}$ and $\mathbf{0}$ are top and respectively bottom elements in the lattice $\mathrm{Con}(A)$).

The mapping $p_i : A_1 \times A_2 \longrightarrow A_i$, $i \in \{1, 2\}$, defined by $p_i((a_1, a_2)) = a_i$ is called the *projection map on the ith coordinate* of $A_1 \times A_2$.

An algebra A is *directly indecomposable* if A is not isomorphic to a direct product of two nontrivial algebras.

A *subdirect representation* of an algebra A with factors A_i is an embedding $f : A \longrightarrow \prod_{i \in I} A_i$ such that each $f_i = p_i \circ f$ is onto A_i for all $i \in I$. A is also called a *subdirect product* of A_i. An algebra A is *subdirectly irreducible* iff it is nontrivial and for any subdirect representation $f : A \longrightarrow \prod_{i \in I} A_i$, there exists a $j \in I$ such that f_j is an isomorphism of A onto A_j.

An algebra A is said to be *simple* if it has a two element congruence lattice.

In what follows we recall some results from [19].

For $i \in \{1, 2\}$, the mapping $p_i : A_1 \times A_2 \longrightarrow A_i$ is a surjective homomorphism from $A = A_1 \times A_2$ to A_i. Furthermore, in $\mathrm{Con}(A_1 \times A_2)$, $\mathrm{Ker}(p_1)$ and $\mathrm{Ker}(p_2)$ are permutable and $\mathrm{Ker}(p_1) \cap \mathrm{Ker}(p_2) = \mathbf{0}$, $\mathrm{Ker}(p_1) \vee \mathrm{Ker}(p_2) = \mathbf{1}$.

If θ_1, θ_2 are complementary factor congruences of an algebra A, then $A \cong A/\theta_1 \times A/\theta_2$. As a consequence, an algebra A is directly indecomposable iff the only complementary factor congruences on A are $\mathbf{0}$ and $\mathbf{1}$.

A subdirectly irreducible algebra is directly indecomposable.

Every algebra is isomorphic to a subdirect product of subdirectly irreducible algebras (Birkhoff's theorem).

Proposition 3.18 *Let A be an FL_w-algebra and $a \in B(A)$. Then the congruences $\theta_{[a)}$ and $\theta_{[a^-)}$ form a pair of complementary factor congruences.*

Proof We proved in Proposition 3.15 that $[a)$ and $[a^-)$ are normal filters of A, so $\theta_{[a)}$ and $\theta_{[a^-)}$ are congruences on A.

Since every FL_w-algebra is congruence permutable, it follows that the congruences $\theta_{[a)}$ and $\theta_{[a^-)}$ are permutable. Hence we have to prove that $\theta_{[a)} \cap \theta_{[a^-)} = \{1\}$ and $\theta_{[a)} \vee \theta_{[a^-)} = A$ with the join defined in the lattice of filters of A. In the proof of Proposition 3.15 we showed that $[a) = \{x \in A \mid x \geq a\}$ and $[a^-) = \{x \in A \mid x \geq a^-\}$. We have $x \in [a) \cap [a^-)$ iff $x \geq a$ and $x \geq a^-$ iff $x \geq a \vee a^- = 1$, so $x = 1$, that is, $[a) \cap [a^-) = \{1\}$.

Since $[a) \vee [a^-) = [\{x \in A \mid x \geq a\} \cup \{x \in A \mid x \geq a^-\})$, applying Proposition 1.36 we get $[a) \vee [a^-) = \{x \in A \mid x\} \geq a \odot a^- = \{x \in A \mid x \geq 0\} = A$.   □

Corollary 3.3 *If A is an FL$_w$-algebra and a $\in$ B(A), then the congruences $\theta_{[a)}$ and $\theta_{[a^\sim)}$ form a pair of complementary factor congruences.*

Proof This follows from the fact that $[a^\sim) = [a^-)$.   □

Theorem 3.5 *A nontrivial FL$_w$-algebra A is directly indecomposable iff B(A) = $\{0, 1\}$.*

Proof As mentioned above, an FL$_w$-algebra A is directly indecomposable iff the only factor congruences on A are **0** and **1**. By Proposition 3.18, the number of pairs of complementary factor congruences coincides with the number of elements of $B(A)$. Thus A is directly indecomposable iff $B(A)$ has only two elements, that is, $B(A) = \{0, 1\}$.   □

Corollary 3.4 *A simple FL$_w$-algebra is subdirectly irreducible and a subdirectly irreducible FL$_w$-algebra is directly indecomposable.*

Proof This follows from the definitions of simple and subdirectly irreducible algebras and applying Theorem 3.5.   □

Example 3.7 If A is the FL$_w$-algebra from Example 3.3, we can see that $B(A) = \{0, 1\}$, so it is directly indecomposable.

Proposition 3.19 *Any linearly ordered FL$_w$-algebra is directly indecomposable.*

Proof Let A be a linearly ordered FL$_w$-algebra and $a \in B(A)$. By Lemma 3.1 we have $a \vee a^- = 1$. Since A is linearly ordered, it follows that $a \leq a^-$ or $a^- \leq a$, hence $a = 1$ or $a^- = 1$. If $a^- = 1$, we get $1 = a^- = a \rightarrow 0$, so $1 \odot a \leq 0$, that is, $a = 0$. Thus $a \in \{0, 1\}$. We conclude that $B(A) = \{0, 1\}$ and according to Theorem 3.5 it follows that A is directly indecomposable.   □

In what follows we give some applications of the above results.

The next two results are proved in a similar way as in [201] for the case of commutative FL$_w$-algebras (FL$_{ew}$-algebras).

Proposition 3.20 *If A is an FL$_w$-algebra and a $\in$ B(A), then a $\odot$ x = x $\odot$ a = a $\wedge$ x for all x $\in$ A.*

Proof According to Birkhoff's theorem, A is isomorphic to a subdirect product of subdirectly irreducible algebras. Consider the isomorphisms f_i ($i \in I$) which define the subdirect representation of A with the factors A_i ($i \in I$). Applying Theorem 3.5 we get that $f_i(a) \in \{0, 1\}$, so $f_i(a) \odot f_i(x) = f_i(a) \wedge f_i(x)$ for all $i \in I$. Hence $f_i(a \odot x) = f_i(a) \odot f_i(x) = f_i(a) \wedge f_i(x) = f_i(a \wedge x)$ for all $i \in I$. Thus $a \odot x = a \wedge x$. Similarly, $x \odot a = x \wedge a = a \wedge x$.   □

Proposition 3.21 *If A is an FL_w-algebra, then $B(A)$ is the universe of a Boolean subalgebra of A.*

Proof We have to prove that $(B(A), \wedge, \vee)$ is distributive and closed under the operations $\wedge, \vee, \odot, \rightarrow$ and $\rightsquigarrow$.

For distributivity, we must prove the identity $a \wedge (b \vee c) = (a \wedge b) \vee (a \wedge c)$ for all $a, b, c \in B(A)$. Applying Proposition 3.20 we get:

$$a \wedge (b \vee c) = a \odot (b \vee c) = (a \odot b) \vee (a \odot c) = (a \wedge b) \vee (a \wedge c).$$

By the hypothesis there exist $a', b' \in A$ such that:

$$a \wedge a' = 0, \qquad a \vee a' = 1, \qquad b \wedge b' = 0, \qquad b \vee b' = 1.$$

We prove that $a' \vee b'$ is the complement of $a \wedge b$:

$$(a \wedge b) \wedge (a' \vee b') = (a \wedge b \wedge a') \vee (a \wedge b \wedge b') = 0 \vee 0 = 0;$$
$$(a \wedge b) \vee (a' \vee b') = (a' \vee b' \vee a) \wedge (a' \vee b' \vee b) = 1 \wedge 1 = 1.$$

Similarly, we prove that $a' \wedge b'$ is the complement of $a \vee b$:

$$(a \vee b) \wedge (a' \wedge b') = (a' \wedge b' \wedge a) \vee (a' \wedge b' \wedge b) = 0 \vee 0 = 0;$$
$$(a \vee b) \vee (a' \wedge b') = (a \vee b \vee a') \wedge (a \vee b \vee b') = 1 \wedge 1 = 1.$$

Thus $a \wedge b, a \vee b \in B(A)$.

If $a, b \in B(A)$, then $a \odot b = a \wedge b \in B(A)$, so $B(A)$ is closed under $\odot$.

If b' is the complement of b, then $b' = b^- = b^\sim$ and applying Corollary 3.2 we have $b^{-\sim} = b^{\sim-} = b$. It follows that:

$$a \rightarrow b = a \rightarrow b^{\sim-} = (a \odot b^\sim)^- \in B(A) \quad \text{and}$$
$$a \rightsquigarrow b = a \rightsquigarrow b^{-\sim} = (b^- \odot a)^\sim \in B(A).$$

Hence $B(A)$ is closed under $\rightarrow$ and $\rightsquigarrow$.

We conclude that $B(A)$ is a Boolean subalgebra of A. $\qquad\qquad\qquad\square$

Corollary 3.5 (De Morgan's laws) $(a \vee b)' = a' \wedge b'$ *and* $(a \wedge b)' = a' \vee b'$.

Corollary 3.6 $(B(A), \wedge, \vee)$ *is a De Morgan lattice.*

Proposition 3.22 *If A is an FL_w-algebra, then the following hold for all $a, b \in B(A)$ and $x, y \in A$:*

(b_1) $(x \rightarrow a) \odot x = x \odot (x \rightsquigarrow a) = a \wedge x$;
(b_2) $(a \rightarrow x) \odot a = a \odot (a \rightsquigarrow x) = a \wedge x$;
(b_3) $(x \rightarrow y) \odot a = [(x \odot a) \rightarrow (y \odot a)] \odot a$;
(b_4) $a \odot (x \rightsquigarrow y) = a \odot [(a \odot x) \rightsquigarrow (a \odot y)]$.

Proof

(b_1) By (*psbck-c25*) we have $(x \to a) \odot x, x \odot (x \leadsto a) \le a \wedge x$.
Since $a \le x \to a, x \leadsto a$, we get

$$a \odot x \le (x \to a) \odot x \quad \text{and} \quad x \odot a \le x \odot (x \leadsto a).$$

Applying Proposition 3.20, it follows that $a \wedge x \le (x \to a) \odot x, x \odot (x \leadsto a)$.
Therefore $(x \to a) \odot x = x \odot (x \leadsto a) = a \wedge x$.

(b_2) Similar to (b_1).

(b_3) By (*psbck-c27*) we have $(x \to y) \odot a \le [(x \odot a) \to (y \odot a)] \odot a$.
Conversely, by (*psbck-c25*) we have:

$$\big[(x \odot a) \to (y \odot a)\big] \odot (x \odot a) \le y \odot a \le y$$

and taking into consideration that $x \odot a = a \odot x$, we get

$$\big[(x \odot a) \to (y \odot a)\big] \odot (a \odot x) \le y.$$

Hence $[(x \odot a) \to (y \odot a)] \odot a \le x \to y$.
By right multiplication with a and applying the fact that $a^2 = a$, we get:

$$\big[(x \odot a) \to (y \odot a)\big] \odot a \le (x \to y) \odot a.$$

Thus $(x \to y) \odot a = [(x \odot a) \to (y \odot a)] \odot a$.

(b_4) Similar to (b_3). $\qquad\qquad\qquad\qquad\qquad\qquad\qquad\qquad\qquad\qquad\qquad\qquad\square$

Proposition 3.23 *If A is an FL_w-algebra, then the following hold for all $a, b \in B(A)$ and $x, y \in A$:*

(b_5) $(a \to b) \odot x = [(x \odot a) \to (x \odot b)] \odot x$;

(b_6) $x \odot (a \leadsto b) = x \odot [(a \odot x) \leadsto (b \odot x)]$;

(b_7) $a \to (x \to y) = (a \to x) \to (a \to y)$;

(b_8) $a \leadsto (x \leadsto y) = (a \leadsto x) \leadsto (a \leadsto y)$;

(b_9) *if $x, y \le a$, then $x \odot (a \leadsto y) = (a \to x) \odot y$.*

Proof

(b_5) Applying (*rl-c3*) we have:

$$\begin{aligned}
\big[(x \odot a) \to (x \odot b)\big] \odot x &= \big[(x \odot a) \to (x \wedge b)\big] \odot x \\
&= \big[(x \odot a \to x) \wedge (x \odot a \to b)\big] \odot x \\
&= (x \odot a \to b) \odot x \\
&= \big[x \to (a \to b)\big] \odot x = (a \to b) \wedge x \\
&= (a \to b) \odot x
\end{aligned}$$

(we also applied (*psbck-c30*), (b_1) and the fact that $a \to b \in B(A)$).

(b_6) Similarly, applying (rl-c_3), ($psbck$-c_{30}) and (b_2) we get:

$$x \odot \left[(a \odot x) \rightsquigarrow (b \odot x) \right] = x \odot \left[(a \odot x) \rightsquigarrow (b \wedge x) \right]$$
$$= x \odot \left[(a \odot x \rightsquigarrow b) \wedge (a \odot x \rightsquigarrow x) \right]$$
$$= x \odot (a \odot x \rightsquigarrow b)$$
$$= x \odot \left[x \rightsquigarrow (a \rightsquigarrow b) \right] = x \wedge (a \rightsquigarrow b)$$
$$= x \odot (a \rightsquigarrow b).$$

(b_7) Applying ($psbck$-c_{30}) and (b_2) we have:

$$a \rightarrow (x \rightarrow y) = (a \odot x) \rightarrow y = a \wedge x \rightarrow y$$
$$= (a \rightarrow x) \odot a \rightarrow y = (a \rightarrow x) \rightarrow (a \rightarrow y).$$

(b_8) Similar to (b_7).

(b_9) Applying (b_1) and (b_2) we get:

$$x \odot (a \rightsquigarrow y) = (a \wedge x) \odot (a \rightsquigarrow y) = a \odot (a \rightsquigarrow x) \odot (a \rightsquigarrow y)$$
$$= (a \rightarrow x) \odot a \odot (a \rightsquigarrow y) = (a \rightarrow x) \odot (a \rightarrow y) \odot a$$
$$= (a \rightarrow x) \odot (a \wedge y) = (a \rightarrow x) \odot y. \qquad \square$$

Proposition 3.24 *If A is an FL_w-algebra, then the following hold for all $a, b \in B(A)$ and $x, y \in A$:*

(b_{10}) $a \vee (x \odot y) = (a \vee x) \odot (a \vee y)$;

(b_{11}) $a \wedge (x \odot y) = (a \wedge x) \odot (a \wedge y)$;

(b_{12}) $(a \, \square_1 \, x) \, \square_2 \, a = a$ *for all* $\square_1, \square_2 \in \{\rightarrow, \rightsquigarrow\}$.

Proof

(b_{10}) Applying (rl-c_2) we get:

$$(a \vee x) \odot (a \vee y) = \left[(a \vee x) \odot a \right] \vee \left[(a \vee x) \odot y \right]$$
$$= \left[(a \vee x) \wedge a \right] \vee \left[(a \odot y) \vee (x \odot y) \right]$$
$$= a \vee (a \odot y) \vee (x \odot y) = a \vee (x \odot y)$$

(since $a \odot y \leq a$, we have $a \vee (a \odot y) = a$).

(b_{11}) Applying the properties of an FL_w-algebra we have:

$$(a \wedge x) \odot (a \wedge y) = (a \odot x) \odot (a \odot y) = a \odot (x \odot a) \odot y$$
$$= a \odot (x \wedge a) \odot y = a \odot (a \wedge x) \odot y = a \odot (a \odot x) \odot y$$
$$= a^2 \odot x \odot y = a \odot (x \odot y) = a \wedge (x \odot y).$$

(b_{12}) Applying (*psbck-c*$_{10}$), (*psbck-c*$_1$) and Corollary 3.1(3) we have:

$$a \to 0 \le a \to x \quad \text{and} \quad (a \to x) \to a \le (a \to 0) \to a = a^- \to a = a.$$

Since by (*psbck-c*$_6$) $a \le (a \to x) \to a$, we get $(a \to x) \to a = a$.
Thus we have proved (b_{12}) for $\Box_1 = \Box_2 = \to$.
The other cases can be proved in the same way. $\Box$

In [105] a bounded $R\ell$-monoid structure was defined on the subintervals $[a, b]$ of
the interval $[0, 1]$ for all $a, b \in [0, 1]$, $a \le b$ (recall that a *bounded $R\ell$-monoid* is an
FL_w-algebra $(A, \wedge, \vee, \odot, \to, \rightsquigarrow, 0, 1)$ satisfying the *pseudo-divisibility* condition:
$(x \to y) \odot x = x \odot (x \rightsquigarrow y) = x \wedge y$ for all $x, y \in A$).

In the case of an FL_w-algebra we can endow the subinterval $[a, 1]$ with the
structure of an FL_w-algebra for all $a \in A$ (Theorem 3.2). We will prove that, if
$a, b \in B(A)$, then the subintervals of the forms $[0, a]$ and $[a, b]$ can also be endowed
with an FL_w-algebra structure.

Theorem 3.6 *Let* $(A, \wedge, \vee, \odot, \to, \rightsquigarrow, 0, 1)$ *be an* FL_w-*algebra,* $a \in B(A)$ *and*
$A_0^a = ([0, a], \wedge, \vee, \odot_0^a, \to_0^a, \rightsquigarrow_0^a, 0, a)$, *where:* $x \odot_0^a y := x \odot (a \rightsquigarrow y)$, $x \to_0^a y :=$
$(x \to y) \odot a$ *and* $x \rightsquigarrow_0^a y := a \odot (x \rightsquigarrow y)$. *Then* A_0^a *is an* FL_w-*algebra.*

Proof We will verify the axioms of an FL_w-algebra.

(A_1) It is clear that $([0, a], \wedge, \vee, 0, a)$ is a bounded lattice with smallest element 0
and greatest element a.

(A_2) Since for all $x \in [0, a]$ we have

$$x \odot_0^a a = x \odot (a \rightsquigarrow a) = x \odot 1 = x \quad \text{and}$$

$$a \odot_0^a x = a \odot (a \rightsquigarrow x) = a \wedge x = x,$$

it follows that a is a unit with respect to $\odot_0^a$.
Associativity can be proved by applying (b_9) and (b_2):

$$x \odot_0^a \left(y \odot_0^a z \right) = (a \to x) \odot \left(y \odot_0^a z \right) = (a \to x) \odot y \odot (a \rightsquigarrow z)$$
$$= \left(x \odot_0^a y \right) \odot (a \rightsquigarrow z) = \left(x \odot_0^a y \right) \odot_0^a z.$$

Thus $([0, a], \odot_0^a, a)$ is a monoid with unit a.

(A_3) Consider $x, y, z \in [0, a]$ such that $x \odot_0^a y \le z$, that is,

$$(a \to x) \odot y \le z \quad \text{and} \quad x \odot (a \rightsquigarrow y) \le z.$$

It follows that $a \to x \le y \to z$ and $a \rightsquigarrow y \le x \rightsquigarrow z$, hence

$$(a \to x) \odot a \le (y \to z) \odot a = y \to_0^a z \quad \text{and}$$

$$a \odot (a \rightsquigarrow y) \le a \odot (x \rightsquigarrow z) = x \rightsquigarrow_0^a z.$$

Thus $x = a \wedge x = (a \to x) \odot a \leq y \to_0^a z$ and $y = a \wedge y = a \odot (a \rightsquigarrow y) \leq x \rightsquigarrow_0^a z$.

Conversely, suppose $x \leq y \to_0^a z = (y \to z) \odot a$. It follows that:

$$x \odot_0^a y = x \odot (a \rightsquigarrow y) \leq (y \to z) \odot a \odot (a \rightsquigarrow y)$$
$$= (y \to z) \odot (a \wedge y) = (y \to z) \odot y \leq y \wedge z \leq z.$$

In a similar way, from $y \leq x \rightsquigarrow_0^a z$ we get $x \odot_0^a y \leq z$.

Thus A_0^a is an FL_w-algebra.   □

Theorem 3.7 *Let* $(A, \wedge, \vee, \odot, \to, \rightsquigarrow, 0, 1)$ *be an* FL_w-*algebra*, $a, b \in B(A)$, $a \leq b$ *and* $A_a^b = ([a, b], \wedge, \vee, \odot_a^b, \to_a^b, \rightsquigarrow_a^b, a, b)$, *where:* $x \odot_a^b y := (x \odot (b \rightsquigarrow y)) \vee a$, $x \to_a^b y := (x \to y) \odot b$ *and* $x \rightsquigarrow_a^b y := b \odot (x \rightsquigarrow y)$. *Then* A_a^b *is an* FL_w-*algebra*.

Proof According to Theorem 3.2, the algebra $([a, 1], \odot_a^1, \wedge, \vee, \to_a^1, \rightsquigarrow_a^1, a, 1)$ with the operations $x \odot_a^1 y = (x \odot y) \vee a$, $x \to_a^1 y = x \to y$ and $x \rightsquigarrow_a^1 y = x \rightsquigarrow y$ is an FL_w-algebra. Let $x, y \in [a, b]$. Since $x \leq b$, by (psbck-c_1) we have $b \to y \leq x \to y$, hence $(x \to y) \odot b \geq (b \to y) \odot b = b \wedge y \geq a$.

Similarly, $b \rightsquigarrow y \leq x \rightsquigarrow y$, so $b \odot (x \rightsquigarrow y) \geq b \odot (b \rightsquigarrow y) = b \wedge y \geq a$.

By Theorem 3.6 it follows that the algebra $([a, b], \wedge, \vee, \odot_a^b, \to_a^b, \rightsquigarrow_a^b, a, b)$ is an FL_w-algebra with the operations:

$$x \odot_a^b y = x \odot_a^1 (b \rightsquigarrow_a^1 y) = x \odot_a^1 (b \rightsquigarrow y) = (x \odot (b \rightsquigarrow y)) \vee a,$$

$$x \to_a^b y = (x \to_a^1 y) \odot_a^1 b = (x \to y) \odot_a^1 b = ((x \to y) \odot b) \vee a = (x \to y) \odot b,$$

$$x \rightsquigarrow_a^b y = b \odot_a^1 (x \rightsquigarrow_a^1 y) = b \odot_a^1 (x \rightsquigarrow y) = (b \odot (x \rightsquigarrow y)) \vee a = b \odot (x \rightsquigarrow y).$$

□

3.5 FL_w-Algebras of Fractions Relative to a Meet-Closed System

In this section we introduce the FL_w-algebra of fractions relative to a meet-closed system. For more on this subject see [24–29, 31–33, 252].

Let $(A, \wedge, \vee, \odot, \to, \rightsquigarrow, 0, 1)$ be an FL_w-algebra.

Definition 3.5 A nonempty subset $S \subseteq A$ is called a *meet-closed system* in A if $1 \in S$ and $x, y \in S$ implies $x \wedge y \in S$.

$S(A)$ will denote the set of all meet-closed systems of A (obviously, $\{1\}$, $A \in S(A)$).

Consider the relation Θ_S on A defined by $(x, y) \in \Theta_S$ iff there exists $e \in S \cap B(A)$ such that $x \wedge e = y \wedge e$.

Lemma 3.2 *The relation* Θ_S *is a congruence on* A.

Proof Reflexivity holds: $(x, x) \in \Theta_S$, since $x \wedge 1 = x \wedge 1$ and $1 \in S \cap B(A)$.

Symmetry is also straightforward: $(x, y) \in \Theta_S$ iff there exists an $e \in S \cap B(A)$ such that $x \wedge e = y \wedge e$ iff $(y, x) \in \Theta_S$.

We prove transitivity:

$(x, y) \in \Theta_S$ iff   there exists an $e \in S \cap B(A)$ such that $x \wedge e = y \wedge e$.

$(y, z) \in \Theta_S$ iff   there exists an $f \in S \cap B(A)$ such that $y \wedge f = z \wedge f$.

Consider $g = e \wedge f$ which belongs to $S \cap B(A)$ and we have:

$$x \wedge g = x \wedge (e \wedge f) = (x \wedge e) \wedge f = (y \wedge e) \wedge f$$
$$= (y \wedge f) \wedge e = (z \wedge f) \wedge e = z \wedge g,$$

which proves transitivity.

We now prove that Θ_S is compatible with the operations: $\wedge, \vee, \odot, \rightarrow, \rightsquigarrow$.

If $(x, y) \in \Theta_S$, $(z, t) \in \Theta_S$, where $x, y, z, t \in A$, then we need to prove that:

$$(x \wedge z, y \wedge t), (x \vee z, y \vee t), (x \odot z, y \odot t), (x \rightarrow z, y \rightarrow t), (x \rightsquigarrow z, y \rightsquigarrow t) \in \Theta_S.$$

By the definition of Θ_S we know that there exist $e, f \in S \cap B(A)$ such that $x \wedge e = y \wedge e$ and $z \wedge f = t \wedge f$.

Let $g = e \wedge f$, so $g \in S \cap B(A)$.

Compatibility with $\wedge$:

$$(x \wedge z) \wedge g = (x \wedge z) \wedge (e \wedge f) = (x \wedge e) \wedge (z \wedge f) = (y \wedge e) \wedge (t \wedge f) = (y \wedge t) \wedge g,$$

so $(x \wedge z, y \wedge t) \in \Theta_S$.

We now prove the compatibility with $\vee$. Applying $(rl\text{-}c_2)$, we get:

$$(x \vee z) \wedge g = (x \vee z) \odot g = g \odot (x \vee z) = (g \odot x) \vee (g \odot z)$$
$$= (g \wedge x) \vee (g \wedge z) = \big[(e \wedge f) \wedge x\big] \vee \big[(e \wedge f) \wedge z\big]$$
$$= (y \wedge g) \vee (t \wedge g) = (y \odot g) \vee (t \odot g) = (y \vee t) \odot g = (y \vee t) \wedge g,$$

so $(x \vee z, y \vee t) \in \Theta_S$.

For the compatibility with $\odot$, applying (b_{11}) we have:

$$(x \odot z) \wedge g = g \wedge (x \odot z) = (g \wedge x) \odot (g \wedge z) = (e \wedge f \wedge x) \odot (e \wedge f \wedge z)$$
$$= (y \wedge e \wedge f) \odot (t \wedge e \wedge f) = (y \wedge g) \odot (t \wedge g) = (y \odot t) \wedge g,$$

so $(x \odot z, y \odot t) \in \Theta_S$.

We check the compatibility with $\rightarrow$, taking into consideration (b_3):

$$(x \rightarrow z) \wedge g = (x \rightarrow z) \odot g = \big[(x \odot g) \rightarrow (z \odot g)\big] \odot g$$
$$= \big[(x \wedge g) \rightarrow (z \wedge g)\big] \odot g = \big[(y \wedge g) \rightarrow (t \wedge g)\big] \odot g$$

$$= \left[(y \odot g) \to (t \odot g)\right] \odot g = (y \to t) \odot g$$
$$= (y \to t) \wedge g,$$

so $(x \to z, y \to t) \in \Theta_S$.

Finally, for the compatibility with $\rightsquigarrow$, applying (b_4) we have:

$$(x \rightsquigarrow z) \wedge g = g \wedge (x \rightsquigarrow z) = g \odot (x \rightsquigarrow z) = g \odot \left[(g \odot x) \rightsquigarrow (g \odot z)\right]$$
$$= g \odot \left[(g \wedge x) \rightsquigarrow (g \wedge z)\right] = g \odot \left[(g \wedge y) \rightsquigarrow (g \wedge t)\right]$$
$$= g \odot \left[(g \odot y) \rightsquigarrow (g \odot t)\right] = g \odot (y \rightsquigarrow t) = g \wedge (y \rightsquigarrow t)$$
$$= (y \rightsquigarrow t) \wedge g,$$

so $(x \rightsquigarrow z, y \rightsquigarrow t) \in \Theta_S$.   □

Remark 3.7 For $x \in A$, denote by x/S the equivalence class of x relative to Θ_S and let $A[S] = A/\Theta_S$. Let $p_S : A \to A[S]$, $p_S(x) := x/S$.

Then $A[S]$ is an FL_w-algebra with $\mathbf{0} = 0/S$, $\mathbf{1} = 1/S$ and for every $x, y \in S$ we have:

$$x/S \vee y/S := (x \vee y)/S,$$
$$x/S \wedge y/S := (x \wedge y)/S,$$
$$x/S \odot y/S := (x \odot y)/S,$$
$$x/S \to y/S := (x \to y)/S,$$
$$x/S \rightsquigarrow y/S := (x \rightsquigarrow y)/S.$$

Hence p_S is an onto morphism of FL_w-algebras.

Remark 3.8

(1) If $0 \in S$, then $A[S] = \mathbf{0}$.
    Indeed, since $0 \in S \cap B(A)$ and $x \wedge 0 = y \wedge 0$, it follows that $(x, y) \in \Theta_S$, for all $x, y \in A$.
(2) If $S \cap B(A) = \{1\}$, then $A[S] = A$.
    Indeed, $(x, y) \in \Theta_S$ iff $x \wedge 1 = y \wedge 1$ iff $x = y$, that is, $A[S] = A$.
(3) $p_S(S \cap B(A)) = \{\mathbf{1}\}$.
    Indeed, $s \wedge s = 1 \wedge s$ for all $s \in S \cap B(A)$, so $s/S = 1/S = \mathbf{1}$.

Definition 3.6 $A[S]$ is called the FL_w-algebra of fractions relative to the meet-closed system S.

Example 3.8 Consider the FL_w-algebra A from Example 3.3. Obviously, $B(A) = \{0, 1\}$.

For any meet-closed system S which contains 0 (for example $S = \{0, a, 1\}$) we have $A[S] = \mathbf{0}$.

For any meet-closed system S such that $0 \notin S$ (for example $S = \{a, b, 1\}$) we have $A[S] = A$.

Theorem 3.8 *Let A' be an FL_w-algebra and let $f : A \rightarrow A'$ be a morphism of FL_w-algebras such that $f(S \cap B(A)) = \{1\}$.*

There exists a unique morphism of FL_w-algebras $f' : A[S] \rightarrow A'$ such that the following diagram is commutative (that is, $f' \circ p_S = f$):

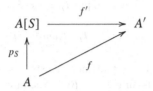

Proof Consider $p_S(x) = p_S(y)$ with $x, y \in A$. It follows that $(x, y) \in \Theta_S$ and thus there exists an $e \in S \cap B(A)$ such that $x \wedge e = y \wedge e$.

Hence $f(x \wedge e) = f(y \wedge e)$ and since f is a morphism of FL_w-algebras we obtain: $f(x) \wedge f(e) = f(y) \wedge f(e)$ and using the fact that $f(e) = 1$ we get $f(x) = f(y)$.

It follows that the map $f' : A[S] \rightarrow A$ defined by $f'(x/S) = f(x)$ for all $x \in A$ is correctly defined. It is also easy to check that f' is a morphism of FL_w-algebras and $f' \circ p_S = f$. The unicity of f' follows from the fact that f is an onto map. $\square$

Remark 3.9 If A is a pseudo-MTL algebra (see Definition 4.1), then $A[S]$ is a pseudo-MTL algebra too. Moreover, if A is a pseudo-BL algebra (see Definition 4.5), then $A[S]$ is also a pseudo-BL algebra. Indeed:

$$(x/S \rightarrow y/S) \vee (y/S \rightarrow x/S) = (x \rightarrow y)/S \vee (y \rightarrow x)/S$$
$$= \left((x \rightarrow y) \vee (y \rightarrow x)\right)/S = 1/S = 1,$$
$$(x/S \rightsquigarrow y/S) \vee (y/S \rightsquigarrow x/S) = (x \rightsquigarrow y)/S \vee (y \rightsquigarrow x)/S$$
$$= \left((x \rightsquigarrow y) \vee (y \rightsquigarrow x)\right)/S = 1/S = 1,$$
$$x/S \wedge y/S = (x \wedge y)/S = \left((x \rightarrow y) \odot x\right)/S$$
$$= (x \rightarrow y)/S \odot x/S = (x/S \rightarrow y/S) \odot x/S,$$
$$x/S \wedge y/S = (x \wedge y)/S = \left(x \odot (x \rightsquigarrow y)\right)/S$$
$$= x/S \odot (x \rightsquigarrow y)/S = x/S \odot (x/S \rightsquigarrow y/S).$$

So $A[S]$ is a pseudo-BL algebra of fractions relative to the meet-closed system. (See Chap. 4 for pseudo-MTL algebras and pseudo-BL algebras.)

Definition 3.7

(1) A subset I of a bounded lattice $(L, \wedge, \vee, 0, 1)$ is called an *ideal* if it satisfies the conditions:

(LI_1) $0 \in I$;
(LI_2) if $a, b \in I$, then $a \vee b \in I$;
(LI_3) if $a \in I$ and $b \leq a$, then $b \in I$.

(2) An ideal I of L is called *prime* if $a \wedge b \in I$ implies $a \in I$ or $b \in I$.

Let $(A, \wedge, \vee, \odot, \rightarrow, \rightsquigarrow, 0, 1)$ be an FL_w-algebra and P be a prime ideal of the underlying lattice $L(A)$. So $P \neq A$ and $S = A \setminus P$ is a meet-closed system in A. We denote by A_P the algebra $A[S]$ and we put $I_P = \{x/S \mid x \in P\}$.

Lemma 3.3 *Let $x \in A$ such that $x/S \in I_P$. Then $x \in P$.*

Proof From $x/S \in I_P$ it follows that there exists a $y \in P$ such that $x/S = y/S$ which means that there exists an $e \in S \cap B(A)$ such that $x \wedge e = y \wedge e \leq y$, hence $x \wedge e \in P$. Since P is prime and $e \in S = A \setminus P$, it follows that $x \in P$. □

Proposition 3.25 *The set I_P is a proper prime ideal of the underlying lattice $L(A_P)$.*

Proof Since $0 \in P$, it follows that $\mathbf{0} \in I_P$.

Let $x, y \in P$, so $x/S \vee y/S = (x \vee y)/S \in I_P$. Now, let $x \in P$, $y \in A$ such that $y/S \leq x/S$. This implies $y/S \rightarrow x/S = 1/S$, thus $(y \rightarrow x)/S = 1/S$, which means there exists an $e \in S \cap B(A)$ such that $e \wedge (y \rightarrow x) = e \wedge 1 = e$, hence $e \leq y \rightarrow x$. Thus $e \odot y \leq x$, which means $e \wedge y \leq x$. It follows that $e \wedge y \in P$, so $y \in P$ and thus $y/S \in I_P$. Hence I_P is an ideal of A_P.

We prove now that $I_P \neq A_P$. Assume $I_P = A_P$ and so $1/S \in I_P$, hence $1 \in P$ by Lemma 3.3. But this implies $P = A$, which is a contradiction. Finally, we prove that I_P is prime. Consider $x, y \in P$ such that $x/S \wedge y/S \in I_P$. It follows that $(x \wedge y)/S \in I_P$, so using Lemma 3.3 we get $x \wedge y \in P$.

Since P is prime, we get $x \in P$ or $y \in P$ so $x/S \in I_P$ or $y/S \in I_P$.

We conclude that I_P is a proper prime ideal of the underlying lattice $L(A_P)$. □

Chapter 4
Other Non-commutative Multiple-Valued Logic Algebras

In this chapter we present some specific properties of other non-commutative multiple-valued logic algebras: pseudo-MTL algebras, bounded $R\ell$-monoids, pseudo-BL algebras and pseudo-MV algebras. As main results, we extend to the case of pseudo-MTL algebras some results regarding the prime filters proved for pseudo-BL algebras.

4.1 Pseudo-MTL Algebras

In order to capture the logic of all left-continuous t-norms and their residua, Esteva and Godo ([117]) introduced the *Monoidal T-norm based Logic* (MTL for short). Jenei and Montagna proved in [197] that MTL is standard complete, i.e. it is complete with respect to the semantics given by the class of all left-continuous t-norms and their residua. Esteva and Godo also developed the algebraic counterpart of this logic, that is, MTL-algebra (bounded integral commutative prelinear residuated lattice). P. Flondor, G. Georgescu and A. Iorgulescu have independently introduced MTL-algebra under the name *weak-BL algebra*. They have also introduced the *weak pseudo-BL algebra* also called a *pseudo-MTL algebra* ([122]) which is an FL_w-algebra satisfying pseudo-prelinearity condition.

In fact, there are two important residuated structures which derive from an FL_w-algebra:

- Pseudo-MTL algebra, which is an FL_w-algebra together with the pseudo-prelinearity condition (studied in [122, 181, 232, 233]);
- Divisible FL_w-algebra or bounded $R\ell$-monoid, which is an FL_w-algebra together with the pseudo-divisibility condition (investigated in [111, 181, 205, 240]).

Therefore, all the properties of an FL_w-algebra hold in a pseudo-MTL algebra, so in this section we will focus on some specific properties of this structure by extending some similar properties proven in the case of pseudo-BL algebras.

We will also give an example of a pseudo-MTL algebra which is not a chain and an example of a locally finite pseudo-MTL algebra.

L.C. Ciungu, *Non-commutative Multiple-Valued Logic Algebras*,
Springer Monographs in Mathematics, DOI 10.1007/978-3-319-01589-7_4,
© Springer International Publishing Switzerland 2014

Definition 4.1 *A pseudo-MTL algebra is an algebra* $\mathcal{A} = (A, \wedge, \vee, \odot, \rightarrow, \rightsquigarrow, 0, 1)$ of type $(2, 2, 2, 2, 2, 0, 0)$ satisfying the following conditions:

$(psMTL_1)$ $(A, \wedge, \vee, 0, 1)$ is a bounded lattice;
$(psMTL_2)$ $(A, \odot, 1)$ is a monoid;
$(psMTL_3)$ $x \odot y \leq z$ iff $x \leq y \rightarrow z$ iff $y \leq x \rightsquigarrow z$ for any $x, y, z \in A$;
$(psMTL_4)$ $(x \rightarrow y) \vee (y \rightarrow x) = (x \rightsquigarrow y) \vee (y \rightsquigarrow x) = 1$.

In other words, a pseudo-MTL algebra is an FL_w-algebra which satisfies the pseudo-prelinearity condition $(psMTL_4)$.

A pseudo-MTL algebra A is *proper* if it does not satisfy the pseudo-divisibility condition, that is, A is not a pseudo-BL algebra (see Sect. 4.3).

Example 4.1 The FL_w-algebra $(A, \wedge, \vee, \odot, \rightarrow, \rightsquigarrow, 0, 1)$ from Example 3.3 is a pseudo-MTL chain.

Remark 4.1 Consider the pseudo-MTL chain A from Example 4.1.

(1) One can easily prove that A is not a pseudo-BL algebra because:

$$(b \rightarrow a) \odot b \neq b \odot (b \rightsquigarrow a).$$

(2) A is a good pseudo-MTL chain.

The following propositions describe some specific properties of pseudo-MTL algebras.

Proposition 4.1 ([186]) *In any pseudo-MTL algebra A the following properties hold*:

$(psmtl\text{-}c_1)$ $(x \wedge y)^- = x^- \vee y^-$ *and* $(x \wedge y)^\sim = x^\sim \vee y^\sim$;
$(psmtl\text{-}c_2)$ $x \vee y = [(x \rightarrow y) \rightsquigarrow y] \wedge [(y \rightarrow x) \rightsquigarrow x]$ *and* $x \vee y = [(x \rightsquigarrow y) \rightarrow y] \wedge [(y \rightsquigarrow x) \rightarrow x]$.

Proof

$(psmtl\text{-}c_1)$ From $(rl\text{-}c_{21})$ we have $x \rightarrow y = x \rightarrow x \wedge y$ and applying $(psbck\text{-}c_{15})$ we get: $x \rightarrow y = x \rightarrow x \wedge y \leq (x \wedge y)^- \rightsquigarrow x^-$.
   Hence $(x \wedge y)^- \odot (x \rightarrow y) \leq x^-$.
   Changing x and y in the above inequality we also get: $(x \wedge y)^- \odot (y \rightarrow x) \leq y^-$.
   Applying $(rl\text{-}c_2)$, it follows that:

$$(x \wedge y)^- = (x \wedge y)^- \odot 1 = (x \wedge y)^- \odot \big[(x \rightarrow y) \vee (y \rightarrow x)\big]$$
$$= \big[(x \wedge y)^- \odot (x \rightarrow y)\big] \vee \big[(x \wedge y)^- \odot (y \rightarrow x)\big] \leq x^- \vee y^-.$$

On the other hand, according to $(psbck\text{-}c_{42})$ we have $x^- \vee y^- \leq (x \wedge y)^-$.
Hence $(x \wedge y)^- = x^- \vee y^-$. Similarly, $(x \wedge y)^\sim = x^\sim \vee y^\sim$.

($psmtl$-c_2) Denote by u the right term of the equality.

By (rl-c_{24}) we have $x \vee y \leq [(x \to y) \rightsquigarrow y] \wedge [(y \to x) \rightsquigarrow x] = u$.

On the other hand, applying (rl-c_2) we have:

$$u = 1 \odot u = [(x \to y) \vee (y \to x)] \odot u = [(x \to y) \odot u] \vee [(y \to x) \odot u].$$

But,

$$(x \to y) \odot u = (x \to y) \odot [[(x \to y) \rightsquigarrow y] \wedge [(y \to x) \rightsquigarrow x]]$$
$$\leq (x \to y) \odot ((x \to y) \rightsquigarrow y) \leq y \quad (\text{by } (rl\text{-}c_5)).$$

Similarly, $(y \to x) \odot u \leq x$.

It follows that $u = [(x \to y) \odot u] \vee [(y \to x) \odot u] \leq y \vee x = x \vee y$.

We conclude that $x \vee y = [(x \to y) \rightsquigarrow y] \wedge [(y \to x) \rightsquigarrow x]$.

Similarly, $x \vee y = [(x \rightsquigarrow y) \to y] \wedge [(y \rightsquigarrow x) \to x]$. □

Proposition 4.2 *In any pseudo-MTL algebra A the following properties hold*:

($psmtl$-c_3) $(x \to y) \to z \leq ((y \to x) \to z) \to z$ *and* $(x \rightsquigarrow y) \rightsquigarrow z \leq ((y \rightsquigarrow x) \rightsquigarrow z) \rightsquigarrow z$;

($psmtl$-c_4) $(x \to y) \to z \leq ((y \to x) \to z) \rightsquigarrow z$ *and* $(x \rightsquigarrow y) \rightsquigarrow z \leq ((y \rightsquigarrow x) \rightsquigarrow z) \to z$;

($psmtl$-c_5) $(x \to y)^n \vee (y \to x)^n = 1$ *and* $(x \rightsquigarrow y)^n \vee (y \rightsquigarrow x)^n = 1,$ *for all* $n \in \mathbb{N},$ $n \geq 1$.

Proof

($psmtl$-c_3) Applying ($psbck$-c_{25}) and (rl-c_4) we get:

$$[(x \to y) \to z] \odot [(y \to x) \to z] \leq [(x \to y) \to z] \wedge [(y \to x) \to z]$$
$$= [(x \to y) \vee (y \to x)] \to z = 1 \to z = z.$$

Thus $(x \to y) \to z \leq ((y \to x) \to z) \to z$.

Similarly, $(x \rightsquigarrow y) \rightsquigarrow z \leq ((y \rightsquigarrow x) \rightsquigarrow z) \rightsquigarrow z$.

($psmtl$-c_4) This can be proved in a similar way as ($psmtl$-c_3).

($psmtl$-c_5) This follows from (rl-c_{22}), since $(x \to y) \vee (y \to x) = 1$ and $(x \rightsquigarrow y) \vee (y \rightsquigarrow x) = 1$. □

We will present some interesting results regarding the locally finite pseudo-MTL algebras and some examples of normal filters of a pseudo-MTL algebra. These examples will be used in later chapters.

Example 4.2

(1) Consider the pseudo-MTL chain from Example 4.1. Since $\text{ord}(c) = \infty$, it follows that A is not locally finite.

(2) Consider the algebra $(A, \wedge, \vee, \odot, \rightarrow, \rightsquigarrow, 0, 1)$ from Example 3.5.
Then A is a pseudo-MTL chain and we have:

$$\mathrm{ord}(0) = 1, \qquad \mathrm{ord}(a) = 2, \qquad \mathrm{ord}(b) = 2, \qquad \mathrm{ord}(c) = 3.$$

Thus A is a locally finite pseudo-MTL chain.

Remark 4.2

(1) In Proposition 39 in [212] it is proved that every locally finite pseudo-MV algebra is commutative.
(2) In Corollary 4.10 in [143] it is proved that every locally finite pseudo-BL algebra is an MV-algebra, so it is commutative.
(3) In Theorem 3.10 in [110] it is proved that every locally finite bounded Rℓ-
monoid is an MV-algebra, so it is commutative.
(4) By the above example we proved that there exist locally finite pseudo-MTL
algebras which are non-commutative.

Remark 4.3 It is known that, if A is a locally finite pseudo-BL algebra A, then
$x^{-\sim} = x^{\sim -} = x$ for all $x \in A$, i.e. condition (pDN) holds (see Proposition 4.9 in
[143]). This result does not hold in the case of pseudo-MTL algebras. Indeed, in the
pseudo-MTL algebra A from Example 3.5 we have $b^{-\sim} = c \neq b$.

Example 4.3 Consider the filter $F = \{c, 1\}$ of the pseudo-MTL chain A from Exam-
ple 4.1. Since $b \rightarrow a = c \in F$ and $b \rightsquigarrow a = b \notin F$, it follows that F is not a normal
filter of A.

Definition 4.2 A proper filter P of A is called *prime* if for all $x, y \in A$, $x \vee y \in P$
implies $x \in P$ or $y \in P$.

The set of all prime filters of A will be denoted by $\mathrm{Spec}(A)$. We also denote by
$\mathrm{Spec}_n(A)$ the set of all prime normal filters of A. Clearly, $\mathrm{Spec}_n(A) \subseteq \mathrm{Spec}(A)$.

Example 4.4 ([182]) Consider $A = \{0, a, b, c, 1\}$ with $0 < a < b < c < 1$ and the
operations $\odot, \rightarrow, \rightsquigarrow$ given by the following tables:

$\odot$	0	a	b	c	1
0	0	0	0	0	0
a	0	a	a	a	a
b	0	a	a	a	b
c	0	a	b	c	c
1	0	a	b	c	1

$\rightarrow$	0	a	b	c	1
0	1	1	1	1	1
a	0	1	1	1	1
b	0	b	1	1	1
c	0	b	b	1	1
1	0	a	b	c	1

$\rightsquigarrow$	0	a	b	c	1
0	1	1	1	1	1
a	0	1	1	1	1
b	0	c	1	1	1
c	0	a	b	1	1
1	0	a	b	c	1

$\mathcal{A} = (A, \wedge, \vee, \odot, \rightarrow, \rightsquigarrow, 0, 1)$ is a proper pseudo-MTL chain. We have:

$$\mathcal{F}(A) = \big\{\{1\}, \{c, 1\}, \{a, b, c, 1\}, A\big\}, \qquad \mathcal{F}_n(A) = \big\{\{1\}, \{a, b, c, 1\}, A\big\},$$

$$\text{Max}(A) = \{\{a, b, c, 1\}\}, \qquad \text{Max}_n(A) = \{\{a, b, c, 1\}\},$$

$$\text{Spec}(A) = \{\{1\}, \{c, 1\}, \{a, b, c, 1\}\}, \qquad \text{Spec}_n(A) = \{\{1\}, \{a, b, c, 1\}\}.$$

Obviously, in this case we have $\text{Max}_n(A) = \text{Max}(A)$.

In the examples of FL_w-algebras and pseudo-MTL algebras that have been presented, all proper filters are prime. The pseudo-MTL algebra in the next example has a proper filter which is not prime.

Example 4.5 Let $A = \{0, a, b, c, 1\}$ with $0 < a < b, c < 1$, but where b and c are incomparable. Consider the operations $\odot, \rightarrow, \rightsquigarrow$ given by the following tables:

$\odot$	0	a	b	c	1
0	0	0	0	0	0
a	0	0	a	0	a
b	0	0	b	0	b
c	0	a	a	c	c
1	0	a	b	c	1

$\rightarrow$	0	a	b	c	1
0	1	1	1	1	1
a	b	1	1	1	1
b	0	c	1	c	1
c	b	b	b	1	1
1	0	a	b	c	1

$\rightsquigarrow$	0	a	b	c	1
0	1	1	1	1	1
a	c	1	1	1	1
b	c	c	1	c	1
c	0	b	b	1	1
1	0	a	b	c	1

Then $(A, \wedge, \vee, \odot, \rightarrow, \rightsquigarrow, 0, 1)$ is a proper pseudo-MTL algebra. Clearly, A is not a pseudo-MTL chain.

Obviously, $P = \{c, 1\}$ is a prime filter of A. Consider the filter $F = \{1\}$ of A. Since $b \vee c = 1 \in F$, but $b, c \notin F$, it follows that F is not a prime filter of A. Thus:

$$\mathcal{F}(A) = \{\{1\}, \{b, 1\}, \{c, 1\}, A\}, \qquad \mathcal{F}_n(A) = \{\{1\}, A\},$$

$$\text{Max}(A) = \{\{b, 1\}, \{c, 1\}\}, \qquad \text{Max}_n(A) = \emptyset,$$

$$\text{Spec}(A) = \{\{b, 1\}, \{c, 1\}\}, \qquad \text{Spec}_n(A) = \emptyset.$$

Proposition 4.3 *If P is a proper filter of A, then the following properties are equivalent:*

(a) *P is prime;*
(b) *for all $x, y \in A$, $x \rightarrow y \in P$ or $y \rightarrow x \in P$;*
(c) *for all $x, y \in A$, $x \rightsquigarrow y \in P$ or $y \rightsquigarrow x \in P$.*

Proof

(a) $\Rightarrow$ (b) This follows by the definition of a prime filter, taking into consideration that $(x \rightarrow y) \vee (y \rightarrow x) = 1 \in P$.

(b) $\Rightarrow$ (a) Assume that $x \vee y \in P$ and for example $x \rightarrow y \in P$.
Since $x \vee y = [(x \rightarrow y) \rightsquigarrow y] \wedge [(y \rightarrow x) \rightsquigarrow x] \in P$ we get $(x \rightarrow y) \rightsquigarrow y \in P$. Thus $y \in P$. Similarly, if $y \rightarrow x \in P$ we get $x \in P$.
We conclude that P is prime.

(a) $\Leftrightarrow$ (c) This follows similarly as (a) $\Leftrightarrow$ (b). $\qquad\square$

Corollary 4.1 *If P is a prime filter and Q is a proper filter such that $P \subseteq Q$, then Q is a prime filter.*

Proof Consider $x, y \in A$. Since P is a prime filter, by Proposition 4.3 we get $x \to y \in P$ or $y \to x \in P$, so $x \to y \in Q$ or $y \to x \in Q$.

Thus Q is also a prime filter. $\square$

The next result was proved in [85] in the case of a pseudo-BL algebra without using the pseudo-divisibility axiom, so the proof is also valid for a pseudo-MTL algebra.

Theorem 4.1 (Prime filter theorem) *Let A be a pseudo-MTL algebra, $F \in \mathcal{F}(A)$ and let I be an ideal of the bounded lattice $(A, \wedge, \vee, 0, 1)$ such that $F \cap I = \emptyset$. Then there exists a prime filter P of A such that $F \subseteq P$ and $P \cap I = \emptyset$.*

Proof Let $\mathcal{H} = \{H \in \mathcal{F}(A) \mid F \subseteq H \text{ and } H \cap I = \emptyset\}$. A routine application of Zorn's lemma shows that $\mathcal{H}$ has a maximal element, P. Suppose that P is not a prime filter of A. Then there are $a, b \in A$ such that $a \to b \notin P$ and $b \to a \notin P$. It follows that the filters $[P \cup \{a \to b\})$ and $[P \cup \{b \to a\})$ are not in $\mathcal{H}$. Hence there exist $c \in I \cap [P \cup \{a \to b\})$ and $d \in I \cap [P \cup \{b \to a\})$.

By Remark 1.13(3), $c \geq (s_1 \odot (a \to b)^{p_1}) \odot \cdots \odot (s_m \odot (a \to b)^{p_m})$, for some $m \geq 1$, $p_1, \ldots, p_m \geq 0$ and $s_1, \ldots, s_m \in P$ and $d \geq (t_1 \odot (b \to a)^{q_1}) \odot \cdots \odot (t_n \odot (b \to a)^{q_n})$, for some $n \geq 1$, $q_1, \ldots, q_n \geq 0$ and $t_1, \ldots, t_n \in P$.

Let $s = s_1 \odot \cdots \odot s_m$ and $t = t_1 \odot \cdots \odot t_n$. It follows that $s, t \in P$.

Let $p = \max\{p_i \mid i = 1, \ldots, m\}$ and $q = \max\{q_i \mid i = 1, \ldots, n\}$.

Then $c \geq \prod_{i=1}^{m}(s \odot (a \to b)^p) = [s \odot (a \to b)^p]^m$ and $d \geq \prod_{i=1}^{n}(t \odot (b \to a)^q) = [t \odot (b \to a)^q]^n$. Now let $u = s \odot t$ and $r = \max\{p, q\}$. It follows that $u \in P$ and $c \geq [u \odot (a \to b)^r]^m$, $d \geq [u \odot (b \to a)^r]^n$.

Applying $(rl$-$c_{26})$, $(rl$-$c_2)$ and $(rl$-$c_{22})$ we get:

$$x = c \vee d \geq \left[u \odot (a \to b)^r\right]^m \vee \left[u \odot (b \to a)^r\right]^n$$
$$\geq \left(\left[u \odot (a \to b)^r\right] \vee \left[u \odot (b \to a)^r\right]\right)^{mn}$$
$$= \left(u \odot \left[(a \to b)^r \vee (b \to a)^r\right]\right)^{mn} = (u \odot 1)^{mn} = u^{mn} \in P.$$

Thus $x \in P$. Since I is an ideal, we have $x \in I$, that is, $P \cap I \neq \emptyset$, a contradiction. We conclude that P is a prime filter. $\square$

Corollary 4.2 *Let F be a filter of A and $a \in A \setminus F$. Then there is a prime filter P of A such that $F \subseteq P$ and $a \notin P$.*

Proof Consider $I = \{x \in A \mid x \leq a\}$.

We prove that I is an ideal of the lattice $\mathcal{L}(A)$ and $F \cap I = \emptyset$.

First we will verify the axioms of a lattice ideal (Definition 3.7):

(LI_1) Since $0 \leq a$, it follows that $0 \in I$.
(LI_2) If $x, y \in I$, we have $x \leq a$ and $y \leq a$, so $x \vee y \leq a$, that is, $x \vee y \in I$.
(LI_3) If $x \in I$ and $y \leq x$, we have $y \leq a$, so $y \in I$.

Suppose that $F \cap I \neq \emptyset$, that is, there is an $x \in F \cap I$. It follows that $x \leq a$ and $x \in F$, so $a \in F$ which is a contradiction. Thus $F \cap I = \emptyset$.

According to the prime filter theorem, there is a prime filter P of A such that $F \subseteq P$ and $P \cap I = \emptyset$. Since $a \in I$, it follows that $a \notin P$.   □

Corollary 4.3 *Let $a \in A$, $a \neq 1$. Then there is a prime filter P of A such that $a \notin P$.*

Proof This follows by Corollary 4.2 where $F = \{1\}$.   □

Corollary 4.4 *Every proper filter F is the intersection of those prime filters which contain F. In particular, $\bigcap \mathrm{Spec}(A) = \{1\}$.*

Proof Since F is a proper filter, then there is an $a \in A \setminus F$ and according to Corollary 4.2 there is a prime filter P of A such that $F \subseteq P$.

Thus F is the intersection of those prime filters which contain F.

If $F = \{1\}$, we obtain $\bigcap \mathrm{Spec}(A) = \{1\}$.   □

Corollary 4.5 $\mathrm{Max}(A) \subseteq \mathrm{Spec}(A)$.

Proof Consider $F \in \mathrm{Max}(A)$. Since F is a proper filter of A, by Corollary 4.2 it follows that there exists a prime filter F of A such that $F \subseteq P$.

Because P is a proper filter, it follows that $F = P$, so F is a prime filter of A, that is, $F \in \mathrm{Spec}(A)$. Thus $\mathrm{Max}(A) \subseteq \mathrm{Spec}(A)$.   □

Proposition 4.4 *The set of proper filters including a prime filter P of A is a chain.*

Proof Consider the proper filters P_1, P_2 of A such that $P \subseteq P_1$ and $P \subseteq P_2$, so $P \subseteq P_1 \cap P_2$. By Corollary 4.1, $P_1 \cap P_2$ is a prime filter of A.

Assume there exist $x \in P_1 \setminus P_2$ and $y \in P_2 \setminus P_1$. Since $x \in P_1$ and $x \leq x \vee y$, it follows that $x \vee y \in P_1$. Similarly, from $y \in P_2$ and $y \leq x \vee y$ we have $x \vee y \in P_2$. Thus $x \vee y \in P_1 \cap P_2$. Hence $x \in P_1 \cap P_2$ or $y \in P_1 \cap P_2$ which is a contradiction.

We conclude that $P_1 \subseteq P_2$ or $P_2 \subseteq P_1$, that is, the set of proper filters including P is a chain.   □

Proposition 4.5 *A pseudo-MTL algebra A is a chain if and only if every proper filter of A is prime.*

Proof Assume A is a chain and let P be a proper filter of A. Consider $x, y \in A$ such that $x \vee y \in P$. Since A is a chain we have $x \leq y$ or $y \leq x$, that is, $x \vee y = x$ or $x \vee y = y$. Hence $x \in P$ or $y \in P$, so P is prime.

Conversely, since $P = \{1\}$ is a proper filter of A, it follows that P is prime.

Let $x, y \in A$. Since $(x \rightarrow y) \vee (y \rightarrow x) = 1 \in P$, we get $x \rightarrow y = 1$ or $y \rightarrow x = 1$. It follows that $x \leq y$ or $y \leq x$, that is, A is a chain. $\square$

Definition 4.3 An element $p < 1$ of a bounded lattice $(A, \wedge, \vee, 0, 1)$ is said to be *meet-irreducible* if $p = x \wedge y$ implies $p = x$ or $p = y$.

Theorem 4.2 *If P is a proper filter of the lattice $\mathcal{F}(A)$, then the following are equivalent*:

(a) *P is prime*;
(b) *P is meet-irreducible in the lattice $\mathcal{F}(A)$*;
(c) *if $x, y \in A$ such that $x \vee y = 1$, then $x \in P$ or $y \in P$*;
(d) *for all $x, y \in A \setminus P$ there is a $z \in A \setminus P$ such that $x \leq z$ and $y \leq z$*;
(e) *if $x, y \in A$ and $[x) \wedge [y) \subseteq P$, then $x \in P$ or $y \in P$*.

Proof We will follow the idea used in [30].

(a) $\Rightarrow$ (b) Let $P_1, P_2 \in \mathcal{F}(A)$ such that $P_1 \wedge P_2 = P$.
    Since $P \subseteq P_1$ and $P \subseteq P_2$, by Proposition 4.4 we have $P_1 \subseteq P_2$ or $P_2 \subseteq P_1$.
    Hence $P = P_1$ or $P = P_2$. Thus P is meet-irreducible.
(b) $\Rightarrow$ (a) Consider $x, y \in A$ such that $x \vee y \in P$. We have:

$$P(x) \cap P(y) = \left(P \vee [x)\right) \cap \left(P \vee [y)\right) = P \vee \left([x) \cap [y)\right) = P \vee [x \vee y) = P.$$

Since P is meet-irreducible, it follows that $P = P(x)$ or $P = P(y)$.
Hence $x \in P$ or $y \in P$, that is, P is a prime filter.
(a) $\Rightarrow$ (c) This is obvious, since $1 \in P$.
(c) $\Rightarrow$ (a) Let $x, y \in A$. Since $(x \rightarrow y) \vee (y \rightarrow x) = 1$, we get $x \rightarrow y \in P$ or $y \rightarrow x \in P$. By Proposition 4.3 it follows that P is prime.
(a) $\Rightarrow$ (d) Let P be a prime filter of A and $x, y \in A \setminus P$. Suppose for every $z \in A$ with $x \leq z$ and $y \leq z$ we have $z \in P$. Then for $z = x \vee y$, since $x, y \leq z$, we get $x \vee y \in P$. Hence $x \in P$ or $y \in P$, a contradiction. Thus there exists a $z \in A \setminus P$ such that $x \leq z$ and $y \leq z$.
(d) $\Rightarrow$ (b) Suppose there exist $F_1, F_2 \in \mathcal{F}(A)$ such that $P = F_1 \cap F_2$ and $P \neq F_1$, $P \neq F_2$. Hence there exist $x \in F_1 \setminus P$ and $y \in F_2 \setminus P$. By hypothesis, there is a $z \in A \setminus P$ such that $x \leq z$ and $y \leq z$. It follows that $z \in F_1 \cap F_2 = P$, a contradiction. Thus P is meet-irreducible in $\mathcal{F}(A)$.
(d) $\Rightarrow$ (e) Consider $x, y \in A$ such that $[x) \cap [y) \subseteq P$ and suppose $x, y \notin P$. By the hypothesis, there is a $z \in A \setminus P$ such that $x \leq z$ and $y \leq z$. Hence $z \in [x) \cap [y) \subseteq P$, so $z \in P$, a contradiction. Thus $x \in P$ or $y \in P$.
(e) $\Rightarrow$ (a) Let $x, y \in A$ such that $x \vee y \in P$. It follows that $[x \vee y) \subseteq P$. Since $[x \vee y) = [x) \cap [y)$, we have $[x) \cap [y) \subseteq P$. By the hypothesis, we get $x \in P$ or $y \in P$, that is, P is a prime filter of A. $\square$

Proposition 4.6 *Any locally finite pseudo-MTL algebra A is a chain.*

Proof Let $x, y \in A$ such that $x \vee y = 1$. Applying (*psmtl-c2*), we get:

$$1 = x \vee y = \left[(x \to y) \rightsquigarrow y\right] \wedge \left[(y \rightsquigarrow x) \to x\right] \leq (x \to y) \rightsquigarrow y,$$

so $(x \to y) \rightsquigarrow y = 1$, that is, $x \to y \leq y$. Taking into consideration that $y \leq x \to y$, we get $x \to y = y$. Suppose that $x \neq 1$. Since A is locally finite, there is an $n \in \mathbb{N}$ such that $x^n = 0$. We have:

$$y = x \to y = x \to (x \to y) = x^2 \to y = \cdots = x^n \to y = 0 \to y = 1.$$

Thus $x \vee y = 1$ iff $x = 1$ or $y = 1$. But, for all $x, y \in A$ we have $(x \to y) \vee (y \to x) = 1$, so, applying the above result we get $x \to y = 1$ or $y \to x = 1$. Hence $x \leq y$ or $y \leq x$. We conclude that A is a chain. $\qquad\square$

Remark 4.4 If H is a maximal and normal filter of a pseudo-MTL algebra A, then it is a natural problem to determine whether A/H is a linearly ordered pseudo-IMTL algebra (involutive pseudo-MTL algebra, i.e. a pseudo-MTL algebra satisfying condition (pDN)).

Consider $x/H, y/H \in A/H$. Since any maximal filter of A is also a prime filter (see Corollary 4.5), by Proposition 4.3 it follows that for $x, y \in A$ we have $x \to y \in H$ or $y \to x \in H$. Thus by Lemma 1.13, $x/H \leq y/H$ or $y/H \leq x/H$, that is, A/H is a linearly ordered pseudo-MTL algebra. Moreover, since $x \to y \in H$ iff $x \rightsquigarrow y \in H$, it follows that $(x \to y)/H = (x \rightsquigarrow y)/H$, so $x/H \to y/H = x/H \rightsquigarrow y/H$ for all $x, y \in A$, that is, A/H is an MTL-algebra. We also have $x^{-\sim}/H = x/H$ iff $(x^{-\sim} \to x) \wedge (x \to x^{-\sim}) \in H$ iff $x^{-\sim} \to x \in H$ (because $x \leq x^{-\sim}$ for all $x \in A$, so $x \to x^{-\sim} = 1$).

Similarly, $x^{\sim-}/H = x/H$ iff $x^{\sim-} \to x \in H$. We conclude that A is a linearly ordered IMTL algebra if and only if A satisfies the property:

$$x^{-\sim} \to x \in H \quad and \quad x^{\sim-} \to x \in H \quad for\ all\ x \in A.$$

In the case of the pseudo-MTL algebra A from Example 4.4 with the maximal and normal filter $H = \{a, b, c, 1\}$, the above condition is satisfied, so A/H is a linearly ordered IMTL algebra, more precisely, $A/H = \{\mathbf{0}, \mathbf{1}\}$, where $\mathbf{0} = 0/H$ and $\mathbf{1} = 1/H$.

4.2 Bounded Residuated Lattice-Ordered Monoids

Bounded residuated lattice-ordered monoids (bounded $R\ell$-monoids) are a common generalization of pseudo-BL algebras and Heyting algebras, i.e. the algebras behind fuzzy and intuitionistic reasoning.

Definition 4.4 A *bounded $R\ell$-monoid* or *divisible residuated lattice* is an algebra $(A, \wedge, \vee, \odot, \to, \rightsquigarrow, 0, 1)$ of type $(2, 2, 2, 2, 2, 0, 0)$ satisfying the following conditions:

$(R\ell_1)$ $(A, \odot, 1)$ is a monoid;

$(R\ell_2)$ $(A, \wedge, \vee, 0, 1)$ is a bounded lattice;

$(R\ell_3)$ $x \odot y \le z$ iff $x \le y \to z$ iff $y \le x \rightsquigarrow z$ for all $x, y, z \in A$;

$(R\ell_4)$ $(x \to y) \odot x = x \odot (x \rightsquigarrow y) = x \wedge y$ for all $x, y \in A$.

In other words, a bounded $R\ell$-monoid is an FL_w-algebra satisfying the pseudo-divisibility condition $(R\ell_4)$. Properties of bounded $R\ell$-monoids have been studied by J. Rachůnek, J. Kühr, A. Iorgulescu and A. Dvurečenskij ([110, 186, 205, 240, 244]). In this section we will recall some properties of bounded $R\ell$-monoids which do not hold in the case of pseudo-MTL algebras, and hence of FL_w-algebras.

In order to present some examples of bounded $R\ell$-monoids (see [176]), we consider the linearly ordered set $L_{n+1} = \{0, 1, 2, \ldots, n\}$, $n \ge 1$, organized as a lattice with $\wedge = \min$ and $\vee = \max$ and organized as a left-MV algebra $\mathcal{L}_{n+1} = (L_{n+1}, \odot, ^-, n)$ with:

$$x \odot y = \max\{0, x + y - n\}, \qquad x \to y = \min\{n, y - x + n\}, \qquad x^- = x \to 0.$$

We also consider the non-linearly ordered MV-algebra:

$$L_{2 \times 2} = \{0, a, b, 1\} \cong L_2 \times L_2 = \{0, 1\} \times \{0, 1\},$$

where $0 < a, b < 1$ and a, b are incomparable whose tables are the following:

$\odot$	0	a	b	1		$\to$	0	a	b	1
0	0	0	0	0		0	1	1	1	1
a	0	a	0	a		a	b	1	b	1
b	0	0	b	b		b	a	a	1	1
1	0	a	b	1		1	0	a	b	1

We recall that a *pseudo-Product algebra* is a pseudo-BL algebra A which satisfies the following conditions for all $x, y, z \in A$ ([181]):

$(pP1)$ $x \wedge x^- = 0$ and $x \wedge x^{\sim} = 0$;

$(pP2)$ $(z^-)^- \odot [(x \odot z) \to (y \odot z)] \le x \to y$ and $(z^{\sim})^{\sim} \odot [(z \odot x) \rightsquigarrow (z \odot y)] \le x \rightsquigarrow y$.

Example 4.6 ([182]) If $\mathcal{A}$ is a non-linearly ordered pseudo-MV algebra or pseudo-Product algebra, then the ordinal sums $\mathcal{A} \oplus \mathcal{L}_2$, $\mathcal{A} \oplus \mathcal{L}_3$, $\mathcal{A} \oplus \mathcal{L}_4$ are proper, good $R\ell$-monoids.

Example 4.7 ([182]) If $\mathcal{A}$ is a proper pseudo-MV algebra or a proper pseudo-BL algebra, then the ordinal sum $\mathcal{L}_{2 \times 2} \oplus \mathcal{A}$ is a proper, good $R\ell$-monoid.

Proposition 4.7 ([186]) *Let* $\mathcal{A} = (A, \wedge, \vee, \odot, \to, \rightsquigarrow, 0, 1)$ *be a bounded $R\ell$-monoid. Then*

$(divrl\text{-}c_1)$ $x \wedge (\bigvee_{i \in I} y_i) = \bigvee_{i \in I} (x \wedge y_i),$

whenever the arbitrary union exists.

Proof Applying (*rl-c₂*) we have:

$$x \wedge \left(\bigvee_{i \in I} y_i\right) = \left(\bigvee_{i \in I} y_i\right) \odot \left(\bigvee_{j \in I} y_j \rightsquigarrow x\right) = \bigvee_{i \in I}\left[y_i \odot \left(\bigvee_{j \in I} y_j \rightsquigarrow x\right)\right].$$

Since $y_i \leq \bigvee_{j \in I} y_j$ for all $i \in I$, applying (*psbck-c₁*) we get $\bigvee_{j \in I} y_j \rightsquigarrow x \leq y_i \rightsquigarrow x$. Hence $y_i \odot (\bigvee_{j \in I} y_j \rightsquigarrow x) \leq y_i \odot (y_i \rightsquigarrow x) = y_i \wedge x$.

It follows that $\bigvee_{i \in I}[y_i \odot (\bigvee_{j \in I} y_j \rightsquigarrow x)] \leq \bigvee_{i \in I}(x \wedge y_i)$.

Thus $x \wedge (\bigvee_{i \in I} y_i) \leq \bigvee_{i \in I}(x \wedge y_i)$.

On the other hand, from $y_i \leq \bigvee_{j \in I} y_j$ we have $x \wedge y_i \leq x \wedge \bigvee_{j \in I} y_j$ for all $i \in I$. Thus $\bigvee_{i \in I}(x \wedge y_i) \leq x \wedge \bigvee_{i \in I} y_i$. We conclude that $x \wedge \bigvee_{i \in I} y_i = \bigvee_{i \in I}(x \wedge y_i)$. $\qquad\square$

Corollary 4.6 ([186]) *If* $(A, \wedge, \vee, \odot, \rightarrow, \rightsquigarrow, 0, 1)$ *is a bounded Rℓ-monoid, then* $\mathcal{L}(A) = (A, \wedge, \vee)$ *is a distributive lattice.*

Proposition 4.8 *In any bounded Rℓ-monoid A the following holds for all $x, y, z \in A$:*

(*divrl-c₂*) $(x \rightarrow y) \rightarrow (x \rightarrow z) = (y \rightarrow x) \rightarrow (y \rightarrow z)$ *and* $(x \rightsquigarrow y) \rightsquigarrow (x \rightsquigarrow z) = (y \rightsquigarrow x) \rightsquigarrow (y \rightsquigarrow z)$.

Proof Applying (*psbck-c₃₀*) we have:

$$(x \rightarrow y) \rightarrow (x \rightarrow z) = (x \rightarrow y) \odot x \rightarrow z = x \wedge y \rightarrow z = y \wedge x \rightarrow z$$
$$= (y \rightarrow x) \odot y \rightarrow z = (y \rightarrow x) \rightarrow (y \rightarrow z).$$

Similarly,

$$(x \rightsquigarrow y) \rightsquigarrow (x \rightsquigarrow z) = x \odot (x \rightsquigarrow y) \rightsquigarrow z = x \wedge y \rightsquigarrow z = y \wedge x \rightsquigarrow z$$
$$= y \odot (y \rightsquigarrow x) \rightsquigarrow z = (y \rightsquigarrow x) \rightsquigarrow (y \rightsquigarrow z). \qquad\square$$

Proposition 4.9 (Lemma 3.7 in [110]) *In every bounded Rℓ-monoid A we have:*

(1) $z \rightarrow x = z \rightarrow y$ *and* $x, y \leq z$ *imply* $x = y$;
(2) $z \rightsquigarrow x = z \rightsquigarrow y$ *and* $x, y \leq z$ *imply* $x = y$;
(3) *If A is linearly ordered, then* $z \rightarrow x = z \rightarrow y \neq 1$ *implies* $x = y$;
(4) *If A is linearly ordered, then* $z \rightsquigarrow x = z \rightsquigarrow y \neq 1$ *implies* $x = y$.

Proof

(1) We have $x = z \wedge x = (z \rightarrow x) \odot z = (z \rightarrow y) \odot z = z \wedge y = y$.
(2) Similar to (1).
(3) If $z \rightarrow x = z \rightarrow y \neq 1$, then $z \not\leq x$, $z \not\leq y$, hence $x < z$, $y < z$.
    Applying (1) we get $x = y$.
(4) Similar to (3). $\qquad\square$

Remark 4.5 (Theorem 11.1.1 in [186]) For any pseudo-BCK(pP) lattice $(A, \wedge, \vee, \rightarrow, \rightsquigarrow, 0, 1)$ the pseudo-divisibility condition is equivalent to properties (1) and (2) from Proposition 4.9.

Proposition 4.10 *In every locally finite bounded $R\ell$-monoid we have*:

(1) *For $y < 1$, $x \odot y = x$ implies $x = 0$.*
(2) *For $y < 1$, $y \odot x = x$ implies $x = 0$.*

Proof This follows from Proposition 1.39, taking into consideration that in a locally finite bounded pseudo-BCK(pP) the only maximal filter is $D = \{1\}$. $\square$

Proposition 4.11 (Proposition 3.11 in [110]) *Let A be a bounded $R\ell$-monoid.*

(1) *If F is a maximal filter of A, then F is prime;*
(2) *If F is a normal and maximal filter of A, then for all $x, y \in A$:*

$$(x \rightarrow y) \vee (y \rightarrow x) \in F \quad and \quad (x \rightsquigarrow y) \vee (y \rightsquigarrow x) \in F.$$

Remark 4.6 In [199] the notion of a *generalized BL-algebra* (GBL-algebra for short) is defined as a residuated lattice satisfying the pseudo-divisibility condition and it is proved that every finite GBL-algebra is commutative. Since a bounded $R\ell$-monoid is a bounded integral generalized BL-algebra, every finite bounded $R\ell$-monoid is commutative.

Proposition 4.12 *Every bounded Wajsberg pseudo-hoop is a bounded $R\ell$-monoid.*

Proof Let $(A, \odot, \rightarrow, \rightsquigarrow, 0, 1)$ be a bounded Wajsberg pseudo-hoop.
   Condition $(R\ell_1)$ follows by $(pshoop\text{-}c_2)$.
   According to Proposition 2.14 for every $x, y \in A$, $x \vee y$ exists, thus $(A, \vee, \wedge, 0, 1)$ is a bounded lattice, hence $(R\ell_2)$ is satisfied.
   Conditions $(R\ell_3)$ and $(R\ell_4)$ follow from the properties $(pshoop\text{-}c_1)$ and $(pshoop\text{-}c_3)$, respectively.
   We conclude that A is a bounded $R\ell$-monoid. $\square$

Corollary 4.7 *Every finite bounded Wajsberg pseudo-hoop is commutative.*

Proof This is a consequence of Remark 4.6 and Proposition 4.12. $\square$

4.3 Pseudo-BL Algebras

G. Georgescu and A. Iorgulescu introduced in [136] the *pseudo-BL algebras* as a natural generalization of BL-algebras for the non-commutative case. A pseudo-BL algebra is an FL_w-algebra which satisfies the pseudo-divisibility and pseudo-prelinearity conditions. Properties of pseudo-BL algebras were deeply investigated by A. Di Nola, G. Georgescu and A. Iorgulescu in [85] and [86]. Some classes of pseudo-BL algebras were investigated in [143] and the corresponding propositional logic was established by Hájek in [158] and [159].

Definition 4.5 A *(left-)pseudo-BL algebra* is an algebra $\mathcal{A} = (A, \wedge, \vee, \odot, \rightarrow, \rightsquigarrow, 0, 1)$ of type $(2, 2, 2, 2, 2, 0, 0)$ satisfying the following conditions:

$(psBL_1)$ $(A, \wedge, \vee, 0, 1)$ is a bounded lattice;
$(psBL_2)$ $(A, \odot, 1)$ is a monoid;
$(psBL_3)$ $x \odot y \leq z$ iff $x \leq y \rightarrow z$ iff $y \leq x \rightsquigarrow z$ for any $x, y, z \in A$;
$(psBL_4)$ $(x \rightarrow y) \odot x = x \odot (x \rightsquigarrow y) = x \wedge y$ for all $x, y \in A$;
$(psBL_5)$ $(x \rightarrow y) \vee (y \rightarrow x) = (x \rightsquigarrow y) \vee (y \rightsquigarrow x) = 1$.

In other words, a pseudo-BL algebra is a divisible residuated lattice (bounded $R\ell$-monoid) satisfying the pseudo-prelinearity axiom as well as a pseudo-MTL algebra satisfying the pseudo-divisibility axiom.

Example 4.8 ([85]) Let $(A, \odot, \oplus, ^-, ^\sim, 0, 1)$ be a left-pseudo-MV algebra (see Definition 4.6). We define two implications corresponding to the two negations:

$$x \rightarrow y = y \oplus x^- = (x \odot y^\sim)^- \quad \text{and} \quad x \rightsquigarrow y = x^\sim \oplus y = (y^- \odot x)^\sim$$

for any $x, y \in A$. Then $(A, \vee, \wedge, \odot, \rightarrow, \rightsquigarrow, 0, 1)$ is a pseudo-BL algebra.

We will see that a pseudo-BL algebra $(A, \wedge, \vee, \odot, \rightarrow, \rightsquigarrow, 0, 1)$ is a pseudo-MV algebra if and only if it satisfies property (pDN).

Example 4.9 ([85]) Consider an arbitrary ℓ-group $(G, \vee, \wedge, +, -, 0)$ and $u \in G$, $u \leq 0$. If we define:

$$x \odot y := (x + y) \vee u,$$

$$x^- := u - x,$$

$$x^\sim := -x + u,$$

$$x \rightarrow y := (y - x) \wedge 0,$$

$$x \rightsquigarrow y := (-x + y) \wedge 0,$$

$$x \oplus y := (x - u + y) \wedge 0,$$

then $([u, 0], \vee, \wedge, \odot, \rightarrow, \rightsquigarrow, \mathbf{0} = u, \mathbf{1} = 0)$ is a pseudo-BL algebra.

Theorem 4.3 ([148]) *Bounded basic pseudo-hoops are termwise equivalent to pseudo-BL algebras.*

Proof Let $(A, \odot, \rightarrow, \rightsquigarrow, 0, 1)$ be a bounded basic pseudo-hoop. According to Proposition 2.16, $\vee$ is defined and $\vee = \cup_1 = \cup_2$, and by Proposition 2.17, $(x \rightarrow$

$y) \vee (y \rightarrow x) = 1$, $(x \rightsquigarrow y) \vee (y \rightsquigarrow x) = 1$ for all $x, y \in A$. Since A is a meet-semilattice, it follows that $(A, \wedge, \vee, \odot, \rightarrow, \rightsquigarrow, 0, 1)$ is a pseudo-BL algebra.

Conversely, if $(A, \wedge, \vee, \odot, \rightarrow, \rightsquigarrow, 0, 1)$ is a pseudo-BL algebra, then conditions $(psHOOP_1)$–$(psHOOP_5)$ are satisfied, thus A is a bounded pseudo-hoop. Applying Proposition 2.17, it follows that A is a bounded basic pseudo-hoop.   □

Since any pseudo-BL algebra is a bounded basic pseudo-hoop, a pseudo-MTL algebra and a divisible FL_w-algebra, all the properties and results presented in the previous sections are also valid in the case of pseudo-BL algebras.

In this section we will mention some specific properties which will be used in later chapters.

Proposition 4.13 ([85]) *In any pseudo-BL algebra we have*:

$(psbl\text{-}c_1)$ *If* $x \vee y = 1$, *then* $x \odot y = x \wedge y$.

$(psbl\text{-}c_2)$ $z \odot (x \wedge y) = (z \odot x) \wedge (z \odot y)$ *and* $(x \wedge y) \odot z = (x \odot z) \wedge (y \odot z)$.

$(psbl\text{-}c_3)$ $z \odot (x_1 \wedge x_2 \wedge \cdots \wedge x_n) = (z \odot x_1) \wedge (z \odot x_2) \wedge \cdots \wedge (z \odot x_n)$ *and* $(x_1 \wedge x_2 \wedge \cdots \wedge x_n) \odot z = (x_1 \odot z) \wedge (x_2 \odot z) \wedge \cdots \wedge (x_n \odot z)$.

$(psbl\text{-}c_4)$ *If* d_1 *and* d_2 *are the two distance functions from Definition 3.2, then*

$$d_1(x, y) = (x \rightarrow y) \odot (y \rightarrow x) \quad and \quad d_2(x, y) = (x \rightsquigarrow y) \odot (y \rightsquigarrow x)$$

for all $x, y \in A$.

Proof

$(psbl\text{-}c_1)$ According to $(psmtl\text{-}c_2)$ we have

$$1 = x \vee y = \left[(x \rightsquigarrow y) \rightarrow y \right] \wedge \left[(y \rightsquigarrow x) \rightarrow x \right].$$

Thus $(x \rightsquigarrow y) \rightarrow y = 1$, hence $x \rightsquigarrow y \leq y$.

It follows that $x \wedge y = x \odot (x \rightsquigarrow y) \leq x \odot y$.

Since $x \odot y \leq x \wedge y$, we conclude that $x \odot y = x \wedge y$.

$(psbl\text{-}c_2)$ Applying $(rl\text{-}c_{21})$, $(psbck\text{-}c_{27})$ and $(psbck\text{-}c_6)$ we get:

$$x \rightsquigarrow y = x \rightsquigarrow x \wedge y \leq z \odot x \rightsquigarrow z \odot (x \wedge y)$$

$$\leq (z \odot x \rightsquigarrow z \odot y) \rightsquigarrow \left[z \odot x \rightsquigarrow z \odot (x \wedge y) \right]$$

and similarly $y \rightsquigarrow x \leq (z \odot y \rightsquigarrow z \odot x) \rightsquigarrow [z \odot y \rightsquigarrow z \odot (x \wedge y)]$.

But according to $(divrl\text{-}c_2)$ we have

$$(z \odot x \rightsquigarrow z \odot y) \rightsquigarrow \left[z \odot x \rightsquigarrow z \odot (x \wedge y) \right]$$

$$= (z \odot y \rightsquigarrow z \odot x) \rightsquigarrow \left[z \odot y \rightsquigarrow z \odot (x \wedge y) \right].$$

It follows that $1 = (x \rightsquigarrow y) \vee (y \rightsquigarrow x) \leq (z \odot x \rightsquigarrow z \odot y) \rightsquigarrow [z \odot x \rightsquigarrow z \odot (x \wedge y)]$.

Thus $(z \odot x \rightsquigarrow z \odot y) \rightsquigarrow [z \odot x \rightsquigarrow z \odot (x \wedge y)] = 1$, that is,

$$z \odot x \rightsquigarrow z \odot y \leq z \odot x \rightsquigarrow z \odot (x \wedge y).$$

Hence $(z \odot x) \odot (z \odot x \rightsquigarrow z \odot y) \leq z \odot (x \wedge y)$, that is $(z \odot x) \wedge (z \odot y) \leq z \odot (x \wedge y)$.

On the other hand, by $(rl\text{-}c_{18})$ we have

$$z \odot (x \wedge y) \leq (z \odot x) \wedge (z \odot y), \quad \text{hence } z \odot (x \wedge y) = (z \odot x) \wedge (z \odot y).$$

Similarly, $(x \wedge y) \odot z = (x \odot z) \wedge (y \odot z)$.

$(psbl\text{-}c_3)$ This follows from $(psbl\text{-}c_2)$ by induction.

$(psbl\text{-}c_4)$ This follows by $(psbl\text{-}c_1)$, taking into consideration $(psBL_5)$. $\square$

Remark 4.7 In [86] it was asked whether every pseudo-BL algebra is good. This problem was solved in [113], by proving that there are uncountable many varieties of pseudo-BL algebras which are not good. We recall that every linearly ordered pseudo-BL algebra is good ([98]) and every linearly ordered pseudo-hoop is good ([99]).

Proposition 4.14 *If A is a good pseudo-BL algebra, then*

$$(x \odot y)^{-\sim} = x^{-\sim} \odot y^{-\sim}.$$

Proof Applying $(psbl\text{-}c_2)$, $(psbck\text{-}c_{37})$, $(pshoop\text{-}c_{11})$ and $(psbck\text{-}c_{36})$ we have:

$$
\begin{aligned}
(x \odot y)^{-\sim} &= (x \odot y)^{-\sim} \wedge y^{-\sim} = \left[y^{-\sim} \to (x \odot y)^{-\sim} \right] \odot y^{-\sim} \\
&= \left[y^{-\sim} \to (y \rightsquigarrow x^{\sim})^{-} \right] \odot y^{-\sim} = \left[y^{-\sim} \to (y^{-\sim} \rightsquigarrow x^{\sim})^{-} \right] \odot y^{-\sim} \\
&= \left[y^{-\sim} \odot (y^{-\sim} \rightsquigarrow x^{\sim}) \right]^{-} \odot y^{-\sim} = (y^{-\sim} \wedge x^{\sim})^{-} \odot y^{-\sim} \\
&= (y^{-\sim-} \vee x^{-\sim}) \odot y^{-\sim} = (y^{-} \vee x^{-\sim}) \odot y^{-\sim} \\
&= (y^{-} \odot y^{-\sim}) \vee (x^{-\sim} \odot y^{-\sim}) = 0 \vee (x^{-\sim} \odot y^{-\sim}) \\
&= x^{-\sim} \odot y^{-\sim}.
\end{aligned}
$$
$\square$

Corollary 4.8 *Every good pseudo-BL algebra is normal.*

Proposition 4.15 *In every locally finite pseudo-BL algebra A we have:*

(1) $x \odot z = y \odot z \neq 0$ *implies* $x = y$;
(2) $z \odot x = z \odot y \neq 0$ *implies* $x = y$.

Proof

(1) Assume $x \neq y$. Then $x \wedge y < y$ or $x \wedge y < x$.

In the first case we have $y \to x = y \to x \wedge y < 1$.
Applying $(psbl\text{-}c_2)$ and by hypothesis we get

$$(x \wedge y) \odot z = (x \odot z) \wedge (y \odot z) = x \odot z.$$

Since $x \wedge y = (y \to x) \odot y$, we have

$$x \odot z = (x \wedge y) \odot z = (y \to x) \odot (y \odot z) = (y \to x) \odot (x \odot z).$$

Taking into consideration that $y \to x < 1$, applying Proposition 4.10 we get $x \odot z = 0$, which is a contradiction. Therefore $x = y$.
For the case $x \wedge y < x$ we get the same result.

(2) Similar to (1).   □

Let A be a pseudo-BL algebra. Recall that ([86]):

- $B(A)$ is the Boolean algebra of all complemented elements in the distributive lattice $L(A) = (A, \wedge, \vee, 0, 1)$ of A;
- $\text{Reg}(A) := \{x \in A \mid x = x^{-\sim} = x^{\sim -}\}$;
- $\text{Id}(A) := \{x \in A \mid x \odot x = x\}$.

Proposition 4.16 *Let A be a pseudo-BL algebra, $x \in \text{Id}(A)$ and $y \in A$. Then the following hold*:

(1) $x \odot y = x \wedge y = y \odot x$;
(2) $x \wedge x^- = x \wedge x^\sim = 0$;
(3) $x \to y = x \rightsquigarrow y$;
(4) $x^- = x^\sim$.

Proof

(1) We have:

$$x \wedge y = x \odot (x \rightsquigarrow y) = x \odot x \odot (x \rightsquigarrow y) = x \odot (x \wedge y) = (x \odot x) \wedge (x \odot y)$$

$$= x \wedge (x \odot y) = x \odot y.$$

Similarly, $x \wedge y = y \odot x$.
(2) Applying (1) we get $x \wedge x^- = x^- \odot x = 0$ and $x \wedge x^\sim = x^\sim \odot x = 0$.
(3) Since $u \leq x \to y$ iff $u \odot x \leq y$ iff $x \odot u \leq y$ iff $u \leq x \rightsquigarrow y$ for any $u \in A$, it follows that $x \to y = x \rightsquigarrow y$.
(4) This follows from (3) where $y = 0$.   □

Proposition 4.17 ([86]) *If A is a pseudo-BL algebra, then $B(A) = \text{Reg}(A) \cap \text{Id}(A)$.*

Remark 4.8 The pseudo-BL algebras presented in this section are in fact left-pseudo-BL algebras. For the sake of completeness we also define the notion of a right-pseudo-BL algebra as a structure $\mathcal{A}_R = (A_R, \vee_R, \wedge_R, \oplus_R, \to_R, \rightsquigarrow_R, 0_R, 1_R)$ of type $(2, 2, 2, 2, 2, 0, 0)$ satisfying the following axioms:

$(rpsBL_1)$ $(A_R, \vee_R, \wedge_R, 0_R, 1_R)$ is a bounded lattice;
$(rpsBL_2)$ $(A_R, \oplus_R, 0_R)$ is a monoid;
$(rpsBL_3)$ $z \leq_R x \oplus_R y$ iff $y \to_R z \leq_R x$ iff $x \rightsquigarrow_R z \leq_R y$ for any $x, y, z \in A_R$;
$(rpsBL_4)$ $(x \to_R y) \oplus_R x = x \oplus_R (x \rightsquigarrow_R y) = x \wedge_R y$ for all $x, y \in A_R$;
$(rpsBL_5)$ $(x \to_R y) \wedge_R (y \to_R x) = (x \rightsquigarrow_R y) \wedge_R (y \rightsquigarrow_R x) = 0_R$.

For other results regarding pseudo-BL algebras we refer the reader to [85, 86] and [143].

4.4 Pseudo-MV Algebras

(Right-)pseudo-MV algebras were introduced by G. Georgescu and A. Iorgulescu in [135] and [137] as a generalization of MV-algebras, that is, a non-commutative extension of MV-algebras. An equivalent definition of pseudo-MV algebras was presented by J. Rachůnek in [241]. The notion of left-pseudo-MV algebras was introduced and studied in [122].

A. Dvurečenskij proved in [97] that any pseudo-MV algebra is isomorphic with some unit interval $\Gamma(G, u) = \{x \in G \mid 0 \leq x \leq u\}$, where (G, u) is an ℓ-group with strong unit u. Then the category of unital ℓ-groups is equivalent to the category of pseudo-MV algebras.

In this section we will point out some basic definitions and results concerning pseudo-MV algebras. For unproven results or unexplained notions we refer the reader to [88, 100, 122, 128, 135, 137, 169, 188, 207, 238, 241].

In the sequel, an ℓu-group will be a pair (G, u) where $(G, \vee, \wedge, +, -, 0)$ is an ℓ-group and u is a strong unit of G.

For each $x \in G$, let $x^+ = x \vee 0$, $x^- = (-x) \vee 0$, and $|x| = x^+ + x^-$. It follows that $x = x^+ - x^-$.

Proposition 4.18 ([9]) *In any ℓ-group G the following hold*:

(1) $|(x \vee z) - (y \vee z)| \leq |x - y|$;
(2) $|(x \wedge z) - (y \wedge z)| \leq |x - y|$;
(3) $|x + y| \leq |x| + |y|$;
(4) $|x^+ - y^+| \leq |x - y|$;
(5) $|x^- - y^-| \leq |x - y|$.

Example 4.10 ([96]) Let G be the group of all matrices of the form

$$A = \begin{pmatrix} \xi & \alpha \\ 0 & 1 \end{pmatrix}$$

where $\xi, \alpha \in \mathbb{R}$, $\xi > 0$ and the group operation is the usual multiplication of matrices. We denote A by (ξ, α).

Then $A^{-1} = (1/\xi, -\alpha/\xi)$ and $(1, 0)$ is the neutral element.

If we define $G^+ := \{(\xi, \alpha) \mid$ (i) $\xi > 1$ or (ii) $\xi = 1$ and $\alpha \geq 0\}$, then G with the positive cone G^+ is a linearly ordered ℓ-group with strong unit $U = (2, 0)$.

Now we will recall an important construction in the theory of ℓ-groups, namely the *lexicographic product* ([9]). If H is a linearly ordered group and G is an ℓ-group, then the lexicographic product of H and G is $H \times_{lex} G := (H \times G, \leq, +, -, (0, 0))$, where $+, -$ and $\leq$ are defined as follows:

$$(x_1, y_1) + (x_2, y_2) := (x_1 + x_2, y_1 + y_2),$$

$$(x_1, y_1) - (x_2, y_2) := (x_1 - x_2, y_1 - y_2),$$

$$(x_1, y_1) \leq (x_2, y_2) \quad \text{iff} \quad \text{(i) } x_1 \leq x_2 \text{ or (ii) } x_1 = x_2 \text{ and } y_1 \leq y_2.$$

Definition 4.6 ([122]) A *left-pseudo-MV algebra* is a structure $(A_L, \odot_L, \oplus_L, ^{-L},$ $^{\sim L}, 0_L, 1_L)$ of type $(2, 2, 1, 1, 0, 0)$ such that the following axioms are satisfied for all $x, y, z \in A_L$:

$(lpsMV_1)$ $x \odot_L (y \odot_L z) = (x \odot_L y) \odot_L z$;
$(lpsMV_2)$ $x \odot_L 1_L = 1_L \odot_L x = x$;
$(lpsMV_3)$ $x \odot_L 0_L = 0_L \odot_L x = 0_L$;
$(lpsMV_4)$ $0_L^{-L} = 1_L, 0_L^{\sim L} = 1_L$;
$(lpsMV_5)$ $(x^{-L} \odot_L y^{-L})^{\sim L} = (x^{\sim L} \odot_L y^{\sim L})^{-L}$;
$(lpsMV_6)$ $x \odot_L (x^{\sim L} \oplus_L y) = y \odot_L (y^{\sim L} \oplus_L x) = (x \oplus_L y^{-L}) \odot_L y = (y \oplus_L x^{-L}) \odot_L x$;
$(lpsMV_7)$ $x \oplus_L (x^{-L} \odot_L y) = (x \odot_L y^{\sim L}) \oplus_L y$;
$(lpsMV_8)$ $(x^{-L})^{\sim L} = x$;

where $x \oplus_L y := (y^{-L} \odot x^{-L})^{\sim L} = (y^{\sim L} \odot x^{\sim L})^{-L}$.

Example 4.11 ([85]) Let us consider an arbitrary ℓ-group $(G, \vee, \wedge, +, -, 0)$ and let $u' \in G, u' \leq 0$. Define:

$$x' \odot_L y' := (x' + y') \vee u',$$

$$x'^{-L} := u' - x', \qquad x'^{\sim L} = -x' + u',$$

$$x' \oplus_L y' := (x' - u' + y') \wedge 0.$$

Then $(A_L = [u', 0], \odot_L, \oplus_L, ^{-L}, ^{\sim L}, 0_L = u', 1_L = 0)$ is a left-pseudo-MV algebra.

If $(A_L, \odot_L, \oplus_L, ^{-L}, ^{\sim L}, 0_L, 1_L)$ is a left-pseudo-MV algebra, we make the definition:

$$x \to_L y := (x \odot_L y^{\sim L})^{-L} = y \oplus_L x^{-L}, \qquad x \leadsto_L y := (y^{-L} \odot_L x)^{\sim L} = x^{\sim L} \oplus_L y.$$

Proposition 4.19 ([122]) *In any left-pseudo-MV algebra A the following properties hold*:

$(lpsmv\text{-}c_1)$ $(x^{-L})^{\sim L} = x = (x^{\sim L})^{-L}$;

$(lpsmv\text{-}c_2)$ $x \wedge_L y = x \odot_L (x^{\sim L} \oplus_L y) = x \odot_L (x \rightsquigarrow_L y) = y \odot_L (y^{\sim L} \oplus_L x) =$
$(x \oplus_L y^{-L}) \odot_L y = (y \oplus_L x^{-L}) \odot_L x = (x \rightarrow_L y) \odot_L x$;

$(lpsmv\text{-}c_3)$ $x \vee_L y = x \oplus_L (x^{-L} \odot_L y) = y \oplus_L (y^{-L} \odot_L x) = (x \odot_L y^{\sim L}) \oplus_L y =$
$(y \odot_L x^{\sim L}) \oplus_L x = (x \rightarrow_L y) \rightsquigarrow_L y = (x \rightsquigarrow_L y) \rightarrow_L y$;

$(lpsmv\text{-}c_4)$ $x \leq_L y$ iff $y^{-L} \odot_L x = 0_L$ iff $x \rightsquigarrow_L y = 1_L$ iff $x^{\sim L} \oplus_L y = 1_L$ iff $x =$
$y \odot_L (y^{\sim L} \oplus_L x)$ iff $y = y \oplus_L (y^{-L} \odot_L x)$ iff $y \odot_L x^{-L} = 1_L$ iff $x \rightarrow_L y = 1_L$
iff $x \oplus_L y^{\sim L} = 0_L$, $0_L \leq x \leq 1_L$;

$(lpsmv\text{-}c_5)$ $(A_L, \wedge_L, \vee_L, 0_L, 1_L)$ is a bounded distributive lattice;

$(lpsmv\text{-}c_6)$ $x \oplus_L 0_L = 0_L \oplus_L x = x$ and $x \oplus_L 1_L = 1_L \oplus_L x = 1_L$;

$(lpsmv\text{-}c_7)$ $x \odot_L y = (y^{-L} \oplus_L x^{-L})^{\sim L} = (y^{\sim L} \oplus_L x^{\sim L})^{-L} = (x \rightarrow_L y^{-L})^{\sim L} =$
$(y \rightsquigarrow_L x^{\sim L})^{-L}$;

$(lpsmv\text{-}c_8)$ $x \oplus_L (y \oplus_L z) = (x \oplus_L y) \oplus_L z$;

$(lpsmv\text{-}c_9)$ $z \leq_L x \oplus_L y$ iff $x^{-L} \odot_L z \leq_L y$ iff $z \odot_L y^{\sim L} \leq_L x$ and $x \odot_L y \leq_L z$ iff
$x \leq_L z \oplus_L y^{-L} = y \rightarrow_L z$ iff $y \leq_L x^{\sim L} \oplus_L z = x \rightsquigarrow_L z$;

$(lpsmv\text{-}c_{10})$ $(x \oplus_L y^{-L}) \vee_L (y \oplus_L x^{-L}) = (x^{\sim L} \oplus_L y) \vee_L (y^{\sim L} \oplus_L x) = 1_L =$
$(y \rightarrow_L x) \vee_L (x \rightarrow_L y) = (y \rightsquigarrow_L x) \vee_L (x \rightsquigarrow_L y)$;

$(lpsmv\text{-}c_{11})$ $(y \odot_L x^{\sim L}) \wedge_L (x \odot_L y^{\sim L}) = (y^{-L} \odot_L x) \wedge_L (x^{-L} \odot_L y) = 0_L$.

Corollary 4.9 If $(A_L, \odot_L, \oplus_L, {}^{-L}, {}^{\sim L}, 0_L, 1_L)$ is a left-pseudo-MV algebra, then
$(A_L, \wedge_L, \vee_L, \odot_L, \rightarrow_L, \rightsquigarrow_L, 0_L, 1_L)$ is a pseudo-BL algebra.

Definition 4.7 A (right-)pseudo-MV algebra is a structure $(A, \oplus, \odot, {}^-, {}^\sim, 0, 1)$ of
type $(2, 2, 1, 1, 0, 0)$ such that the following axioms hold for all $x, y, z \in A$:

$(psMV_1)$ $x \oplus (y \oplus z) = (x \oplus y) \oplus z$;

$(psMV_2)$ $x \oplus 0 = 0 \oplus x = x$;

$(psMV_3)$ $x \oplus 1 = 1 \oplus x = 1$;

$(psMV_4)$ $1^- = 0, 1^\sim = 0$;

$(psMV_5)$ $(x^- \oplus y^-)^\sim = (x^\sim \oplus y^\sim)^-$;

$(psMV_6)$ $x \oplus (x^\sim \odot y) = y \oplus (y^\sim \odot x) = (x \odot y^-) \oplus y = (y \odot x^-) \oplus x$;

$(psMV_7)$ $x \odot (x^- \oplus y) = (x \oplus y^\sim) \odot y$;

$(psMV_8)$ $(x^-)^\sim = x$;

where $x \odot y := (y^- \oplus x^-)^\sim$.

We consider that the operation $\odot$ has priority over the operation $\oplus$.

In the sequel by a pseudo-MV algebra we mean a right-pseudo-MV algebra.

Example 4.12 ([137]) Consider an arbitrary ℓ-group G and let $u \in G$, $u \geq 0$. If we
put by definition:

$$A_R = \Gamma(G, u) := \{x \in G \mid 0 \leq x \leq u\},$$

$$x \oplus y := (x + y) \wedge u,$$

$$x^- := u - x,$$

$$x^\sim := -x + u,$$

$$x \odot y := (x - u + y) \vee 0,$$

then $(A_R = \Gamma(G, u), \oplus, ^-, ^\sim, 0, u)$ is a pseudo-MV algebra.

Example 4.13 ([116]) Let $A \subset \mathbb{R} \times \mathbb{R}$, $A = \{(1, y) \mid y \geq 0\} \cup \{(2, y) \mid y \leq 0\}$ and $0 = (1, 0), 1 = (2, 0)$. For all $(a, b), (c, d) \in A$ define:

$$(a, b) \oplus (c, d) := \begin{cases} (1, b + d) & \text{if } a = c = 1 \\ (2, ad + b) & \text{if } ac = 2 \text{ and } ad + b \leq 0 \\ (2, 0) & \text{otherwise,} \end{cases}$$

$$(a, b)^- := \left(\frac{2}{a}, -\frac{2b}{a}\right), \qquad (a, b)^\sim = \left(\frac{2}{a}, -\frac{b}{2a}\right).$$

Then $(A, \oplus, \odot, ^-, ^\sim, 0, 1)$ is a pseudo-MV algebra, where $(a, b) \odot (c, d) := ((c, d)^- \oplus (a, b)^-)^\sim$.

Example 4.14 ([96]) Let $G = (\mathbb{Z} \times \mathbb{Z} \times \mathbb{Z}, +, (0, 0, 0), \leq)$ be the *Scrimger 2-group*. The group operation $+$ is defined by:

$$(k_1, m_1, n_1) + (k_2, m_2, n_2) := \begin{cases} (k_1 + m_2, m_1 + k_2, n_1 + n_2) & \text{if } n_2 \text{ is odd} \\ (k_1 + k_2, m_1 + m_2, n_1 + n_2) & \text{if } n_2 \text{ is even.} \end{cases}$$

The neutral element is $0 = (0, 0, 0)$ and

$$-(k, m, n) := \begin{cases} (-m, -k, -n) & \text{if } n \text{ is odd} \\ (-k, -m, -n) & \text{if } n \text{ is even.} \end{cases}$$

The order relation on G is $(k_1, m_1, n_1) \leq (k_2, m_2, n_2)$ iff

(i) $n_1 < n_2$ or   (ii) $n_1 = n_2, k_1 \leq k_2, m_1 \leq m_2$.

Then

$$(k_1, m_1, n_1) \vee (k_2, m_2, n_2) = \begin{cases} (k_1, k_2, n_1) & \text{if } n_1 > n_2 \\ (k_1 \vee k_2, m_1 \vee m_2, n_1 \vee n_2) & \text{if } n_1 = n_2 \\ (k_2, m_2, n_2) & \text{if } n_1 < n_2. \end{cases}$$

One can check that G is a non-Abelian ℓ-group which is not linearly ordered and that $u = (1, 1, 1)$ is a strong unit of G. The positive cone of G is

$$G^+ = \mathbb{Z} \times \mathbb{Z} \times \mathbb{Z}_{>0}^+ \cup \mathbb{Z}^+ \times \mathbb{Z}^+ \times \{0\}.$$

The corresponding pseudo-MV algebra has the form

$$A = \Gamma(G, u) = \mathbb{Z}^+ \times \mathbb{Z}^+ \times \{0\} \cup \mathbb{Z}_{\leq 1} \times \mathbb{Z}_{\leq 1} \times \{1\}$$

with:

$$(k, m, 0)^- := (1 - k, 1 - m, 1),$$
$$(k, m, 0)^\sim := (1 - m, 1 - k, 1),$$
$$(k, m, 1)^- := (1 - m, 1 - k, 0),$$
$$(k, m, 1)^\sim := (1 - k, 1 - m, 0),$$
$$(k_1, m_1, 0) \oplus (k_2, m_2, 0) := (k_1 + k_2, m_1 + m_2, 0),$$
$$(k_1, m_1, 0) \oplus (k_2, m_2, 1) := ((m_1 + k_2) \wedge 1, (m_2 + k_1) \wedge 1, 1),$$
$$(k_1, m_1, 1) \oplus (k_2, m_2, 0) := ((k_1 + k_2) \wedge 1, (m_1 + m_2) \wedge 1, 1),$$
$$(k_1, m_1, 1) \oplus (k_2, m_2, 1) := (1, 1, 1).$$

Proposition 4.20 ([137]) *In any pseudo-MV algebra the following properties hold*:

($psmv$-c_1) $x \odot y = (y^\sim \oplus x^\sim)^-$;
($psmv$-c_2) $(x^\sim)^- = x$;
($psmv$-c_3) $x \oplus x^\sim = 1,\ x^- \oplus x = 1$;
($psmv$-c_4) $(x \oplus y)^- = y^- \odot x^-,\ (x \oplus y)^\sim = y^\sim \odot x^\sim$;
($psmv$-c_5) $(x \odot y)^- = y^- \oplus x^-,\ (x \odot y)^\sim = y^\sim \oplus x^\sim$;
($psmv$-c_6) $x \oplus y = (y^- \odot x^-)^\sim = (y^\sim \odot x^\sim)^-$;
($psmv$-c_7) $x^\sim \odot y \oplus y^\sim = y^\sim \odot x \oplus x^\sim$.

Proposition 4.21 ([137]) *In a pseudo-MV algebra A the following are equivalent*:

(a) $x^- \oplus y = 1$;
(b) $y^\sim \odot x = 0$;
(c) $y = x \oplus x^\sim \odot y$;
(d) $x = x \odot (x^- \oplus y)$;
(e) *there is an $a \in A$ such that $y = x \oplus a$*;
(f) $x \odot y^- = 0$;
(g) $y \oplus x^\sim = 1$.

We define $x \le y$ iff one of the above equivalent conditions holds and "$\le$" defines an order relation on A ([137]). Moreover, A is a distributive lattice with the lattice operations defined as below:

$$x \vee y := x \oplus x^\sim \odot y = y \oplus y^\sim \odot x = x \odot y^- \oplus y = y \odot x^- \oplus x,$$
$$x \wedge y := x \odot (x^- \oplus y) = y \odot (y^- \oplus x) = (x \oplus y^\sim) \odot y = (y \oplus x^\sim) \odot x.$$

Proposition 4.22 ([137]) *In a pseudo-MV algebra A the following hold*:

($psmv$-c_8) $x \le y$ implies $a \oplus x \le a \oplus y,\ x \oplus a \le y \oplus a$;
($psmv$-c_9) $x \odot y \le z$ iff $y \le x^- \oplus z$ iff $x \le z \oplus y^\sim$;
($psmv$-c_{10}) $x \odot y \le x \wedge y \le x \oplus y \le x \vee y$.

In the sequel we will use the notation:

$$0x := 0, \qquad (n+1)x := nx \oplus x \quad \text{for } n \geq 1.$$

Proposition 4.23 ([137]) *In a pseudo-MV algebra A the following hold*:

$(psmv\text{-}c_{11})$ $x \oplus (y \wedge z) = (x \oplus y) \wedge (x \oplus z)$, $(y \wedge z) \oplus x = (y \oplus x) \wedge (z \oplus x)$;

$(psmv\text{-}c_{12})$ $x \oplus (y \vee z) = (x \oplus y) \vee (x \oplus z)$, $(y \vee z) \oplus x = (y \oplus x) \vee (z \oplus x)$;

$(psmv\text{-}c_{13})$ $x \odot y^- \wedge y \odot x^- = 0$;

$(psmv\text{-}c_{14})$ $x^\sim \odot y \wedge y^\sim \odot x = 0$;

$(psmv\text{-}c_{15})$ *if* $x \oplus y = x \oplus z$ *and* $x \odot y = x \odot z$, *then* $y = z$;

$(psmv\text{-}c_{16})$ *if* $y \oplus x = z \oplus x$ *and* $y \odot x = z \odot x$, *then* $y = z$;

$(psmv\text{-}c_{17})$ *if* $x \wedge y = 0$, *then* $nx \wedge ny = 0$ *for all* $n \in \mathbb{N}, n \geq 1$.

Remark 4.9 ([122])

(1) If $(A_L, \odot_L, \oplus_L, ^{-L}, ^{\sim_L}, 0_L, 1_L)$ is a left-pseudo-MV algebra, then $(A_L, \oplus_L, \odot_L, ^{\sim_L}, ^{-L}, 0_L, 1_L)$ is a right-pseudo-MV algebra.

(2) If $(A_R, \oplus_R, \odot_R, ^{-R}, ^{\sim_R}, 0_R, 1_R)$ is a right-pseudo-MV algebra, then $(A_R, \odot_R, \oplus_R, ^{\sim_R}, ^{-R}, 0_R, 1_R)$ is a left-pseudo-MV algebra.

Remark 4.10 ([85]) Let $(A_R, \oplus_R, \odot_R, ^{-R}, ^{\sim_R}, 0_R, 1_R)$ be a right-pseudo-MV algebra and let $\rightarrow_R, \rightsquigarrow_R$ be two implications defined by:

$$x \rightarrow_R y := y \odot_R x^{-R} \quad \text{and} \quad x \rightsquigarrow_R y := x^{\sim_R} \odot_R y.$$

Then $(A_R, \vee_R, \wedge_R, \oplus_R, \rightarrow_R, \rightsquigarrow_R, 0_R, 1_R)$ is a right-pseudo-BL algebra.

Remark 4.11

(1) Every locally finite bounded $R\ell$-monoid is a linearly ordered MV-algebra (Theorem 3.10 in [110]).

(2) If H is a maximal and normal filter of a bounded $R\ell$-monoid A, then A/H is a linearly ordered MV-algebra (Theorem 3.12 in [110]).

Proposition 4.24 ([85]) *Let* $(A, \wedge, \vee, \odot, \rightarrow, \rightsquigarrow, 0, 1)$ *be a pseudo-BL(pDN) algebra (i.e.* $x^{-\sim} = x^{\sim -} = x$ *for all* $x \in A$*). Define the operation* $\oplus$ *on A by*

$$x \oplus y := \left(y^- \odot x^-\right)^\sim = \left(y^\sim \odot x^\sim\right)^- = x^- \rightsquigarrow y = y^\sim \rightarrow x.$$

Then $(A, \odot, \oplus, ^-, ^\sim, 0, 1)$ *is a pseudo-MV algebra.*

Corollary 4.10 ([85]) *A pseudo-BL algebra A is a pseudo-MV algebra if and only if A has condition* (pDN).

Remark 4.12 To be more precise, we have:

(1) A left-pseudo-BL algebra $(A_L, \wedge_L, \vee_L, \odot_L, \rightarrow_L, \rightsquigarrow_L, 0_L, 1_L)$ is a left-pseudo-MV algebra iff $(x^{-L})^{\sim L} = (x^{\sim L})^{-L} = x$ for all $x \in A_L$, where $x^{-L} = x \rightarrow 0_L$ and $x^{\sim L} = x \rightsquigarrow 0_L$.

(2) A right-pseudo-BL algebra $(A_R, \vee_R, \wedge_R, \oplus_R, \rightarrow_R, \rightsquigarrow_R, 0_R, 1_R)$ is a right-pseudo-MV algebra iff $(x^{-R})^{\sim R} = (x^{\sim R})^{-R} = x$ for all $x \in A_R$, where $x^{-R} = x \rightarrow 1_R$ and $x^{\sim R} = x \rightsquigarrow 1_R$.

Proposition 4.25 *Any locally finite pseudo-BL algebra is a locally finite MV-algebra.*

Proof According to Proposition 2.1, any locally finite pseudo-BL algebra satisfies (pDN), so it is a pseudo-MV algebra.   □

Proposition 4.26 *Let A be a pseudo-BL algebra and H a normal filter of A. The following are equivalent:*

(a) *H is a maximal filter;*
(b) *A/H is a locally finite MV-algebra.*

Proof This follows from Propositions 1.42 and 4.25.   □

Theorem 4.4 *Pseudo-MV algebras are termwise equivalent to bounded Wajsberg pseudo-hoops.*

Proof According to Corollary 4.9 and Remark 4.10 every pseudo-MV algebra A is a pseudo-BL algebra satisfying $x \vee y = (y \rightarrow x) \rightsquigarrow x = (y \rightsquigarrow x) \rightarrow x$ for all $x, y \in A$. (Since $x \vee y = y \vee x$, we have $x \vee y = (x \rightarrow y) \rightsquigarrow y = (x \rightsquigarrow y) \rightarrow y$ for all $x, y \in A$.) Conversely, if a pseudo-BL algebra A satisfies the above condition, then for $y = 0$ we get $x^{-\sim} = x^{\sim -} = x$ for all $x \in A$, hence, by Proposition 4.24, A is a pseudo-MV algebra. Applying Theorem 4.3 it follows that pseudo-MV algebras are termwise equivalent to bounded basic pseudo-hoops satisfying $x \vee y = (y \rightarrow x) \rightsquigarrow x = (y \rightsquigarrow x) \rightarrow x$.

Obviously, any pseudo-hoop satisfying $x \vee y = (y \rightarrow x) \rightsquigarrow x = (y \rightsquigarrow x) \rightarrow x$ is a Wajsberg pseudo-hoop and by Proposition 2.19 any Wajsberg pseudo-hoop is a basic pseudo-hoop.

We conclude that pseudo-MV algebras are termwise equivalent to bounded Wajsberg pseudo-hoops.   □

Dvurečenskij proved that any pseudo-MV algebra is isomorphic to some unit interval $\Gamma(G, u)$ defined by the formulas given in Example 4.12 and the category of unital ℓ-groups is categorically equivalent to the category of pseudo-MV algebras.

Theorem 4.5 (Theorem 3.9, Theorem 6.9 in [97]) *Let $(A, \oplus, \odot, ^-, ^\sim, 0, 1)$ be a pseudo-MV algebra. Then there exists an ℓ-group G with strong unit u such that A and $\Gamma(G, u)$ are isomorphic pseudo-MV algebras. The category of unital ℓ-groups is categorically equivalent to the category of pseudo-MV algebras.*

Let $(A, \oplus, \odot, ^-, ^\sim, 0, 1)$ be a pseudo-MV algebra. Dvurečenskij defined a partial addition $+$ on A: $x + y$ is defined iff $x \le y^-$, and in this case $x + y := x \oplus y$ (see [95, 108]). Obviously, $x + y$ is defined iff $x \le y^-$ iff $y \le x^\sim$.

Proposition 4.27 (Proposition 6.4.2 in [108]) *The following properties hold in any pseudo-MV algebra A*:

(1) $x + 0 = x = 0 + x$ *for any $x \in A$*;
(2) $x + x^\sim = 1 = x^- + x$ *for any $x \in A$*;
(3) *if $x^- + y = 1$, then $y = x$*;
(4) *if $x + y = 1$, then $y = x^\sim$ and $x = y^-$*.

On a pseudo-MV algebra A we define two distance functions $\bar{d}_1, \bar{d}_2 : A \times A \to A$ as follows:

$$\bar{d}_1(x, y) := x \odot y^- \oplus y \odot x^-, \qquad \bar{d}_2(x, y) := x^\sim \odot y \oplus y^\sim \odot x.$$

Proposition 4.28 ([137]) *In a pseudo-MV algebra A the following hold*:

$$\bar{d}_1(x, y) = x \odot y^- \vee y \odot x^-, \qquad \bar{d}_2(x, y) = x^\sim \odot y \vee y^\sim \odot x.$$

In [85] the following connections between the distances $\bar{d}_1, \bar{d}_2$ on the pseudo-MV algebra $(A, \oplus, \odot, ^-, ^\sim, 0, 1)$ and the distances d_1, d_2 on the corresponding pseudo-BL algebra are proved:

$$\bar{d}_1(x, y) = \left(d_1(x, y)\right)^- \quad \text{and} \quad \bar{d}_2(x, y) = \left(d_2(x, y)\right)^\sim$$

for all $x, y \in A$.

Definition 4.8 ([212]) The *order* of an element x, denoted ord(x), is the least $n \in \mathbb{N}$ such that $nx = 1$, if such n exists, and ∞ otherwise.

Proposition 4.29 ([212]) *In any pseudo-MV algebra the following hold*:

(1) ord$(x^-) =$ ord$(x^\sim)$;
(2) ord$(x) =$ ord$(x^{--}) =$ ord$(x^{\sim\sim})$.

Definition 4.9 ([137]) A nonempty subset I of a pseudo-MV algebra A is called an *ideal* if the following conditions are satisfied:

(psmv-I_1) if $x \in I$ and $y \in A$, $y \le x$, then $y \in I$;
(psmv-I_2) if $x, y \in I$, then $x \oplus y \in I$.
    An ideal I is *normal* if the following condition holds:
(psmv-I_3) for every $x, y \in A$, $y \odot x^- \in I$ iff $x^\sim \odot y \in I$ (or, equivalently, $x \rightarrow_R$ $y \in I$ iff $x \rightsquigarrow_R y \in I$).

An ideal I of A is called *proper* if $I \neq A$. An ideal I of A is called *maximal* if it is proper and for each ideal $J \neq I$, if $I \subseteq J$, then $J = A$.

An ideal I of A is called *principal* if there is an element $a \in A$ such that I is the ideal *generated* by $\{a\}$, i.e. $I = (a] = \{x \in A \mid x \leq na \text{ for some } n \in \mathbb{N}\}$.

Associated with any normal ideal I of A is a congruence defined by:

$$x \equiv_I y \quad \text{iff} \quad \bar{d}_1(x, y) \in I \quad \text{iff} \quad \bar{d}_2(x, y) \in I.$$

Denote by x/I the congruence class of an element $x \in A$ and by A/I the set of congruence classes of A. Then A/I becomes a pseudo-MV algebra with the operations induced by those of A. We notice that $x/I = 0/I$ iff $x \in I$.

Definition 4.10 A nonempty subset I of a pseudo-MV algebra A is called a *filter* if the following conditions are satisfied:

$(psmv\text{-}F_1)$ if $x \in I$ and $y \in A$, $y \geq x$, then $y \in I$;
$(psmv\text{-}F_2)$ if $x, y \in I$ then $x \odot y \in I$.

Remark 4.13 J. Rachůnek introduced in [241] the so called *non-commutative MV-algebras* which are in fact equivalent to pseudo-MV algebras. The equivalence is given by the fact that if $(A, \oplus, \odot, ^-, ^\sim, 0, 1)$ is a pseudo-MV algebra, then $(A, \oplus, \odot', ^-, ^\sim, 0, 1)$, where $x \odot' y = y \odot x$, is a non-commutative MV-algebra.

Conversely, if $(A, \oplus, \odot, ^-, ^\sim, 0, 1)$ is a non-commutative MV-algebra, then $(A, \oplus, \odot', ^-, ^\sim, 0, 1)$, where $x \odot' y = y \odot x$, is a pseudo-MV algebra.

Remark 4.14 The notions of "right-" and "left-" algebras are connected with the right-continuity of a pseudo-t-conorm and with the left-continuity of a pseudo-t-norm, respectively (see [122]). The statement "$\oplus$ is a pseudo-t-conorm on the poset $(A, \leq, 0)$ with smallest element 0" is equivalent to the statement "the algebra $(A, \leq, \oplus, 0)$ is a partially ordered, integral right-monoid". Dually, the statement "$\odot$ is a pseudo-t-norm on the poset $(A, \geq, 1)$ with greatest element 1" is equivalent to the statement "the algebra $(A, \geq, \odot, 1)$ is a partially ordered, integral left-monoid" (see [179, 181, 186]). Initially, pseudo-MV algebras were defined as "right" algebras, while pseudo-BL algebras were defined as "left" algebras. For this reason, in the present book by pseudo-MV algebras we mean right-pseudo-MV algebras, while in the case of pseudo-BL algebras, pseudo-MTL algebras, FL_w-algebras, pseudo-hoops and pseudo-BCK algebras we choose to work with "left" algebras.

4.5 The Glivenko Property

Glivenko proved that a proposition is classically demonstrable if and only if its double negation is intuitionistically demonstrable ([154]). In other words, classical propositional logic can be interpreted in intuitionistic propositional logic.

Versions of both the logical and algebraic formulations of Glivenko's theorem have been intensively studied ([127, 129, 254]).

In this section we define the Glivenko property for multiple-valued logic algebras.

For a bounded pseudo-BCK algebra $(A, \rightarrow, \rightsquigarrow, 0, 1)$ we define

$$\text{Reg}(A) := \left\{ a \in A \mid a^{-\sim} = a^{\sim-} = a \right\}.$$

Obviously, $0, 1 \in \text{Reg}(A)$.

One can easily see that A satisfies the (pDN) condition if and only if $A = \text{Reg}(A)$.

Moreover, if A is good, then $\text{Reg}(A) = \{a \in A \mid a^{-\sim} = a\}$ and $a^{-\sim}, a^{\sim-} \in \text{Reg}(A)$.

Proposition 4.30 *If $(A, \rightarrow, \rightsquigarrow, 0, 1)$ is a good pseudo-BCK algebra, then:*

(1) $x \oplus y \in \text{Reg}(A)$ *for all $x, y \in A$;*
(2) $x \vee_1 y, x \vee_2 y \in \text{Reg}(A)$ *for all $x, y \in \text{Reg}(A)$;*
(3) $x \rightarrow y, x \rightsquigarrow y \in \text{Reg}(A)$ *for all $x, y \in \text{Reg}(A)$;*
(4) $(\text{Reg}(A), \rightarrow, \rightsquigarrow, 0, 1)$ *is a subalgebra of A.*

If A is a good pseudo-hoop, then

(5) $x \wedge y \in \text{Reg}(A)$ *for all $x, y \in \text{Reg}(A)$.*

If A is a good pseudo-MTL algebra, then

(6) $x \vee y \in \text{Reg}(A)$ *for all $x, y \in \text{Reg}(A)$.*

If A is a good pseudo-BL algebra, then:

(7) $x \odot y \in \text{Reg}(A)$ *for all $x, y \in \text{Reg}(A)$;*
(8) $(\text{Reg}(A), \wedge, \vee, \odot, \rightarrow, \rightsquigarrow, 0, 1)$ *is a subalgebra of A.*

Proof

(1) This follows from Proposition 1.24(5).
(2) This follows from Proposition 1.6(3).
(3) Applying $(psbck\text{-}c_{21})$, if $x, y \in \text{Reg}(A)$ then:

$$x \rightarrow y = x \rightarrow y^{\sim-} = \left(x \rightarrow y^{\sim-} \right)^{\sim-} = (x \rightarrow y)^{\sim-},$$

$$x \rightsquigarrow y = x \rightsquigarrow y^{-\sim} = \left(x \rightsquigarrow y^{-\sim} \right)^{-\sim} = (x \rightsquigarrow y)^{-\sim}.$$

(4) This follows by (3) and from the fact that $0, 1 \in \text{Reg}(A)$.
(5) By $(pshoop\text{-}c_{10})$.
(6) By $(psbck\text{-}c_{41})$ and $(psmtl\text{-}c_1)$.
(7) This follows from Proposition 4.14.
(8) This follows by (3), (5), (6), (7) and from the fact that $0, 1 \in \text{Reg}(A)$. $\square$

Theorem 4.6 *In any good pseudo-BCK algebra A the following are equivalent:*

(a) $(x^{-\sim} \rightarrow x)^{-\sim} = (x^{-\sim} \rightsquigarrow x)^{-\sim} = 1$ *for all $x \in A$;*

(b) $(x^{-\sim} \to x)^- = (x^{-\sim} \rightsquigarrow x)^\sim = 0$;

(c) $(x \to y)^{-\sim} = x \to y^{-\sim}$ and $(x \rightsquigarrow y)^{-\sim} = x \rightsquigarrow y^{-\sim}$ for all $x, y \in A$;

(d) *the mapping* $f : A \longrightarrow \mathrm{Reg}(A)$ *defined by* $f(x) = x^{-\sim}$ *is a surjective morphism of pseudo-BCK algebras.*

Proof

(a) $\Leftrightarrow$ (b) This is obvious.

(a) $\Rightarrow$ (c) Applying *(psbck-c$_{15}$)* and *(psbck-c$_{19}$)* we have:

$$x \to y \le y^- \rightsquigarrow x^- = x \to y^{-\sim}.$$

Hence, by *(psbck-c$_{21}$)* we get:

$$(x \to y)^{-\sim} \le \left(x \to y^{-\sim}\right)^{-\sim} = x \to y^{-\sim}.$$

Conversely, from *(psbck-c$_5$)* we have:

$$y^{-\sim} \to y \le \left(x \to y^{-\sim}\right) \to (x \to y).$$

Taking into consideration that $x \to y \le (x \to y)^{-\sim}$ and *(psbck-c$_{10}$)* we have:

$$\left(x \to y^{-\sim}\right) \to (x \to y) \le \left(x \to y^{-\sim}\right) \to (x \to y)^{-\sim}.$$

Applying (a) and *(psbck-c$_{21}$)*, we get:

$$1 = \left(y^{-\sim} \to y\right)^{-\sim} \le \left[\left(x \to y^{-\sim}\right) \to (x \to y)^{-\sim}\right]^{-\sim}$$
$$= \left(x \to y^{-\sim}\right) \to (x \to y)^{-\sim}.$$

It follows that $(x \to y^{-\sim}) \to (x \to y)^{-\sim} = 1$, so $x \to y^{-\sim} \le (x \to y)^{-\sim}$.
We conclude that $(x \to y)^{-\sim} = x \to y^{-\sim}$.
Similarly, $(x \rightsquigarrow y)^{-\sim} = x \rightsquigarrow y^{-\sim}$.

(c) $\Rightarrow$ (a) Applying (c) we have:

$$\left(x^{-\sim} \to x\right)^{-\sim} = x^{-\sim} \to x^{-\sim} = 1 \quad \text{and}$$

$$\left(x^{-\sim} \rightsquigarrow x\right)^{-\sim} = x^{-\sim} \rightsquigarrow x^{-\sim} = 1.$$

(c) $\Rightarrow$ (d) From (c) and applying *(psbck-c$_{19}$)*, we have:

$$f(x \to y) = (x \to y)^{-\sim} = x \to y^{-\sim} = x^{-\sim} \to y^{-\sim} = f(x) \to f(y),$$

$$f(x \rightsquigarrow y) = (x \rightsquigarrow y)^{-\sim} = x \rightsquigarrow y^{-\sim} = x^{-\sim} \rightsquigarrow y^{-\sim} = f(x) \rightsquigarrow f(y).$$

Hence f is a morphism from A to $\mathrm{Reg}(A)$.
Moreover, for an arbitrary $y \in \mathrm{Reg}(A)$, taking $x = y^{-\sim}$, we have: $f(x) = (y^{-\sim})^{-\sim} = y^{-\sim} = y$, thus f is a surjective morphism.

(d) $\Rightarrow$ (c) If $f : A \longrightarrow \mathrm{Reg}(A)$ defined by $f(x) = x^{-\sim}$ is a morphism, we have:
$f(x \rightarrow y) = f(x) \rightarrow f(y)$, that is, $(x \rightarrow y)^{-\sim} = x^{-\sim} \rightarrow y^{-\sim}$.
Since by $(psbck\text{-}c_{19})$ we have $x^{-\sim} \rightarrow y^{-\sim} = x \rightarrow y^{-\sim}$, it follows that $(x \rightarrow y)^{-\sim} = x \rightarrow y^{-\sim}$. Similarly, $(x \rightsquigarrow y)^{-\sim} = x \rightsquigarrow y^{-\sim}$. $\square$

Definition 4.11 We say that a good pseudo-BCK algebra A has the *Glivenko property* if it satisfies one of the equivalent conditions from Theorem 4.6.

Remark 4.15 By $(psbck\text{-}c_{19})$, in any good pseudo-BCK algebra A satisfying the Glivenko property the following hold:

$$(x \rightarrow y)^{-\sim} = x^{-\sim} \rightarrow y^{-\sim},$$
$$(x \rightsquigarrow y)^{-\sim} = x^{-\sim} \rightsquigarrow y^{-\sim}$$

for all $x, y \in A$.

Example 4.15

(1) The good FL_w-algebra A from Example 4.4 has the Glivenko property.
(2) The good FL_w-algebra A from Example 4.1 does not satisfy the Glivenko property. Indeed, $(b \rightsquigarrow a)^{-\sim} = b \neq 1 = b \rightsquigarrow a^{-\sim}$.
(3) Any pseudo-BCK(pDN) algebra has the Glivenko property.

Remark 4.16

(1) Any good pseudo-hoop has the Glivenko property.
    Indeed, if A is a good pseudo-hoop, then according to $(pshoop\text{-}c_9)$, the following hold for all $x, y \in A$:

$$(x \rightarrow y)^{-\sim} = x^{-\sim} \rightarrow y^{-\sim}, \qquad (x \rightsquigarrow y)^{-\sim} = x^{-\sim} \rightsquigarrow y^{-\sim}.$$

Applying $(psbck\text{-}c_{19})$ it follows that

$$(x \rightarrow y)^{-\sim} = x \rightarrow y^{-\sim}, \qquad (x \rightsquigarrow y)^{-\sim} = x \rightsquigarrow y^{-\sim}.$$

Thus any good pseudo-hoop satisfies the Glivenko property.
(2) Any good $R\ell$-monoid has the Glivenko property.

Remark 4.17

(1) It is obvious that the property $(pshoop\text{-}c_9)$ holds in any good FL_w-algebra with (pDN).
(2) The property $(pshoop\text{-}c_9)$ does not hold in any good FL_w-algebra. Indeed, if A is the good FL_w-algebra from Example 4.1, we have $(b \rightsquigarrow a)^{-\sim} = b \neq 1 = b^{-\sim} \rightsquigarrow a^{-\sim}$.

Chapter 5
Classes of Non-commutative Residuated Structures

In this chapter we study special classes of non-commutative residuated structures: local, perfect and Archimedean structures. The local bounded pseudo-BCK(pP) algebras are characterized in terms of primary deductive systems, while the perfect pseudo-BCK(pP) algebras are characterized in terms of perfect deductive systems. One of the main results consists of proving that the radical of a bounded pseudo-BCK(pP) algebra is normal. We also prove that any linearly ordered pseudo-BCK(pP) algebra and any locally finite pseudo-BCK(pP) algebra are local. Other results state that any local FL_w-algebra and any locally finite FL_w-algebra are directly indecomposable. The classes of Archimedean and hyperarchimedean FL_w-algebras are introduced and it is proved that any locally finite FL_w-algebra is hyperarchimedean and any hyperarchimedean FL_w-algebra is Archimedean.

5.1 Local Pseudo-BCK Algebras with Pseudo-product

Definition 5.1 A pseudo-BCK(pP) algebra is called *local* if it has a unique maximal deductive system.

In this section by a pseudo-BCK(pP) algebra we mean a bounded pseudo-BCK(pP) algebra, even though some notions and properties are also valid for unbounded pseudo-BCK(pP) algebras.

We define:

$$D(A) := \{x \in A \mid \mathrm{ord}(x) = \infty\} \quad \text{and} \quad D(A)^* := \{x \in A \mid \mathrm{ord}(x) < \infty\}.$$

Obviously, $D(A) \cap D(A)^* = \emptyset$ and $D(A) \cup D(A)^* = A$.
We also remark that $1 \in D(A)$ and $0 \in D(A)^*$.
Let A be a pseudo-BCK(pP) algebra and $D \in \mathcal{DS}(A)$. We will use the following notation:

$$D_-^* := \{x \in A \mid x \leq y^- \text{ for some } y \in D\},$$

$$D_\sim^* := \{x \in A \mid x \leq y^\sim \text{ for some } y \in D\}.$$

L.C. Ciungu, *Non-commutative Multiple-Valued Logic Algebras*,
Springer Monographs in Mathematics, DOI 10.1007/978-3-319-01589-7_5,
© Springer International Publishing Switzerland 2014

Proposition 5.1 *Let A be a local pseudo-BCK(pP) algebra. Then:*

(1) *any proper deductive system of A is included in the unique maximal deductive system of A;*
(2) $A_0^- := \{x \in A \mid x^- = 0\}$ *and* $A_0^{\sim} := \{x \in A \mid x^{\sim} = 0\}$ *are included in the unique maximal deductive system of A.*

Proof

(1) This is an immediate consequence of Zorn's lemma.
(2) This follows from Proposition 1.30 and (1). □

Theorem 5.1 *Let A be a pseudo-BCK(pP) algebra. Then the following are equivalent:*

(a) $D(A)$ *is a deductive system of A;*
(b) $D(A)$ *is a proper deductive system of A;*
(c) *A is local;*
(d) $D(A)$ *is the unique maximal deductive system of A;*
(e) *for all* $x, y \in A$, $\mathrm{ord}(x \odot y) < \infty$ *implies* $\mathrm{ord}(x) < \infty$ *or* $\mathrm{ord}(y) < \infty$.

Proof

(a) ⇒ (b) Since $\mathrm{ord}(0) = 1$, we have $0 \notin D(A)$, so $D(A)$ is a proper deductive system of A.
(b) ⇒ (a) This is obvious.
(a) ⇒ (e) Consider $x, y \in A$ such that $\mathrm{ord}(x \odot y) < \infty$, so $x \odot y \notin D(A)$.
   Since $D(A)$ is a deductive system of A, it follows that $x \notin D(A)$ or $y \notin D(A)$.
   Hence $\mathrm{ord}(x) < \infty$ or $\mathrm{ord}(y) < \infty$.
(e) ⇒ (a) Because $1 \in D(A)$ it follows that $D(A)$ is nonempty. Consider $x, y \in D(A)$, that is, $\mathrm{ord}(x) = \infty$ and $\mathrm{ord}(y) = \infty$. By (e) we get $\mathrm{ord}(x \odot y) = \infty$, so $x \odot y \in D(A)$. Consider $x \in D(A)$ and $y \in A$ such that $x \leq y$. It follows that $x^n > 0$ for all $n \in \mathbb{N}$. Since $x^n \leq y^n$ we get $y^n > 0$ for all $n \in \mathbb{N}$, so $\mathrm{ord}(y) = \infty$, that is, $y \in D(A)$. Thus $D(A)$ is a deductive system of A.
(d) ⇒ (c) This follows by the definition of a local pseudo-BCK(pP) algebra.
(c) ⇒ (d) If M is the unique maximal deductive system of A, then by Lemma 1.8 and Proposition 5.1 we have $x \in M$ iff $[x) \subseteq M$ iff $[x)$ is proper iff $\mathrm{ord}(x) = \infty$ iff $x \in D(A)$. Hence $M = D(A)$.
(d) ⇒ (a) This is obvious.
(a) ⇒ (d) Since $0 \notin D(A)$, it follows that $D(A)$ is a proper deductive system of A. Let F be a proper deductive system of A. Consider $x \in F$. Since $[x) \subseteq F$, we have that $[x)$ is a proper deductive system of A, so by Lemma 1.8 it follows that $\mathrm{ord}(x) = \infty$. Hence $x \in D(A)$, so $F \subseteq D(A)$. Thus $D(A)$ is the unique maximal deductive system of A. □

Corollary 5.1 *If A is a local pseudo-BCK(pP) algebra, then:*

(1) *for any* $x \in A$, $\mathrm{ord}(x) < \infty$ *or* $[\mathrm{ord}(x^-) < \infty$ *and* $\mathrm{ord}(x^{\sim}) < \infty]$;

(2) $D(A)^*_- \subseteq D(A)^*$ and $D(A)^*_\sim \subseteq D(A)^*$;
(3) $D(A) \cap D(A)^*_- = D(A) \cap D(A)^*_\sim = \emptyset$.

Proof

(1) Let $x \in A$. Since $x \odot x^- = x^\sim \odot x = 0$, it follows that $\text{ord}(x \odot x^-) = \text{ord}(x^\sim \odot x) = \text{ord}(0) = 1 < \infty$. By Theorem 5.1(e) we get (1).
(2) Let $x \in D(A)^*_-$. This means that there is a $y \in D(A)$ such that $x \le y^-$. Since $\text{ord}(y) = \infty$, by (1) we get that $\text{ord}(y^-) < \infty$. Hence $\text{ord}(x) < \infty$, so $x \in D(A)^*$. Thus $D(A)^*_- \subseteq D(A)^*$. Similarly, $D(A)^*_\sim \subseteq D(A)^*$.
(3) This follows from (2), taking into consideration that $D(A) \cap D(A)^* = \emptyset$. $\square$

Example 5.1 Consider the pseudo-BCK(pP) algebra A from Example 1.9. One can easily check that $D(A) = \{a_2, s, a, b, n, c, d, m, 1\}$ and it is a deductive system of A, so A is a local pseudo-BCK(pP) algebra.

Proposition 5.2 *Any linearly ordered pseudo-BCK(pP) algebra is local.*

Proof Assume that A is a linearly ordered pseudo-BCK(pP) algebra and consider $x, y \in A$ such that $\text{ord}(x \odot y) < \infty$. Since A is linearly ordered, we have $x \le y$ or $y \le x$. Assume that $x \le y$. It follows that $x \odot x \le x \odot y$, so $\text{ord}(x \odot x) < \infty$. Hence $\text{ord}(x) < \infty$. Similarly, from $y \le x$ we get $\text{ord}(y) < \infty$. Thus according to Theorem 5.1(e), A is a local pseudo-BCK(pP) algebra. $\square$

Proposition 5.3 *Any locally finite pseudo-BCK(pP) algebra is local.*

Proof We have $D(A) = \{1\}$, so $D(A)$ is a deductive system of A. Applying Theorem 5.1, it follows that A is local. $\square$

Theorem 5.2 *Any local FL_w-algebra is directly indecomposable.*

Proof Let A be a local FL_w-algebra and $a \in B(A)$. Since $a^- = a^\sim$, according to Corollary 5.1 we get that $\text{ord}(a) < \infty$ or $\text{ord}(a^-) < \infty$. It follows that there exists an $n \in \mathbb{N}$, $n \ge 1$, such that $a^n = 0$ or $(a^-)^n = 0$ (in fact, there exist $n_1, n_2 \in \mathbb{N}$, $n_1, n_2 \ge 1$, such that $a^{n_1} = 0$ or $(a^-)^{n_2} = 0$, and we take $n = \min\{n_1, n_2\}$). Since by Proposition 3.13 and Corollary 3.1 we have $a^n = a$ and $(a^-)^n = a^-$, it follows that $a = 0$ or $a^- = 0$. If $a^- = 0$, then $a^{-\sim} = 1$ and by Corollary 3.2 we get $a = a^{-\sim} = 1$. Thus $a \in \{0, 1\}$, hence $B(A) = \{0, 1\}$. By Theorem 3.5 it follows that A is directly indecomposable. $\square$

Theorem 5.3 *Any locally finite FL_w-algebra is directly indecomposable.*

Proof Let A be a locally finite FL_w-algebra. According to Proposition 5.3, it follows that A is local and applying Theorem 5.2 we conclude that A is directly indecomposable. $\square$

Proposition 5.4 *If P is a proper normal deductive system of a pseudo-BCK(pP) algebra A, then the following are equivalent:*

(a) *P is primary;*
(b) *A/P is a local pseudo-BCK(pP) algebra;*
(c) *P is contained in a unique maximal deductive system of A.*

Proof

(a) $\Leftrightarrow$ (b) Applying Theorem 5.1(e) and Lemma 1.13(2), we have: A/P is local iff [for all $x, y \in A$, $\mathrm{ord}(x/P \odot y/P) < \infty$ implies $\mathrm{ord}(x/P) < \infty$ or
$\mathrm{ord}(y/P) < \infty$] iff [for all $x, y \in A$, $(x/P \odot y/P)^n = 0/P$ for some $n \in \mathbb{N}$
implies $(x/P)^m = 0/P$ or $(y/P)^m = 0/P$ for some $m \in \mathbb{N}$] iff [for all $x, y \in A$,
$(x \odot y)^n/P = 0/P$ for some $n \in \mathbb{N}$ implies $x^m/P = 0/P$ or $y^m/P = 0/P$ for
some $m \in \mathbb{N}$] iff [for all $x, y \in A$, $((x \odot y)^n)^- \in P$ for some $n \in \mathbb{N}$ implies
$(x^m)^- \in P$ or $(y^m)^- \in P$ for some $m \in \mathbb{N}$] iff P is primary.
(a) $\Leftrightarrow$ (c) By (a) $\Leftrightarrow$ (b), P is primary iff A/P is local iff A/P has a unique maximal
deductive system. By Corollary 1.10 there is a bijection between $\mathrm{Max}(A/P)$ and
$\{D \mid D \in \mathrm{Max}(A), P \subseteq D\}$. It follows that P is primary if and only if there is a
unique maximal deductive system of A containing P. $\square$

Theorem 5.4 *If A is a pseudo-BCK(pP) algebra, then the following are equivalent:*

(a) *A is local;*
(b) *any proper normal deductive system of A is primary;*
(c) *$\{1\}$ is a primary deductive system of A.*

Proof

(a) $\Rightarrow$ (b) Let H be a proper normal deductive system of A. By Theorem 5.1(d),
$D(A)$ is the unique maximal deductive system of A. Hence $H \subseteq D(A)$ and
according to Proposition 5.4 it follows that H is primary.
(b) $\Rightarrow$ (c) Since $\{1\}$ is a proper normal deductive system of A, by (b) we get that
$\{1\}$ is primary.
(c) $\Rightarrow$ (a) Since $\{1\}$ is primary, applying Proposition 5.4 it follows that $A/\{1\}$ is
local. Taking into consideration that $A \cong A/\{1\}$, it follows that A is local. $\square$

5.2 Perfect Residuated Structures

5.2.1 Perfect Pseudo-BCK Algebras with Pseudo-product

Definition 5.2 A bounded pseudo-BCK(pP) algebra A is called *perfect* if it satisfies
the following conditions:

(1) A is a local good pseudo-BCK(pP) algebra;
(2) for any $x \in A$, $\mathrm{ord}(x) < \infty$ iff [$\mathrm{ord}(x^-) = \infty$ and $\mathrm{ord}(x^\sim) = \infty$].

Proposition 5.5 *Let A be a local good pseudo-BCK(pP) algebra. Then the following are equivalent:*

(a) *A is perfect;*
(b) $D(A)^*_- = D(A)^*_\sim = D(A)^*$.

Proof

(a) $\Rightarrow$ (b) Since A is a local pseudo-BCK(pP) algebra, applying Corollary 5.1(2) we get $D(A)^*_- \subseteq D(A)^*$ and $D(A)^*_\sim \subseteq D(A)^*$.

Conversely, consider $x \in D(A)^*$, that is, $\text{ord}(x) < \infty$. By the definition of a perfect pseudo-BCK(pP) algebra we get $\text{ord}(x^-) = \infty$ and $\text{ord}(x^\sim) = \infty$, that is, $x^-, x^\sim \in D(A)$. Applying the properties $x \leq x^{\sim-}$ and $x \leq x^{-\sim}$ we get $x \in D(A)^*_-$ and $x \in D(a)^*_\sim$. It follows that $D(A)^* \subseteq D(A)^*_-$ and $D(A)^* \subseteq D(A)^*_\sim$, respectively. Thus $D(A)^*_- = D(A)^*$ and $D(A)^*_\sim = D(A)^*$.

(b) $\Rightarrow$ (a) Consider $x \in A$ such that $\text{ord}(x) < \infty$, that is, $x \in D(A)^*$. Since $D(A)^*_- = D(A)^*$, there exists a $y \in D(A)$ such that $x \leq y^-$, so $y^{-\sim} \leq x^\sim$. By $y \leq y^{-\sim}$ and $\text{ord}(y) = \infty$, we get $\text{ord}(y^{-\sim}) = \infty$. From $y^{-\sim} \leq x^\sim$ we get $\text{ord}(x^\sim) = \infty$. Since $D(A)^*_\sim = D(A)^*$, there exists a $y \in D(A)$ such that $x \leq y^\sim$, so $y^{\sim-} \leq x^-$. By $y \leq y^{\sim-}$ and $\text{ord}(y) = \infty$, we get $\text{ord}(y^{\sim-}) = \infty$. From $y^{\sim-} \leq x^-$ we get $\text{ord}(x^-) = \infty$.

Conversely, consider $x \in A$ such that $\text{ord}(x^-) = \infty$ and $\text{ord}(x^\sim) = \infty$. Since A is local, by Corollary 5.1(1) it follows that $\text{ord}(x) < \infty$. Thus A is a perfect pseudo-BCK(pP) algebra. $\qquad\square$

Corollary 5.2 *If A is a perfect pseudo-BCK(pP) algebra, then*

$$D(A)^* = \left\{ x^- \mid x \in D(A) \right\} = \left\{ x^\sim \mid x \in D(A) \right\}.$$

Example 5.2

(1) Consider the pseudo-BCK(pP) algebra A from Example 1.9. Since A is not good, it follows that it is not a perfect pseudo-BCK(pP) algebra.
(2) If A_1 is the good pseudo-BCK(pP) algebra from Example 1.17, we have $D(A_1) = \{a_1, a_2, b_2, s, a, b, n, c, d, m, 1\}$ and $D(A_1)^* = \{0\}$. It is easy to see that $D(A_1)$ is a deductive system of A_1. Since $\text{ord}(0^-) = \text{ord}(0^\sim) = \infty$, it follows that A_1 is a perfect pseudo-BCK(pP) algebra.

Proposition 5.6 *Let A be a good pseudo-BCK(pP) algebra and P a proper normal deductive system of A. Then the following are equivalent:*

(a) *P is a perfect deductive system of A;*
(b) *A/P is a perfect pseudo-BCK(pP) algebra.*

Proof By Proposition 5.4, A/P is local iff P is primary. Also, A/P is perfect iff the following condition is satisfied:

$$\text{ord}(x/P) < \infty \quad \text{iff} \quad \left[\text{ord}\left((x/P)^-\right) = \infty \text{ and } \text{ord}\left((x/P)^\sim\right) = \infty\right].$$

But, applying Lemma 1.13, we have:

$$\text{ord}(x/P) < \infty \quad \text{iff} \quad (x/P)^n = 0/P \text{ for some } n \in \mathbb{N} \quad \text{iff}$$
$$\left[(x^n)^- \in P \text{ for some } n \in \mathbb{N} \text{ and } (x^n)^\sim \in P \text{ for some } n \in \mathbb{N}\right].$$

We also have:

$$\text{ord}\left((x/P)^-\right) = \infty \quad \text{iff} \quad \left((x/P)^-\right)^m \neq 0/P \text{ for all } m \in \mathbb{N} \quad \text{iff}$$
$$\left((x^-)^m\right)^- \notin P \text{ for all } m \in \mathbb{N}.$$

Taking into consideration the definition of a perfect deductive system it follows that (a) $\Leftrightarrow$ (b). $\qquad\square$

Theorem 5.5 *If A is a local good pseudo-BCK(pP) algebra, then the following are equivalent:*

(a) *A is perfect;*
(b) *any proper normal deductive system of A is perfect;*
(c) *$\{1\}$ is a perfect deductive system of A.*

Proof

(a) $\Rightarrow$ (b) Let D be a proper normal deductive system of A. Since A is local, by Theorem 5.4 it follows that D is primary. Let $x \in A$ such that $(x^n)^- \in D$ for some $n \in \mathbb{N}$ and suppose that $((x^-)^m)^- \in D$ for some $m \in \mathbb{N}$. Since D is proper, $[(x^n)^-), [((x^-)^m)^-) \subseteq D$ are also proper deductive systems of A. By Lemma 1.8(1) it follows that $\text{ord}((x^n)^-) = \text{ord}(((x^-)^m)^-) = \infty$. Since A is perfect, $\text{ord}(x^n) < \infty$ and $\text{ord}((x^-)^m) < \infty$, hence $\text{ord}(x) < \infty$ and $\text{ord}(x^-) < \infty$, contradicting the fact that A is perfect.
Thus $(x^n)^- \in D$ for $n \in \mathbb{N}$ implies $((x^-)^m)^- \notin D$ for all $m \in \mathbb{N}$, that is, D is perfect.

(b) $\Rightarrow$ (c) This is obvious, since $\{1\}$ is a proper normal deductive system of A.

(c) $\Rightarrow$ (a) Since $\{1\}$ is a perfect deductive system of A, applying Proposition 5.6 it follows that $A/\{1\}$ is perfect. Taking into consideration that $A \cong A/\{1\}$ we get that A is perfect. $\qquad\square$

Definition 5.3 Let A be a pseudo-BCK(pP) algebra. The intersection of all maximal deductive systems of A is called the *radical* of A and is denoted by $\text{Rad}(A)$. The intersection of all maximal normal deductive systems of A is called the *normal radical* of A and is denoted by $\text{Rad}_n(A)$.

It is obvious that $\text{Rad}(A)$ and $\text{Rad}_n(A)$ are proper deductive systems of A and $\text{Rad}(A) \subseteq \text{Rad}_n(A)$.

Proposition 5.7 *If A is a local pseudo-BCK(pP) algebra, then $\text{Rad}(A) = D(A)$.*

Proof By Theorem 5.1 it follows that $D(A)$ is the unique maximal deductive system of A, so $\text{Rad}(A) = D(A)$. $\qquad\square$

Example 5.3 Consider the perfect pseudo-BCK(pP) A_1 from Example 1.17. One can easily check that

$$\text{Rad}(A_1) = \text{Rad}_n(A_1) = D(A_1) = \{a_1, a_2, b_2, s, a, b, n, c, d, m, 1\}.$$

Lemma 5.1 *Let A be a local pseudo-BCK(pP) algebra and $x \in \text{Rad}(A)^*$, $y \in A$.*

(1) *If $y \le x$, then $y \in \text{Rad}(A)^*$;*
(2) *$x \odot y \in \text{Rad}(A)^*$.*

Proof

(1) Since $x \in \text{Rad}(A)^*$, there exists an $n \in \mathbb{N}$ such that $x^n = 0$. But $y \le x$, so $y^n \le x^n$. It follows that $y^n = 0$, that is, $\text{ord}(y) < \infty$. Hence $y \in \text{Rad}(A)^*$.
(2) This follows from (1), taking into consideration that $x \odot y \le x$. $\qquad\square$

Lemma 5.2 *Let A be a pseudo-BCK(pP) algebra.*

(1) *If A is local, then $x \in \text{Rad}(A)$ implies $x^-, x^\sim \in \text{Rad}(A)^*$;*
(2) *If A is perfect, then $x \in \text{Rad}(A)^*$ implies $x^-, x^\sim \in \text{Rad}(A)$.* $\qquad\square$

Proof

(1) Let $x \in \text{Rad}(A)$. If $x^- \in \text{Rad}(A)$ or $x^\sim \in \text{Rad}(A)$, then $x^- \odot x \in \text{Rad}(A)$ or $x \odot x^\sim \in \text{Rad}(A)$, respectively. It follows that $0 \in \text{Rad}(A)$, a contradiction. Hence $x^-, x^\sim \in \text{Rad}(A)^*$.
(2) Let $x \in \text{Rad}(A)^*$. If $x^- \in \text{Rad}(A)^*$ or $x^\sim \in \text{Rad}(A)^*$, then $\text{ord}(x^-) < \infty$ or $\text{ord}(x^\sim) < \infty$, respectively. By definition of a perfect pseudo-BCK(pP) algebra it follows that $\text{ord}(x) = \infty$, a contradiction. Thus $x^-, x^\sim \in \text{Rad}(A)$. $\qquad\square$

Corollary 5.3 *If A is a perfect pseudo-BCK(pP) algebra, then $x \in \text{Rad}(A)$ implies $x^{--}, x^{-\sim}, x^{\sim-}, x^{\sim\sim} \in \text{Rad}(A)$.*

Theorem 5.6 *If A is a perfect pseudo-BCK(pP) algebra, then $\text{Rad}(A)$ is a normal deductive system of A.*

Proof We have to prove that $x \to y \in \text{Rad}(A)$ iff $x \rightsquigarrow y \in \text{Rad}(A)$ for all $x, y \in A$. Consider $x, y \in A$ such that $x \to y \in \text{Rad}(A)$ and suppose $x \rightsquigarrow y \notin \text{Rad}(A)$.

From $y \le y^{-\sim}$ we get $x \to y \le x \to y^{-\sim}$ (by $(psbck\text{-}c_{14})$ and $(psbck\text{-}c_{10})$). Since $\text{Rad}(A)$ is a deductive system of A, it follows that $x \to y^{-\sim} \in \text{Rad}(A)$, that is, $(x \odot y^\sim)^- \in \text{Rad}(A)$ (by $(psbck\text{-}c_{37})$ and from the fact that A is good).

Hence $x \odot y^\sim \in \text{Rad}(A)^*$.

On the other hand, from $x \rightsquigarrow y \notin \text{Rad}(A)$, it follows that $x \rightsquigarrow y \in \text{Rad}(A)^*$.

Since $x \le x^{-\sim}$, by $(psbck\text{-}c_1)$ we get $x^{-\sim} \rightsquigarrow y \le x \rightsquigarrow y$, so $x^{-\sim} \rightsquigarrow y \in \text{Rad}(A)^*$ (by Lemma 5.1). By $(psbck\text{-}c_{40})$ we have $x^\sim \le x^{\sim-} \rightsquigarrow y$, so $x^\sim \in$

$\text{Rad}(A)^*$, that is, $x \in \text{Rad}(A)$. But $y \leq x \rightsquigarrow y$, so $y \in \text{Rad}(A)^*$, that is, $y^\sim \in \text{Rad}(A)$. Since $\text{Rad}(A)$ is a deductive system of A and $x, y^\sim \in \text{Rad}(A)$, we get $x \odot y^\sim \in \text{Rad}(A)$ which is a contradiction. Thus $x \to y \in \text{Rad}(A)$ implies $x \rightsquigarrow y \in \text{Rad}(A)$.

Similarly, $x \rightsquigarrow y \in \text{Rad}(A)$ implies $x \to y \in \text{Rad}(A)$ and we conclude that $\text{Rad}(A)$ is a normal deductive system of A. $\square$

Corollary 5.4 *If A is a perfect pseudo-BCK(pP) algebra, then $\text{Rad}(A) = \text{Rad}_n(A)$.*

Corollary 5.5 *If A is a perfect pseudo-BCK(pP) algebra, then $A/\text{Rad}(A)$ is perfect too.*

Proof By Theorem 5.6, $\text{Rad}(A)$ is a proper normal deductive system of A and by Theorem 5.5 it follows that $\text{Rad}(A)$ is perfect. Applying Proposition 5.6 we get that $A/\text{Rad}(A)$ is a perfect pseudo-BCK(pP) algebra. $\square$

Remark 5.1 If the pseudo-BCK(pP) algebra A is not perfect, then the result proved in Theorem 5.6 is not always valid. Indeed, consider the pseudo-BCK(pP) algebra A from Example 1.9. Since A is not good, it is not a perfect pseudo-BCK(pP) algebra. Moreover, $D(A) = \{a_2, s, a, b, n, c, d, 1\}$ is the unique maximal deductive system of A, so $\text{Rad}(A) = D(A)$. But $D(A)$ is not a normal deductive system ($a_1 \rightsquigarrow 0 = a_2 \in D(A)$, while $a_1 \to 0 = a_1 \notin D(A)$).

For a pseudo-BCK(pP) algebra A we define $A^+ := \{x \in A \mid x > x^- \vee x^\sim\}$.

Proposition 5.8 *If A is a linearly ordered pseudo-BCK(pP) algebra, then $\text{Rad}(A) \subseteq A^+$.*

Proof Consider $x \in \text{Rad}(A)$ and suppose that $x \notin A^+$, that is, $x \not> x^- \vee x^\sim$. Since A is a chain, it follows that $x \leq x^- \vee x^\sim$. Taking into consideration that $\text{Rad}(A)$ is a deductive system of A, we get that $x^- \vee x^\sim \in \text{Rad}(A)$, so $x \odot (x^- \vee x^\sim) \in \text{Rad}(A)$. Applying $(rl\text{-}c_2)$ we have $(x \odot x^-) \vee (x \odot x^\sim) \in \text{Rad}(A)$ and since $x \odot x^\sim = 0$, we get $x \odot x^- \in \text{Rad}(A)$. It follows that $(x \odot x^-) \odot x \in \text{Rad}(A)$, hence $0 \in \text{Rad}(A)$, a contradiction. Thus $x \in A^+$ for all $x \in \text{Rad}(A)$, so $\text{Rad}(A) \subseteq A^+$. $\square$

Corollary 5.6 *A^+ is a deductive system of the linearly ordered pseudo-BCK(pP) algebra A iff $A^+ = \text{Rad}(A)$.*

Theorem 5.7 *If A is a linearly ordered pseudo-BCK(pP) algebra, then $\text{Rad}(A) = \{x \in A \mid x^n > x^- \vee x^\sim \text{ for all } n \in \mathbb{N}, n \geq 1\}$.*

Proof Consider $x \in \text{Rad}(A)$. Then $x^n \in \text{Rad}(A) \subseteq A^+$, that is, $x^n > (x^n)^- \vee (x^n)^\sim$ for all $n \in \mathbb{N}, n \geq 1$. From $x^n \leq x$ we get $x^- \leq (x^n)^-$ and $x^\sim \leq (x^n)^\sim$, so $x^- \vee x^\sim \leq (x^n)^- \vee (x^n)^\sim < x^n$.

Conversely, consider $x \in A$ such that $x^n > x^- \vee x^\sim$ for all $n \in \mathbb{N}$, $n \geq 1$. We have $x^n \neq 0$ for all $n \in \mathbb{N}$ (if $x^n = 0$, we get $0 = x^n > x^- \vee x^\sim$, a contradiction). Therefore $[x^n)$ is a proper deductive system of A. Since A is linearly ordered, the union of all its proper deductive systems is the unique maximal deductive system $\mathrm{Rad}(A)$ of A. Thus $x \in [x^n) \subseteq \mathrm{Rad}(A)$. $\qquad\square$

Remark 5.2 ([232]) If A is an MTL-algebra, then $\mathrm{Rad}(A) = \{x \in A \mid x^n > x^-$ for all $n \in \mathbb{N}, n \geq 1\}$.

Remark 5.3 Let A be a pseudo-BCK(pP) algebra.
  Then $\mathrm{Rad}(A) \cap \mathrm{Rad}(A)\text{-}\mathrm{Div}(A) = \emptyset$. Indeed, suppose that there exists an $x \in \mathrm{Rad}(A) \cap \mathrm{Rad}(A)\text{-}\mathrm{Div}(A)$. It follows that there exist $y_1, y_2 \in \mathrm{Rad}(A)$ such that $x \odot y_1 = y_2 \odot x = 0$. Hence $0 \in \mathrm{Rad}(A)$, so $\mathrm{Rad}(A)$ is not proper, a contradiction.

Definition 5.4 A pseudo-BCK(pP) algebra is said to be *relatively free of zero divisors* if $\mathrm{Rad}(A)\text{-}\mathrm{Div}(A) = \{0\}$.

Lemma 5.3 *If the perfect pseudo-BCK(pP) algebra A is relatively free of zero divisors and $x \in \mathrm{Rad}(A)$, $y \in \mathrm{Rad}(A)^*$, then $x \perp y$ or $y \perp x$ if and only if $y = 0$.*

Proof Suppose that there exist $x \in \mathrm{Rad}(A)$ and $y \in \mathrm{Rad}(A)^*$ such that $x \perp y$. According to Proposition 1.28 we have $y \odot x = 0$. Since A is relatively free of zero divisors, it follows that $y = 0$. Similarly, if $y \perp x$ we have $x \odot y = 0$, so $y = 0$. On the other hand $x \perp 0$ and $0 \perp x$, hence the assertion is completely proved. $\qquad\square$

Proposition 5.9 *If a perfect pseudo-BCK(pP) algebra A has no nontrivial zero divisors, then $\mathrm{Rad}(A) = A \setminus \{0\}$.*

Proof Obviously, $\mathrm{Rad}(A) = A \setminus \{0\}$ is equivalent to $\mathrm{Rad}(A)^* = \{0\}$. Consider $x \in \mathrm{Rad}(A)^*$, that is, $\mathrm{ord}(x) < \infty$. Let $n \in \mathbb{N}$, $n > 1$ be the smallest integer such that $x^n = 0$. Then $x \odot x^{n-1} = 0$ with $x^{n-1} \neq 0$. Since A has no nontrivial zero divisors, it follows that $x = 0$. Thus $\mathrm{Rad}(A)^* = \{0\}$ and we are done. $\qquad\square$

Example 5.4

(1) The perfect pseudo-BCK(pP) algebra A from Example 4.4 is relatively free of zero divisors.
(2) The non-perfect pseudo-BCK(pP) algebra A from Example 4.5 is relatively free of zero divisors.
(3) Consider the pseudo-BCK(pP) algebra A from Example 4.5 and $X = \{b, c, 1\} \subseteq A$. Since $a \odot c = 0$ with $c \in X$, it follows that a is a X-left zero divisor of A. From $b \odot a = 0$ with $b \in X$ we get that a is an X-right zero divisor of A.

Lemma 5.4 *If A is a perfect pseudo-BCK(pP) algebra and $x, y \in \mathrm{Rad}(A)$, then $x \not\perp y$.*

Proof Suppose that there exist $x, y \in \text{Rad}(A)$ such that $x \perp y$. According to Proposition 1.27 we have $y^{-\sim} \odot x^{-\sim} = 0$. Since $y^{-\sim}, x^{-\sim} \in \text{Rad}(A)$ and $\text{Rad}(A)$ is a deductive system of A, it follows that $y^{-\sim} \odot x^{-\sim} \in \text{Rad}(A)$, so $0 \in \text{Rad}(A)$, a contradiction. Thus $x \not\perp y$. $\square$

Proposition 5.10 *If A is a perfect pseudo-BCK(pP) algebra, then* $\text{Rad}(A)^*$ *is an ideal of A.*

Proof We verify the axioms from Definition 1.17:

(I_1) If $x, y \in \text{Rad}(A)^*$, then according to Proposition 1.25 we have $x \oplus y = (y^- \odot x^-)^\sim$. Since $y^-, x^- \in \text{Rad}(A)$ and $\text{Rad}(A)$ is a deductive system of A, it follows that $y^- \odot x^- \in \text{Rad}(A)$. Thus $(y^- \odot x^-)^\sim \in \text{Rad}(A)^*$, that is, $x \oplus y \in \text{Rad}(A)^*$.

(I_2) Consider $y \in A$ and $x \in \text{Rad}(A)^*$ such that $y \le x$.
From $y \le x$ we get $x^- \le y^-$. Since $x^- \in \text{Rad}(A)$ and $\text{Rad}(A)$ is a deductive system of A we have $y^- \in \text{Rad}(A)$, that is, $y \in \text{Rad}(A)^*$. $\square$

Recall that, if A is a pseudo-BCK(pP) algebra, then a subset $B \subseteq A$ is a subalgebra of A if B is closed under the operations of A.

Proposition 5.11 *Let A be a pseudo-BCK(pP) algebra and B be a subalgebra of A. Then* $\text{Rad}_n(B) \subseteq \text{Rad}_n(A) \cap B$.

Proof First, we prove that if $H \in \text{Max}_n(A)$, then $H \cap B \in \text{Max}_n(B)$.

One can easily prove that $H \cap B$ is a deductive system of B. Since $0 \notin H$, it follows that $H \cap B$ is proper. For any $x \in A$, applying Theorem 1.12 we have that $x \notin H$ iff $(x^n)^- \in H$ for some $n \in \mathbb{N}$. Then for any $x \in B$, since $(x^n)^- \in B$ for all $n \in \mathbb{N}$, we have $x \notin H \cap B$ iff $(x^n)^- \in H \cap B$ for some $n \in \mathbb{N}$. Thus $H \cap B \in \text{Max}_n(B)$.

It follows that $\text{Rad}_n(B) \subseteq \bigcap \{H \cap B \mid H \in \text{Max}_n(A)\} = \text{Rad}_n(A) \cap B$. $\square$

Corollary 5.7 *If A is a perfect pseudo-BCK(pP) algebra and B a perfect subalgebra of A, then* $\text{Rad}(B) \subseteq \text{Rad}(A) \cap B$.

Remark 5.4 If A is an MTL-algebra and B a subalgebra of A, it was proved in [232] that $\text{Rad}(B) = \text{Rad}(A) \cap B$.

This result does not hold for pseudo-MTL algebras, so it does not hold for pseudo-BCK(pP) algebras (FL_w-algebras).

Indeed, consider the pseudo-MTL algebra A in Example 4.4. According to Theorem 3.2, $(B = [b, 1], \odot_b^1, \to, \leadsto, b, 1)$ is a subalgebra of A (where $x \odot_b^1 y = (x \odot y) \vee b$ for all $x, y \in [b, 1]$). One can easily check that $\text{Rad}(A) = \{a, b, c, 1\}$ and $\text{Rad}(B) = \{c, 1\}$, hence $\text{Rad}(A) \cap B = \{b, c, 1\} \ne \text{Rad}(B)$.

Moreover, A and B are perfect pseudo-BCK(pP) algebras, so $\text{Rad}(A) = \text{Rad}_n(A)$ and $\text{Rad}(B) = \text{Rad}_n(B)$. Hence $\text{Rad}_n(B) \ne \text{Rad}_n(A) \cap B$.

Remark 5.5 The proof of the above mentioned result is based on the *congruence extension property* (CEP, for short). A class $\mathbb{A}$ of algebras of the same type has CEP if for every algebra A in $\mathbb{A}$, every subalgebra B of A and every congruence θ on B, there exists a congruence Θ on A such that $\Theta \cap B^2 = \theta$.

It was proved in [129] that the class of FL_w-algebras does not satisfy CEP.

In order to prove this, the FL_w-algebra $(A = \{0, a, b, c, 1\}, \wedge, \vee, \odot, \rightarrow, \rightsquigarrow, 0, 1)$ with $0 < a < b < c < 1$ and the operations $\odot, \rightarrow, \rightsquigarrow$ given by the following tables was considered:

$\odot$	0	a	b	c	1
0	0	0	0	0	0
a	0	0	0	0	a
b	0	a	b	b	b
c	0	a	b	c	c
1	0	a	b	c	1

$\rightarrow$	0	a	b	c	1
0	1	1	1	1	1
a	a	1	1	1	1
b	a	a	1	1	1
c	a	a	b	1	1
1	0	a	b	c	1

$\rightsquigarrow$	0	a	b	c	1
0	1	1	1	1	1
a	c	1	1	1	1
b	0	a	1	1	1
c	0	a	b	1	1
1	0	a	b	c	1

One can easily check that $(B = \{b, c, 1\}, \odot, \rightarrow, \rightsquigarrow, b, 1)$ is a subalgebra of A and that B has the non-trivial congruence $\theta = \{\{c, 1\}, \{b\}\}$. Suppose that there exists a non-trivial congruence Θ on A such that $\Theta \cap B^2 = \theta$, so $c\,\Theta\,1$. It follows that $(a \rightarrow a \odot c)\,\Theta\,(a \rightarrow a \odot 1)$, that is, $a\,\Theta\,1$. Hence $a \odot c\,\Theta\,1 \odot 1$, so $0\,\Theta\,1$. Thus Θ is the trivial congruence on A and we can conclude that the class of FL_w-algebras does not satisfy CEP.

By contrast, it was proved in [129] that the class of FL_{ew}-algebras satisfies CEP, hence the class of MTL-algebras has this property too. More precisely, if a variety of the class $\mathbb{A}$ of FL_w-algebras $(A, \wedge, \vee, \odot, \rightarrow, \rightsquigarrow, 0, 1)$ satisfies the identity $(x \wedge e)^k \odot y = y \odot (x \wedge e)^k$ for some $k \in \mathbb{N}$, $k \geq 1$, then $\mathbb{A}$ satisfies CEP. This is the case for the variety $\mathbb{CRL}$ of FL_{ew}-algebras.

Since the FL_w-algebra from the above example is a pseudo-MTL algebra, it follows that the class of these algebras also does not satisfy CEP.

5.2.2 Perfect Pseudo-MTL Algebras

Since a pseudo-MTL algebra is a pseudo-BCK(pP) algebra, a perfect pseudo-MTL algebra is defined in the same way as a perfect pseudo-BCK(pP) algebra. All properties and results presented in the previous section for perfect pseudo-BCK(pP) algebras are also valid in the case of perfect pseudo-MTL algebras. In this section we will investigate some specific results for perfect pseudo-MTL algebras.

Proposition 5.12 *If A is a perfect pseudo-MTL algebra, then for any $x, y \in$ Rad(A), $x^- \odot y^- = x^\sim \odot y^\sim = 0$.*

Proof We will follow the idea used in [30] for the case of pseudo-BL algebras.

Let $x, y \in$ Rad(A). Proving $x^- \odot y^- = 0$ is equivalent to proving that $(x^- \odot y^-)^\sim = 1$. Suppose $(x^- \odot y^-)^\sim \neq 1$. By Corollary 4.3, there exists a prime deductive system P such that $(x^- \odot y^-)^\sim \notin P$, that is, $y^- \rightsquigarrow x^{-\sim} \notin P$. Since

P is prime, by Proposition 4.3 it follows that $x^{-\sim} \leadsto y^- \in P$, or equivalently $x^{-\sim} \leadsto y^{-\sim} \in P$. Hence $x^{-\sim} \leadsto y^{--\sim} \in P$, so $(y^{--} \odot x^{-\sim})^\sim \in P$.

By Proposition 1.32, there is a maximal deductive system M such that $P \subseteq M$.

Since A is perfect, we have $M = \mathrm{Rad}(A)$. Obviously, $u = y^{--} \odot x^{-\sim} \notin P$ (if $u \in P$, since $u^\sim \in P$, we get $0 = u \odot u^\sim \in P$, hence P is not a proper deductive system, a contradiction). Thus $y^{--} \odot x^{-\sim} \notin \mathrm{Rad}(A)$.

By Theorem 5.6 we have that $\mathrm{Rad}(A)$ is a normal deductive system, so applying Theorem 1.12, there is an $n \geq 1$ such that $[(y^{--} \odot x^{-\sim})^n]^- \in \mathrm{Rad}(A)$. Let $z = (y^{--} \odot x^{-\sim})^n$, so $z^- \in \mathrm{Rad}(A)$. Since $x, y \in \mathrm{Rad}(A)$, by Corollary 5.3 we have $y^{--}, x^{-\sim} \in \mathrm{Rad}(A)$, so $y^{--}, x^{-\sim} \in \mathrm{Rad}(A)$, hence $z \in \mathrm{Rad}(A)$. Thus $z, z^- \in \mathrm{Rad}(A)$ and $0 = z^- \odot z \in \mathrm{Rad}(A)$, that is, $\mathrm{Rad}(A)$ is not a proper deductive system, a contradiction. We conclude that $x^- \odot y^- = 0$.

Similarly, $x^\sim \odot y^\sim = 0$.   □

Corollary 5.8 *Let A be a perfect pseudo-MTL algebra.*

(1) *If $x, y \in \mathrm{Rad}(A)^*$, then $x \odot y = y \odot x = 0$;*
(2) *If $x \in \mathrm{Rad}(A)$ and $y \in \mathrm{Rad}(A)^*$, then $x^- \leq y^-$ and $x^\sim \leq y^\sim$;*
(3) *If $x \in \mathrm{Rad}(A)$, then $x^- \leq x^{--} \wedge x^{-\sim}$ and $x^\sim \leq x^{-\sim} \wedge x^{\sim\sim}$.*

Proof

(1) Since $x, y \in \mathrm{Rad}(A)^*$, it follows that $x^-, y^- \in \mathrm{Rad}(A)$ and applying Proposition 5.12 we get $x^{-\sim} \odot y^{-\sim} = 0$. Taking into consideration the fact that $x \leq x^{-\sim}$ and $y \leq y^{-\sim}$, we get $x \odot y \leq x^{-\sim} \odot y^{-\sim} = 0$. Hence $x \odot y = 0$. Similarly, $y \odot x = 0$;

(2) From $x, y^\sim \in \mathrm{Rad}(A)$, by Proposition 5.12 we get $x^- \odot y^{\sim-} = 0$.
 Because $y \leq y^{\sim-}$, we have $x^- \odot y \leq x^- \odot y^{\sim-} = 0$, so $x^- \odot y = 0$. Hence $x^- \leq y^-$. Similarly, $x^\sim \leq y^\sim$;

(3) From $x \in \mathrm{Rad}(A)$, it follows that $x^-, x^\sim \in \mathrm{Rad}(A)^*$. Since A is good, we get from (2) that $x^- \leq x^{--}, x^{-\sim}$ and $x^\sim \leq x^{-\sim}, x^{\sim\sim}$. It follows that $x^- \leq x^{--} \wedge x^{-\sim}$ and $x^\sim \leq x^{-\sim} \wedge x^{\sim\sim}$.   □

Proposition 5.13 *If A is a perfect pseudo-MTL algebra and $x, y \in \mathrm{Rad}(A)^*$, then $x \perp y$ and $y \perp x$.*

Proof Since $x, y \in \mathrm{Rad}(A)^*$, it follows that $y^-, x^- \in \mathrm{Rad}(A)$. Hence $y^{-\sim} \odot x^{-\sim} = 0$. By Proposition 1.27, we get $x \perp y$. Similarly, $y \perp x$.   □

Proposition 5.14 *Let A be a perfect pseudo-MTL algebra, $x \in \mathrm{Rad}(A)$ and $y \in \mathrm{Rad}(A)^*$. Then:*

(1) *if $x \perp y$ or $y \perp x$, then $y^2 = 0$;*
(2) *if $x \perp y$, then $y \in \mathrm{Rad}(A)_l\text{-}\mathrm{Div}(A)$;*
(3) *if $y \perp x$, then $y \in \mathrm{Rad}(A)_r\text{-}\mathrm{Div}(A)$.*

Proof

(1) According to Corollary 5.8 we have $x^- \leq y^-$. On the other hand, since $x \perp y$, we get $y \odot x = 0$. It follows that $y \leq x^- \leq y^-$, that is, $y^2 = 0$.

Similarly, from $x^\sim \leq y^\sim$ and $x \odot y = 0$ we get $y \leq x^\sim \leq y^\sim$, so $y^2 = 0$.

(2) Obviously, $x \neq 0$. Since $x \perp y$, we get $y \odot x = 0$, that is, $y \in \mathrm{Rad}(A)_l\text{-}\mathrm{Div}(A)$.

(3) Similarly, from $y \perp x$ we have $x \odot y = 0$, so $y \in \mathrm{Rad}(A)_r\text{-}\mathrm{Div}(A)$. $\square$

5.2.3 Perfect Pseudo-MV Algebras

In this subsection we recall some notions and results regarding the local and perfect pseudo-MV algebras (see [212]).

If A is a pseudo-MV algebra, we denote by $D(A)$ the set $\{x \in A \,|\, \mathrm{ord}(x) = \infty\}$. Obviously, $0 \in D(A)$.

Lemma 5.5 *If A is a pseudo-MV algebra, then:*

$$\{x \in A | x \geq y^- \text{ for some } y \in D(A)\} = \{x \in A | x \geq y^\sim \text{ for some } y \in D(A)\}.$$

We define

$$D(A)^* := \{x \in A | x \geq y^- \text{ for some } y \in D(A)\}$$
$$= \{x \in A | x \geq y^\sim \text{ for some } y \in D(A)\}.$$

Obviously, $1 \in D(A)^*$.

Definition 5.5 A pseudo-MV algebra is called *local* if for every $x, y \in A$ the following condition holds: $\mathrm{ord}(x \oplus y) < \infty$ implies $\mathrm{ord}(x) < \infty$ or $\mathrm{ord}(y) < \infty$.

Remark 5.6 If A is local, then $\mathrm{ord}(x) < \infty$ or $\mathrm{ord}(x^-) < \infty$.

Proposition 5.15 *Let A be a pseudo-MV algebra. The following are equivalent:*

(a) *A is local;*
(b) *$D(A)$ is an ideal of A;*
(c) *$D(A)$ is a proper ideal of A;*
(d) *$D(A)$ is the unique maximal ideal of A.*

Proposition 5.16 *If A is a local pseudo-MV algebra, then:*

(1) *$D(A)^*$ is a filter of A;*
(2) *$D(A) \cap D(A)^* = \emptyset$.*

Definition 5.6 A local pseudo-MV algebra A is called *perfect* if for any $x \in A$, $\mathrm{ord}(x) < \infty$ iff $[\mathrm{ord}(x^-) = \infty$ and $\mathrm{ord}(x^\sim) = \infty]$.

Proposition 5.17 *If A is a local pseudo-MV algebra, then the following are equivalent*:

(a) *A is perfect*;
(b) *$A = D(A) \cup D(A)^*$.*

Definition 5.7 The intersection of all maximal ideals of a pseudo-MV algebra is denoted by $\mathrm{Rad}(A)$ and it is called the *radical* of A.

Corollary 5.9

(1) *If A is a local pseudo-MV algebra, then $\mathrm{Rad}(A) = D(A)$*;
(2) *A local pseudo-MV algebra A is perfect iff $A = \mathrm{Rad}(A) \cup \mathrm{Rad}(A)^*$.*

Proposition 5.18 *If A is a perfect pseudo-MV algebra, then*:

(1)

$$\mathrm{Rad}(A)^* = \{x \in A \mid \mathrm{ord}(x) < \infty\} = \{x^- \mid x \in \mathrm{Rad}(A)\}$$
$$= \{x^\sim \mid x \in \mathrm{Rad}(A)\};$$

(2) *$x \odot y = 0$ for all $x, y \in \mathrm{Rad}(A)$*;
(3) *if $x \in \mathrm{Rad}(A)$ and $y \in \mathrm{Rad}(A)^*$, then $x \leq y$*;
(4) *$(\mathrm{Rad}(A), \oplus, 0)$ is a cancellative monoid.*

Proposition 5.19 *If A is a perfect pseudo-MV algebra, then $\mathrm{Rad}(A)$ is a normal ideal of A.*

Proof We have to prove that $y \odot x^- \in \mathrm{Rad}(A)$ iff $x^\sim \odot y \in \mathrm{Rad}(A)$.

Assume that $y \odot x^- \in \mathrm{Rad}(A)$. If $x^\sim \odot y \notin \mathrm{Rad}(A)$, then $x^\sim \odot y \in \mathrm{Rad}(A)^*$.
Hence $(x^\sim \odot y)^- \in \mathrm{Rad}(A)$, so $y^- \oplus x \in \mathrm{Rad}(A)$.

Since $x, y^- \leq y^- \oplus x$, we get $x, y^- \in \mathrm{Rad}(A)$, so $x^-, y \in \mathrm{Rad}(A)^*$. Because $\mathrm{Rad}(A)^*$ is a filter, it follows that $y \odot x^- \in \mathrm{Rad}(A)^*$ which is a contradiction, so $x^\sim \odot y \in \mathrm{Rad}(A)$.

Similarly, $x^\sim \odot y \in \mathrm{Rad}(A)$ implies $y \odot x^- \in \mathrm{Rad}(A)$.

Thus $\mathrm{Rad}(A)$ is a normal ideal of A. $\square$

Example 5.5 Let $\mathbb{Z}$ be the Abelian ℓ-group of integers, G an ℓ-group, $g_0 \in G^+$ and $A = \Gamma(\mathbb{Z} \times_{lex} G, (1, g_0))$ where $\mathbb{Z} \times_{lex} G$ is the lexicographic product of $\mathbb{Z}$ and G. One can easily prove that

$$A = \{(0, g) \mid g \in G^+\} \cup \{(1, g) \mid g \leq g_0\} \quad \text{and} \quad D(A) = \{(0, g) \mid g \in G^+\}.$$

Obviously, $D(A)$ is an ideal of A, so A is a local pseudo-MV algebra.

Because $\mathrm{ord}((0, g)) = \infty$ for all $g \in G^+$ and $\mathrm{ord}((1, g)) = 2$ for all $g \leq g_0$, we conclude that A is a perfect pseudo-MV algebra.

5.3 Archimedean Residuated Structures

The Archimedean property was stated by Archimedes in the following form: "... the following lemma is assumed: that the excess by which the greater of (two) unequal areas exceeds the less can, by being added to itself, be made to exceed any given finite area" ([164]).

This is one of the most beautiful axioms of classical arithmetic.

In the case of the field of real numbers, the Archimedean property can be formulated as follows: for any real numbers x and y with $x > 0$, there exists an $n \in \mathbb{N}$ such that $nx > y$.

The aim of this section is to extend the Archimedean property to the case of FL_w-algebras. We will also define the hyperarchimedean FL_w-algebras, proving that every hyperarchimedean FL_w-algebra is Archimedean and every locally FL_w-algebra is hyperarchimedean.

We recall the notions of Archimedean ℓ-groups, MV-algebras and pseudo-MV algebras.

An ℓ-group G is *Archimedean* if for $0 \leq x, y \in G$, $nx \leq y$ for all $n \in \mathbb{N}$ implies $x = 0$.

Proposition 5.20 (Theorem 6.1.32 in [108]) *In any MV-algebra A the following conditions are equivalent*:

(a) $\mathrm{Rad}(A) = \{0\}$;
(b) $nx \leq x^-$ *for all* $n \in \mathbb{N}$ *implies* $x = 0$;
(c) $nx \leq y^-$ *for all* $n \in \mathbb{N}$ *implies* $x \wedge y = 0$;
(d) $nx \leq y$ *for all* $n \in \mathbb{N}$ *implies* $x \odot y = x$,

where $nx := x_1 \oplus \cdots \oplus x_n$ *with* $x_1 = \cdots = x_n = x$.

Definition 5.8 An MV-algebra A is *Archimedean* in Dvurečenskij's sense [108] if it satisfies condition (b) of Proposition 5.20 and A is *Archimedean* in Belluce's sense [6] if it satisfies condition (d) of Proposition 5.20.

By Proposition 5.20 the two definitions of Archimedean MV-algebras are equivalent.

Definition 5.9 ([97]) A pseudo-MV algebra A is said to be *Archimedean* if the existence of $na := a_1 + a_2 + \cdots + a_n$, where $a_1 = a_2 = \cdots = a_n = a \in A$, for any integer $n \geq 1$, entails that $a = 0$.

Theorem 5.8 (Theorem 4.2 in [97]) *Any Archimedean pseudo-MV algebra is commutative, that is, an MV-algebra.*

Proposition 5.21 *In any FL_w-algebra the following are equivalent*:

(a) $x^n \geq x^- \vee x^\sim$ *for any* $n \in \mathbb{N}$ *implies* $x = 1$;
(b) $x^n \geq y^- \vee y^\sim$ *for any* $n \in \mathbb{N}$ *implies* $x \vee y = 1$;

(c) $x^n \geq y^- \vee y^\sim$ for any $n \in \mathbb{N}$ implies $x \to y = x \rightsquigarrow y = y$ and $y \to x = y \rightsquigarrow x = x$.

Proof

(a) $\Rightarrow$ (b) Take $x, y \in A$ such that $x^n \geq y^- \vee y^\sim$ for any $n \in \mathbb{N}$.
By the properties of FL_w-algebras and by hypothesis we have:

$$(x \vee y)^- = x^- \wedge y^- \leq y^- \leq y^- \vee y^\sim \leq x^n \leq (x \vee y)^n \quad \text{and}$$
$$(x \vee y)^\sim = x^\sim \wedge y^\sim \leq y^\sim \leq y^- \vee y^\sim \leq x^n \leq (x \vee y)^n,$$

hence $(x \vee y)^n \geq (x \vee y)^- \vee (x \vee y)^\sim$ for any $n \in \mathbb{N}$.
Thus by (a), we get $x \vee y = 1$.

(b) $\Rightarrow$ (a) Consider $x \in A$ such that $x^n \geq x^- \vee x^\sim$ for any $n \in \mathbb{N}$.
Applying (b) for $y = x$ we get $x \vee x = 1$, hence $x = 1$.

(b) $\Rightarrow$ (c) Assume that $x^n \geq y^- \vee y^\sim$ for any $n \in \mathbb{N}$.
Applying (b), it follows that $x \vee y = 1$. But, for $x, y \in A$ we have:

$$x \vee y \leq \left[(x \to y) \rightsquigarrow y\right] \wedge \left[(y \to x) \rightsquigarrow x\right],$$
$$x \vee y \leq \left[(x \rightsquigarrow y) \to y\right] \wedge \left[(y \rightsquigarrow x) \to x\right].$$

Since $x \vee y = 1$, it follows that:

$$\left[(x \to y) \rightsquigarrow y\right] \wedge \left[(y \to x) \rightsquigarrow x\right] = 1 \quad \text{and}$$
$$\left[(x \rightsquigarrow y) \to y\right] \wedge \left[(y \rightsquigarrow x) \to x\right] = 1,$$

hence $(x \to y) \rightsquigarrow y = 1$ and $(x \rightsquigarrow y) \to y = 1$.
From $(x \to y) \rightsquigarrow y = 1$ we have $x \to y \leq y$ and taking into consideration that $y \leq x \to y$, we obtain $x \to y = y$. Similarly, $x \rightsquigarrow y = y$.
In a similar way we can prove that $y \to x = y \rightsquigarrow x = x$.

(c) $\Rightarrow$ (a) Consider $x \in A$ such that $x^n \geq x^- \vee x^\sim$ for any $n \in \mathbb{N}$.
Applying (c) for $y = x$ we get $x \to x = x$, hence $x = 1$. $\qquad \square$

Definition 5.10 An FL_w-algebra is called *Archimedean* if one of the equivalent conditions from Proposition 5.21 is satisfied.

Example 5.6 Take $A = \{0, a, b, c, 1\}$ where $0 < a < b, c < 1$ and b, c are incomparable. Consider the operations $\odot, \to, \rightsquigarrow$ given by the following tables:

$\odot$	0	a	b	c	1
0	0	0	0	0	0
a	0	0	0	a	a
b	0	a	b	a	b
c	0	0	0	c	c
1	0	a	b	c	1

$\to$	0	a	b	c	1
0	1	1	1	1	1
a	c	1	1	1	1
b	c	c	1	c	1
c	0	b	b	1	1
1	0	a	b	c	1

$\rightsquigarrow$	0	a	b	c	1
0	1	1	1	1	1
a	b	1	1	1	1
b	0	c	1	c	1
c	b	b	b	1	1
1	0	a	b	c	1

Then $(A, \wedge, \vee, \odot, \rightarrow, \rightsquigarrow, 0, 1)$ is an FL_w-algebra.
   We have:

$$0^n = 0 \not\geq 0^- \vee 0^\sim = 1 \vee 1 = 1, \quad n \geq 1,$$
$$a^n = 0 \not\geq a^- \vee a^\sim = c \vee b = 1, \quad n \geq 2,$$
$$b^n = b \not\geq b^- \vee b^\sim = c \vee 0 = c, \quad n \geq 1,$$
$$c^n = c \not\geq c^- \vee c^\sim = 0 \vee b = b, \quad n \geq 1,$$
$$1^n = 1 \geq 1^- \vee 1^\sim = 0 \vee 0 = 0, \quad n \geq 1.$$

Obviously: $a \not\geq a^- \vee a^\sim = c \vee b = 1$.
   We conclude that, $x^n \geq x^- \vee x^\sim$ for all $n \in \mathbb{N}$, $n \geq 1$, implies $x = 1$. Hence A is
an Archimedean FL_w-algebra.

Definition 5.11 If A is an FL_w-algebra, then an element $x \in A$ is called
Archimedean if there is an $n \in \mathbb{N}$, $n \geq 1$, such that $x^n \in B(A)$. An FL_w-algebra
A is called *hyperarchimedean* if all its elements are Archimedean.

Example 5.7 Consider the FL_w-algebra A from Example 3.4.
   Since $a_1^2 = 0 \in B(A)$, it follows that a_1 is Archimedean.
   By contrast, $a_2^n = a_2 \notin B(A)$ for all $n \in \mathbb{N}$, $n \geq 1$, so a_2 is not Archimedean.
   Thus A is not a hyperarchimedean FL_w-algebra.

Proposition 5.22 *Any locally finite FL_w-algebra is hyperarchimedean.*

Proof Let A be a locally finite FL_w-algebra and $x \in A$. Hence there exists an $n \in \mathbb{N}$,
$n \geq 1$, such that $x^n = 0 \in B(A)$. It follows that any element x of A is Archimedean,
so A is hyperarchimedean. $\qquad\qquad\qquad\qquad\qquad\qquad\qquad\qquad\qquad\qquad\square$

Proposition 5.23 *Any hyperarchimedean FL_w-algebra is Archimedean.*

Proof Let A be a hyperarchimedean FL_w-algebra and $x \in A$ such that $x^n \geq x^- \vee$
$x^\sim$ for any $n \in \mathbb{N}$. Since A is hyperarchimedean, there exists an $m \in \mathbb{N}$, $m \geq 1$,
such that $x^m \in B(A)$. According to Proposition 3.17 it follows that $x = 1$, so A is
Archimedean. $\qquad\qquad\qquad\qquad\qquad\qquad\qquad\qquad\qquad\qquad\qquad\qquad\quad\square$

Corollary 5.10 *Any locally finite FL_w-algebra is Archimedean.*

Proof This follows from Propositions 5.22 and 5.23. $\qquad\qquad\qquad\qquad\qquad\qquad\square$

Example 5.8 Consider again the pseudo-MTL algebra A from Example 4.1. Since
$c^n = c \geq 0 = c^- \vee c^\sim$ for all $n \geq 1$ and $c \neq 1$, it follows that A is not an
Archimedean pseudo-MTL algebra.

Example 5.9 The pseudo-MTL algebra A from Example 3.5 is locally finite, so it is Archimedean.

We give an example of an Archimedean pseudo-MTL algebra which is not a chain and is not a hyperarchimedean pseudo-MTL algebra.

Example 5.10 Consider the pseudo-MTL algebra A from Example 4.5.
Since $a^2 = 0 \in B(A)$, it follows that a is Archimedean.
By contrast, $b^n = b \notin B(A)$ for all $n \in \mathbb{N}$, $n \geq 1$, so b is not Archimedean.
Thus A is not a hyperarchimedean pseudo-MTL algebra.
We have:

$$0^n = 0 \ngeq 0^- \vee 0^\sim = 1 \vee 1 = 1, \quad n \geq 1,$$
$$a^n = 0 \ngeq a^- \vee a^\sim = b \vee c = 1, \quad n \geq 2,$$
$$b^n = b \ngeq b^- \vee b^\sim = 0 \vee c = c, \quad n \geq 1,$$
$$c^n = c \ngeq c^- \vee c^\sim = b \vee 0 = b, \quad n \geq 1$$
$$1^n = 1 \geq 1^- \vee 1^\sim = 0 \vee 0 = 0, \quad n \geq 1.$$

We conclude that, if $x^n \geq x^- \vee x^\sim$ for all $n \in \mathbb{N}$, $n \geq 1$, then $x = 1$.
Hence A is an Archimedean pseudo-MTL algebra.

Remark 5.7 By Examples 5.9 and 5.10 we have proved that, in general, an Archimedean FL_w-algebra is not commutative. This result seems to be important, taking into consideration the known results in the case of other structures: any Archimedean ℓ-group is Abelian (Theorem 10.19 in [12]) and any Archimedean pseudo-MV algebra is an MV-algebra, so it is commutative (Theorem 4.2 in [97]).

Remark 5.8 Obviously, an FL_{ew}-algebra is Archimedean if it satisfies one of the following equivalent conditions:

(a) $x^n \geq x^-$ for any $n \in \mathbb{N}$ implies $x = 1$;
(b) $x^n \geq y^-$ for any $n \in \mathbb{N}$ implies $x \vee y = 1$;
(c) $x^n \geq y^-$ for any $n \in \mathbb{N}$ implies $x \to y = y$ and $y \to x = x$.

We will give below an example of Archimedean FL_{ew}-algebra.

Example 5.11 (Example 19.3.1 in [178]) Consider $A = \{0, a, b, c, d, 1\}$ with $0 < a < b, c < d < 1$ and b, c incomparable. Define the operations $\odot, \to$ by the following tables:

$\odot$	0	a	b	c	d	1
0	0	0	0	0	0	0
a	0	0	0	0	0	a
b	0	0	0	0	0	b
c	0	0	0	0	0	c
d	0	0	0	0	0	d
1	0	a	b	c	1	1

$\rightarrow$	0	a	b	c	d	1
0	1	1	1	1	1	1
a	d	1	1	1	1	1
b	d	d	1	d	1	1
c	d	d	d	1	1	1
d	d	d	d	d	1	1
1	0	a	b	c	d	1

Then $(A, \wedge, \vee, \odot, \rightarrow, 0, 1)$ is a proper bounded FL_{ew}-algebra.

Indeed, since $(b \rightarrow c) \odot b = 0 \neq a = b \wedge c$, it follows that the condition (B_4) is not satisfied, so A is neither a BL-algebra nor a divisible FL_w-algebra.

Moreover, $(b \rightarrow c) \vee (c \rightarrow b) = d \neq 1$, so A is not an MTL-algebra.

In fact, A is a bounded FL_{ew}-algebra with *weak nilpotent minimum* (WNM) and $(C_\vee)$ conditions:

$$(\text{WNM}): \quad (x \odot y)^- \vee \left[(x \wedge y) \rightarrow (x \odot y)\right] = 1,$$

$$(C_\vee): \quad x \vee y = \left[(x \rightarrow y) \rightarrow y\right] \wedge \left[(y \rightarrow x) \rightarrow x\right].$$

We have:

$$0^n = 0 \not\geq 0^- = 1, \quad n \geq 1,$$

$$a^n = 0 \not\geq a^- = d, \quad n \geq 2,$$

$$b^n = 0 \not\geq b^- = d, \quad n \geq 2,$$

$$c^n = 0 \not\geq c^- = d, \quad n \geq 2,$$

$$d^n = 0 \not\geq d^- = d, \quad n \geq 2,$$

$$1^n = 1 \geq 1^- = 0, \quad n \geq 1.$$

We conclude that, if $x^n \geq x^-$ for all $n \in \mathbb{N}$, $n \geq 1$, then $x = 1$. Hence A is an Archimedean FL_{ew}-algebra.

Chapter 6
States on Multiple-Valued Logic Algebras

If a trial is governed by the laws of classical logic, then the set of its associated events has the structure of Boolean algebra and based on this principle classical probability theory was developed. A *probability* (*Boolean state*) on a Boolean algebra $(B, \wedge, \vee, ^-, 0, 1)$ is a function $m : B \longrightarrow [0, 1]$ such that:

$$m(x \vee y) = m(x) + m(y), \quad \text{if } x \wedge y = 0;$$
$$m(1) = 1.$$

In order to develop a probability theory starting from a logical system, it is necessary to solve two problems:

- to establish an algebraic structure on the set of events;
- to define an appropriate notion of probability (state).

The sets of events will always have the structure of the Lindenbaum-Tarski algebra associated to the logical system, however establishing a sound notion of probability is more difficult. In the classical case the definition of probability is based on the existence of a binary additive operation: the Boolean join $\vee$ in the case of probabilities defined on Boolean algebras and the MV-sum $\oplus$ in the case of probabilities on MV-algebras. If the algebras of events are not endowed with this kind of sum, then we can approach the problem in two ways:

(i) by defining a pseudo-sum which satisfies some of the usual properties of the sum, and by defining the probability by the "additivity" property;
(ii) by stipulating some rules which determine how the probability behaves relative to certain event operations (such as implication).

Approach (i) was taken by Riečan in [245] and this led to the Riečan states on pseudo-BL algebras ([131]), bounded $R\ell$-monoids ([110, 111]), FL_w-algebras ([49]) and bounded pseudo-BCK algebras ([52, 67]).

Approach (ii) was used in the above mentioned papers to define the Bosbach states based on the implication operations on pseudo-BL algebras, bounded $R\ell$-monoids, FL_w-algebras and bounded pseudo-BCK algebras, respectively.

L.C. Ciungu, *Non-commutative Multiple-Valued Logic Algebras*,
Springer Monographs in Mathematics, DOI 10.1007/978-3-319-01589-7_6,
© Springer International Publishing Switzerland 2014

The states on ℓ-groups with strong unit are studied in [155].

A *state* on an ℓ-group (G, u) with strong unit u is a function $s : G \longrightarrow \mathbb{R}$ such that:

$$s(g_1 + g_2) = s(g_1) + s(g_2) \quad \text{for all } g_1, g_2 \in G;$$

$$s(g) \geq 0 \quad \text{for all } g \in G^+;$$

$$g(1) = 1.$$

It is known that any Abelian ℓ-group (G, u) with strong unit u possesses at least one state (Sect. 4 in [155]).

The states on an MV-algebra $(A, \oplus, ^-, 0)$ were first introduced by D. Mundici in [230] as an averaging of the truth-value in Łukasiewicz logic. A state on an MV-algebra is defined as a function $s : A \longrightarrow [0, 1]$ satisfying the conditions:

$$s(1) = 1 \quad (\textit{normality});$$

$$s(x \oplus y) = s(x) + s(y), \quad \text{if } x \odot y = 0 \quad (\textit{additivity}),$$

where $x \odot y = (x^- \oplus y^-)^-$.

There exists a bijective correspondence between the set of states on an MV-algebra and the set of states on (G, u), where (G, u) is an ℓu-group such that $\Gamma(G, u) = A$ (Theorem 15.2.10 in [89]). As a consequence, every MV-algebra admits at least one state.

The states on MV-algebras have been entirely studied in [203, 249, 250].

In the case of non-commutative structures, the states were first introduced for pseudo-MV algebras in [96] and it was proved that any linearly ordered pseudo-MV algebra possesses a unique state and that there exists a pseudo-MV algebra having no states on it.

The notion of a *Bosbach state* on pseudo-BL algebras was defined in [131] using an identity studied by Bosbach in residuation groupoids ([17]). The *Riečan state* was also defined on a good pseudo-BL algebra in [131], which extends the additive measures introduced by Riečan for BL-algebras in [245]. It was proved that every Bosbach state is a Riečan state, but the converse was an open question. It was also proved in [131] that the existence of a state-morphism on a pseudo-BL algebra is equivalent to the existence of a maximal filter which is normal. Based on this result, in [98] it was proved that every linearly ordered pseudo-BL algebra admits a state. The notion of state was extended in [111] to the case of bounded $R\ell$-monoids and it was shown that the Bosbach and Riečan states on good $R\ell$-monoids coincide. In [208] the states on a pseudo-BCK semilattice were defined and it was proved that any Bosbach state on a good pseudo-BCK semilattice is a Riečan state, but the converse is not true.

In this chapter we will present the notion of a state for pseudo-BCK algebras ([52, 67]). One of the main results consists of proving that any Bosbach state on a good pseudo-BCK algebra is a Riečan state, but the converse turns out not to be true. Some conditions are given for a Riečan state on a good pseudo-BCK algebra to be a Bosbach state. In contrast to the case of pseudo-BL algebras, we show that there exist linearly ordered pseudo-BCK algebras having no Bosbach states and that there exist pseudo-BCK algebras having normal deductive systems which are maximal, but having no Bosbach states.

Some specific properties of states on FL_w-algebras, pseudo-MTL algebras, bounded $R\ell$-monoids and subinterval algebras of pseudo-hoops are proved.

A special section is dedicated to the existence of states on the residuated structures, showing that every perfect FL_w-algebra admits at least a Bosbach state and every perfect pseudo-BL algebra has a unique state-morphism.

Finally, we introduce the notion of a local state on a perfect pseudo-MTL algebra and we prove that every local state can be extended to a Riečan state.

Many results in this and the following chapters are based on the notion of the standard MV-algebra.

The *standard MV-algebra*, denoted $[0, 1]_Ł$, is the interval $[0, 1]$ of reals, equipped with the operations:

$$x \oplus_Ł y := \min\{x + y, 1\}, \qquad x \odot_Ł y := \max\{x + y - 1, 0\},$$

$$x \wedge y := \min\{x, y\}, \qquad x \vee y := \max\{x, y\},$$

$$x \rightarrow_Ł y := x^- \oplus_Ł y = \min\{1 - x + y, 1\} \quad \text{(Łukasiewicz implication)},$$

$$x^- : x \rightarrow_Ł 0 = 1 - x,$$

for all $x, y \in [0, 1]$.

6.1 States on Bounded Pseudo-BCK Algebras

Definition 6.1 A *Bosbach state* on a bounded pseudo-BCK algebra A is a function $s : A \longrightarrow [0, 1]$ such that the following conditions hold for any $x, y \in A$:

(B_1) $s(x) + s(x \rightarrow y) = s(y) + s(y \rightarrow x)$;
(B_2) $s(x) + s(x \rightsquigarrow y) = s(y) + s(y \rightsquigarrow x)$;
(B_3) $s(0) = 0$ and $s(1) = 1$,

where $+$ is the usual addition of real numbers.

Example 6.1 Consider the bounded pseudo-BCK lattice A_1 from Example 1.16. The function $s : A_1 \longrightarrow [0, 1]$ defined by: $s(0) := 0$, $s(a) = s(b) = s(c) = s(d) = s(1) := 1$ is a unique Bosbach state on A_1.

Now we present an example of a bounded linearly ordered pseudo-BCK algebra having a unique Bosbach state. On the other hand, not every linearly ordered pseudo-BCK algebra admits a Bosbach state (see Example 6.6).

Example 6.2 Consider the bounded pseudo-BCK lattice A from Example 4.4. The function $s : A \longrightarrow [0, 1]$ defined by: $s(0) := 0$, $s(a) = s(b) = s(c) = s(1) := 1$ is a unique Bosbach state on A.

Not every bounded pseudo-BCK algebra has Bosbach states.

Example 6.3 Consider the bounded pseudo-BCK lattice A from Example 1.3. One can prove that A has no Bosbach states. Indeed, assume that A admits a Bosbach state s such that $s(0) = 0$, $s(a) = \alpha$, $s(b) = \beta$, $s(c) = \gamma$, $s(1) = 1$. From $s(x) + s(x \to y) = s(y) + s(y \to x)$, taking $x = a$, $y = 0$, $x = b$, $y = 0$, and $x = c$, $y = 0$, respectively, we get $\alpha = 1$, $\beta = 0$, $\gamma = 1$.

On the other hand, taking $x = b$, $y = 0$ in $s(x) + s(x \rightsquigarrow y) = s(y) + s(y \rightsquigarrow x)$ we get $\beta + 0 = 0 + 1$, so $0 = 1$ which is a contradiction. Hence A does not admit a Bosbach state.

Proposition 6.1 *Let A be a bounded pseudo-BCK algebra and s be a Bosbach state on A. For all $x, y \in A$, the following properties hold*:

(1) $s(y \to x) = 1 + s(x) - s(y) = s(y \rightsquigarrow x)$ *and* $s(x) \le s(y)$ *whenever* $x \le y$;
(2) $s(x \vee_1 y) = s(y \vee_1 x)$ *and* $s(x \vee_2 y) = s(y \vee_2 x)$;
(3) $s(x \vee_1 y^{-\sim}) = s(x^{-\sim} \vee_1 y^{-\sim})$ *and* $s(x \vee_2 y^{\sim-}) = s(x^{\sim-} \vee_2 y^{\sim-})$;
(4) $s(x^{-\sim} \vee_1 y) = s(x \vee_1 y^{-\sim})$ *and* $s(x^{\sim-} \vee_2 y) = s(x \vee_2 y^{\sim-})$;
(5) $s(x^{-\sim}) = s(x) = s(x^{\sim-})$;
(6) $s(x^-) = 1 - s(x) = s(x^\sim)$.

Proof

(1) This is straightforward.
(2) By Proposition 1.7 and property (1), we have $s(x \to y) = s(x \vee_1 y \to y) = 1 + s(y) - s(x \vee_1 y)$ and $s(y \to x) = s(y \vee_1 x \to x) = 1 + s(x) - s(y \vee_1 x)$.
    Then $s(x \to y) = s(y) + s(y \to x) - s(x) = s(y) + 1 + s(x) - s(y \vee_1 x) - s(x)$ proving $s(x \vee_1 y) = s(y \vee_1 x)$. Similarly, $s(x \vee_2 y) = s(y \vee_2 x)$.
(3) This follows from Proposition 1.6.
(4) This follows from (2) and (3).
(5) Since $x^{-\sim} = x \vee_1 0$, by (2) we have

$$s\left(x^{-\sim}\right) = s(x \vee_1 0) = s(0 \vee_1 x) = s\big((0 \to x) \rightsquigarrow x\big) = s(x).$$

In a similar way, we have $s(x) = s(x^{\sim-})$.
(6) $s(x^-) = s(x \to 0) = s(0) - s(x) + s(0 \to x) = 1 - s(x)$.
    Similarly, $s(x^\sim) = 1 - s(x)$. $\square$

Proposition 6.2 *Let A be a bounded pseudo-BCK algebra and let $s : A \longrightarrow [0, 1]$ be a function such that $s(0) = 0$, $s(x \vee_1 y) = s(y \vee_1 x)$ and $s(x \vee_2 y) = s(y \vee_2 x)$ for all $x, y \in A$. Then the following are equivalent*:

(a) *s is a Bosbach state on A*;
(b) *for all $x, y \in A$, $y \le x$ implies $s(x \to y) = s(x \rightsquigarrow y) = 1 - s(x) + s(y)$*;
(c) *for all $x, y \in A$, $s(x \to y) = 1 - s(x \vee_1 y) + s(y)$ and $s(x \rightsquigarrow y) = 1 - s(x \vee_2 y) + s(y)$*.

Proof

(a) $\Rightarrow$ (b) This follows from Proposition 6.1(1).
(b) $\Rightarrow$ (c) This follows from the inequalities $x, y \leq x \vee_1 y$ and $x, y \leq x \vee_2 y$.
(c) $\Rightarrow$ (a) Using (c), we get:

$$s(x) + s(x \to y) = s(x) + 1 - s(x \vee_1 y) + s(y)$$
$$= 1 - s(y \vee_1 x) + s(x) + s(y) = s(y) + s(y \to x).$$

Similarly,

$$s(x) + s(x \rightsquigarrow y) = s(x) + 1 - s(x \vee_2 y) + s(y)$$
$$= 1 - s(y \vee_2 x) + s(x) + s(y) = s(y) + s(y \rightsquigarrow x).$$

Moreover, by (c) we have:

$$s(1) = s(x \to x) = 1 - s(x) + s(x) = 1.$$

Thus s is a Bosbach state on A.   □

Proposition 6.3 *Let A be a bounded pseudo-BCK algebra and s be a Bosbach state on A. For all $x, y \in A$, the following properties hold:*

(1) $s(x^{-\sim} \to y) = s(x \to y^{-\sim})$ *and* $s(x^{\sim-} \rightsquigarrow y) = s(x \rightsquigarrow y^{\sim-})$;
(2) $s(x \to y^{-\sim}) = s(y^- \rightsquigarrow x^-) = s(x^{-\sim} \to y^{-\sim}) = s(x^{-\sim} \to y)$ *and* $s(x \rightsquigarrow y^{\sim-}) = s(y^\sim \to x^\sim) = s(x^{\sim-} \rightsquigarrow y^{\sim-}) = s(x^{\sim-} \rightsquigarrow y)$;
(3) $s(x^\sim \to y^{-\sim}) = s(x^\sim \to y)$ *and* $s(x^- \rightsquigarrow y^{\sim-}) = s(x^- \rightsquigarrow y)$.

Proof

(1) Using Propositions 6.2(c) and 6.1(4), we have:

$$s\left(x^{-\sim} \to y\right) = 1 - s\left(x^{-\sim} \vee_1 y\right) + s(y)$$
$$= 1 - s\left(x \vee_1 y^{-\sim}\right) + s\left(y^{-\sim}\right) = s\left(x \to y^{-\sim}\right).$$

Similarly, $s(x^{\sim-} \rightsquigarrow y) = s(x \rightsquigarrow y^{\sim-})$.
(2) This follows by (*psbck-c*$_{19}$) and (1).
(3) Applying Proposition 6.1(4) we get:

$$s\left(x^\sim \to y^{-\sim}\right) = 1 - s\left(x^\sim \vee_1 y^{-\sim}\right) + s\left(y^{-\sim}\right) = 1 - s\left(x^{\sim-\sim} \vee_1 y\right) + s(y)$$
$$= 1 - s\left(x^\sim \vee_1 y\right) + s(y) = s\left(x^\sim \to y\right).$$

Similarly, $s(x^- \rightsquigarrow y^{\sim-}) = s(x^- \rightsquigarrow y)$.   □

Proposition 6.4 *Let s be a Bosbach state on a bounded pseudo-BCK algebra A. Then for all $x, y \in A$, we have:*

(1) $s(x \vee_1 y) = s(x \vee_2 y)$;
(2) $s(x \rightarrow y) = s(x \rightsquigarrow y)$.

Proof

(1) First we prove the equality for $y \leq x$. Applying Propositions 6.1(2) and 1.5(4), we have $s(x \vee_1 y) = s(y \vee_1 x) = s(x)$ and $s(x \vee_2 y) = s(y \vee_2 x) = s(x)$, that is, $s(x \vee_1 y) = s(x \vee_2 y)$.

Assume now that x and y are arbitrary elements of A. Using Proposition 6.1(2) again and the first part of the proof, we have

$$s(x \vee_1 y) = s\big(x \vee_1 (x \vee_1 y)\big) = s\big((x \vee_1 y) \vee_1 x\big)$$
$$= s\big((x \vee_1 y) \vee_2 x\big) \geq s(y \vee_2 x)$$
$$= s(x \vee_2 y) = s\big(y \vee_2 (x \vee_2 y)\big)$$
$$= s\big((x \vee_2 y) \vee_1 y\big) \geq s(x \vee_1 y).$$

(2) This follows immediately from Proposition 6.2(c) and the first equation. ☐

Proposition 6.5 *Let s be a Bosbach state on a bounded pseudo-BCK algebra A. Then for all $x, y \in A$, we have:*

(1) $s(x^- \rightarrow y^-) = s(y^{-\sim} \rightarrow x^{-\sim})$;
(2) $s(x^\sim \rightarrow y^\sim) = s(y^{\sim-} \rightarrow x^{\sim-})$.

Proof

(1) By (*psbck-c15*) we have $x^- \rightsquigarrow y^- \leq y^{-\sim} \rightarrow x^{-\sim}$, so by Proposition 6.4(2) and Proposition 6.1(1) it follows that:

$$s\big(x^- \rightarrow y^-\big) = s\big(x^- \rightsquigarrow y^-\big) \leq s\big(y^{-\sim} \rightarrow x^{-\sim}\big) \leq s\big(x^{-\sim-} \rightsquigarrow y^{-\sim-}\big)$$
$$= s\big(x^- \rightsquigarrow y^-\big) = s\big(x^- \rightarrow y^-\big).$$

Thus $s(x^- \rightarrow y^-) = s(y^{-\sim} \rightarrow x^{-\sim})$.
(2) Similar to (1). ☐

Proposition 6.6 *Let s be a Bosbach state on a bounded pseudo-BCK algebra A. Then for all $x, y \in A$, we have:*

(1) $s(x \rightarrow y) = s(x^{-\sim} \rightarrow y^{-\sim})$;
(2) $s(x \rightsquigarrow y) = s(x^{\sim-} \rightsquigarrow y^{\sim-})$. ☐

Proof

(1) By Proposition 6.3(2) we have $s(x \rightarrow y^{-\sim}) = s(x^{-\sim} \rightarrow y^{-\sim})$.
From $y \leq y^{-\sim}$ and (*psbck-c10*) we get $x \rightarrow y \leq x \rightarrow y^{-\sim}$, hence

$$s(x \rightarrow y) \leq s\big(x \rightarrow y^{-\sim}\big) = s\big(x^{-\sim} \rightarrow y^{-\sim}\big).$$

On the other hand, by $x \leq x^{-\sim}$ and $(psbck\text{-}c_1)$ we have $x^{-\sim} \rightarrow y \leq x \rightarrow y$, so using Proposition 6.3(2) again we get

$$s\left(x^{-\sim} \rightarrow y^{-\sim}\right) = s\left(x^{-\sim} \rightarrow y\right) \leq s(x \rightarrow y).$$

We conclude that $s(x \rightarrow y) = s(x^{-\sim} \rightarrow y^{-\sim})$.

(2) Similar to (1). $\qquad \square$

Definition 6.2 Let A be a bounded pseudo-BCK algebra. A *state-morphism* on A is a function $m : A \longrightarrow [0, 1]$ such that:

(SM_1) $m(0) = 0$;
(SM_2) $m(x \rightarrow y) = m(x \rightsquigarrow y) = m(x) \rightarrow_{\mathrm{L}} m(y)$ for all $x, y \in A$.

Proposition 6.7 *Every state-morphism on a bounded pseudo-BCK algebra A is a Bosbach state on A.*

Proof It is obvious that $m(1) = m(x \rightarrow x) = m(x) \rightarrow_{\mathrm{L}} m(x) = 1$.
We also have:

$$
\begin{aligned}
m(x) + m(x \rightarrow y) &= m(x) + m(x) \rightarrow_{\mathrm{L}} m(y) = m(x) + \min\{1 - m(x) + m(y), 1\} \\
&= \min\{1 + m(y), 1 + m(x)\} \\
&= m(y) + \min\{1 - m(y) + m(x), 1\} \\
&= m(y) + m(y) \rightarrow_{\mathrm{L}} m(x) = m(y) + m(y \rightarrow x).
\end{aligned}
$$

Similarly, $m(x) + m(x \rightsquigarrow y) = m(y) + m(y \rightsquigarrow x)$.
Thus s is a Bosbach state on A. $\qquad \square$

Proposition 6.8 *Let A be a bounded pseudo-BCK algebra. A Bosbach state m on A is a state-morphism if and only if:*

$$m(x \vee_1 y) = \max\{m(x), m(y)\}$$

for all $x, y \in A$, or equivalently,

$$m(x \vee_2 y) = \max\{m(x), m(y)\}$$

for all $x, y \in A$.

Proof In view of Proposition 6.4, the two equations are equivalent. If m is a state-morphism on A, then by Proposition 6.7, m is a Bosbach state.
Using the relation $m(x \rightarrow y) = 1 - m(x \vee_1 y) + m(y)$, we obtain:

$$
\begin{aligned}
m(x \vee_1 y) &= 1 + m(y) - m(x \rightarrow y) = 1 + m(y) - \left(m(x) \rightarrow_{\mathrm{L}} m(y)\right) \\
&= 1 + m(y) - \min\{1 - m(x) + m(y), 1\} \\
&= 1 + m(y) + \max\{-1 + m(x) - m(y), -1\} = \max\{m(x), m(y)\}.
\end{aligned}
$$

For the converse, assume that m is a Bosbach state on A such that

$$m(x \vee_1 y) = \max\{m(x), m(y)\} \quad \text{for all } x, y \in A.$$

Then, again using the relation $m(x \to y) = 1 - m(x \vee_1 y) + m(y)$, we have:

$$\begin{aligned} m(x \to y) &= 1 + m(y) - \max\{m(x), m(y)\} \\ &= 1 + m(y) + \min\{-m(x), -m(y)\} \\ &= \min\{1 - m(x) + m(y), 1\} = m(x) \to_L m(y). \end{aligned}$$

Similarly, $m(x \rightsquigarrow y) = m(x) \to_L m(y)$.

Thus m is a state-morphism on A.   □

Example 6.4 Consider the pseudo-BCK algebra $A = \{0, a, b, c, 1\}$ from Example 1.11. The function $m : A \longrightarrow [0, 1]$ defined by:

$$m(0) := 0, \qquad m(a) = m(b) = m(c) = m(1) := 1$$

is a unique Bosbach state on A.

In addition, $m(x \vee_1 y) = m(x \vee_2 y) = \max\{m(x), m(y)\}$ for all $x, y \in A$, hence m is also a state-morphism on A.

The set $\text{Ker}(s) := \{a \in A \mid s(a) = 1\}$ is called the *kernel* of the Bosbach state s on A.

Proposition 6.9 *Let A be a bounded pseudo-BCK algebra and let s be a Bosbach state on A. Then $\text{Ker}(s)$ is a proper and normal deductive system of A.*

Proof Obviously, $1 \in \text{Ker}(s)$ and $0 \notin \text{Ker}(s)$.

Assume that $a, a \to b \in \text{Ker}(s)$. We have $1 = s(a) \le s(a \vee_1 b)$, so $s(a \vee_1 b) = 1$.

It follows that $1 = s(a \to b) = 1 - s(a \vee_1 b) + s(b) = s(b)$.

Hence $b \in \text{Ker}(s)$, so $\text{Ker}(s)$ is a proper deductive system of A.

By Proposition 6.4(2), $s(a \rightsquigarrow b) = s(a \to b)$, and this proves that $\text{Ker}(s)$ is normal.   □

Lemma 6.1 *Let s be a Bosbach state on a bounded pseudo-BCK algebra A and $K = \text{Ker}(s)$. In the bounded quotient pseudo-BCK algebra $(A/K, \le, \to, \rightsquigarrow, 0/K, 1/K)$ we have:*

(1) *$a/K \le b/K$ iff $s(a \to b) = 1$ iff $s(a \vee_1 b) = s(b)$ iff $s(a \vee_2 b) = s(b)$;*

(2) *$a/K = b/K$ iff $s(a \to b) = s(b \to a) = 1$ iff $s(a) = s(b) = s(a \vee_1 b)$ iff $s(a \rightsquigarrow b) = s(b \rightsquigarrow a) = 1$ iff $s(a) = s(b) = s(a \vee_2 b)$.*

Moreover, the mapping $\hat{s} : A/K \to [0, 1]$ defined by $\hat{s}(a/K) := s(a)$ $(a \in A)$ is a Bosbach state on A/K.

Proof

(1) It follows easily that: $a/K \leq b/K$ iff $(a \to b)/K = a/K \to b/K = 1/K = K$
    iff $a \to b \in K$ iff $s(a \to b) = 1$.
    As $s(a \to b) = 1 - s(a \vee_1 b) + s(b)$, we get $a/K \leq b/K$ iff $s(a \vee_1 b) = s(b)$.
    Similarly, $a/K \leq b/K$ iff $(a \rightsquigarrow b)/K = a/K \rightsquigarrow b/K = 1/K = K$ iff $a \rightsquigarrow$
    $b \in K$ iff $s(a \rightsquigarrow b) = 1$.
    As $s(a \rightsquigarrow b) = 1 - s(a \vee_2 b) + s(b)$, we get $a/K \leq b/K$ iff $s(a \vee_2 b) = s(b)$.
(2) This follows easily from (1).

The fact that $\hat{s}$ is a well-defined Bosbach state on A/K is now straightforward.
□

Proposition 6.10 *Let s be a Bosbach state on a bounded pseudo-BCK algebra A
and let $K = \mathrm{Ker}(s)$. For every element $x \in A$, we have*

$$x^{-\sim}/K = x/K = x^{\sim-}/K,$$

that is, A/K satisfies the (pDN) *condition.*

Proof We have $x \leq x^{-\sim}$. By the definition of a Bosbach state and Proposition 6.1(5), we have

$$s\left(x^{-\sim} \to x\right) = s(x) + s\left(x \to x^{-\sim}\right) - s\left(x^{-\sim}\right) = s\left(x \to x^{-\sim}\right) = s(1) = 1.$$

Hence $x^{-\sim}/K = x/K$. In a similar way, we prove the second equality.   □

Remark 6.1 Let s be a Bosbach state on a bounded pseudo-BCK algebra A. According to the proof of Proposition 6.10, we have

$$s\left(x^{-\sim} \to x\right) = 1 = s\left(x^{\sim-} \to x\right) \quad \text{and} \quad s\left(x^{-\sim} \rightsquigarrow x\right) = 1 = s\left(x^{\sim-} \rightsquigarrow x\right).$$

Proposition 6.11 *Let s be a Bosbach state on a bounded pseudo-BCK algebra A.
Then A/K is $\vee_1$-commutative as well as $\vee_2$-commutative, where $K = \mathrm{Ker}(s)$. In
addition, A/K is a $\vee$-semilattice and good.*

Proof Since s is a Bosbach state, A/K is a pseudo-BCK algebra.
    We denote x/K by $\bar{x}$, $x \in A$. Then $\hat{s}$, defined by $\hat{s}(a) := s(a)$ $(a \in A)$, is a
Bosbach state on A/K.
(1) We show that if $\bar{x} \leq \bar{y}$, then $\bar{x} \vee_1 \bar{y} = \bar{y} = \bar{y} \vee_1 \bar{x}$.
    By Proposition 1.5(4), we have $\bar{x} \vee_1 \bar{y} = \bar{y}$.
    We have to show that $s((y \vee_1 x) \to y) = 1$.
    By Proposition 6.1(1), we have

$$s(y \vee_1 x) = s\left((y \to x) \rightsquigarrow x\right) = \hat{s}\left((\bar{y} \to \bar{x}) \rightsquigarrow \bar{x}\right) = 1 + \hat{s}(\bar{x}) - \hat{s}(\bar{y} \to \bar{x})$$
$$= 1 + \hat{s}(\bar{x}) - \left[1 + \hat{s}(\bar{x}) - \hat{s}(\bar{y})\right] = \hat{s}(\bar{y}) = s(y).$$

Therefore, using Proposition 6.1(2),

$$s\big((y \vee_1 x) \to y\big) = \hat{s}\big((\bar{y} \vee_1 \bar{x}) \to \bar{y}\big) = 1 + \hat{s}(\bar{y}) - \hat{s}(\bar{y} \vee_1 \bar{x})$$
$$= 1 + \hat{s}(\bar{y}) - \hat{s}(\bar{x} \vee_1 \bar{y}) = 1 + \hat{s}(\bar{y}) - \hat{s}(\bar{y}) = 1.$$

Hence $\bar{x} \vee_1 \bar{y} = \bar{y} = \bar{y} \vee_1 \bar{x}$ holds for $\bar{x} \leq \bar{y}$.

(2) Now we show that $\bar{x} \vee_1 \bar{y} = \bar{y} \vee_1 \bar{x}$ holds for all $x, y \in A$.

By (1), we have

$$\bar{x} \vee_1 \bar{y} = \bar{x} \vee_1 (\bar{x} \vee_1 \bar{y}) = (\bar{x} \vee_1 \bar{y}) \vee_1 \bar{x} \geq \bar{y} \vee_1 \bar{x} = \bar{y} \vee_1 (\bar{y} \vee_1 \bar{x})$$
$$= (\bar{y} \vee_1 \bar{x}) \vee_1 \bar{y} \geq \bar{x} \vee_1 \bar{y}.$$

Hence $\bar{x} \vee_1 \bar{y} = \bar{y} \vee_1 \bar{x}$ for all $x, y \in A$. This implies that A/K is $\vee_1$-commutative. In a similar way we prove that A/K is $\vee_2$-commutative, that is, A/K is sup-commutative. By Theorem 1.2, A/K is a $\vee$-semilattice. Moreover, according to Proposition 6.10, A/K is good. $\square$

An *MV-state* on an MV-algebra A is a mapping $s : A \to [0, 1]$ such that $s(1) = 1$ and $s(a \oplus b) = s(a) + s(b)$ whenever $a \odot b = 0$. Every MV-algebra admits at least one MV-state, and due to [131], every MV-state on A coincides with a Bosbach state on the BCK algebra A and vice versa.

We note that the *radical*, Rad(A), of an MV-algebra A is the intersection of all maximal ideals of A ([41]).

Theorem 6.1 *Let s be a Bosbach state on a bounded pseudo-BCK algebra A and let $K = \mathrm{Ker}(s)$. Then $(A/K, \oplus, ^-, 0/K)$, where*

$$a/K \oplus b/K := \big(a^\sim \to b\big)/K \quad and \quad (a/K)^- := a^-/K,$$

is an Archimedean MV-algebra and the map $\hat{s}(a/K) := s(a)$ is an MV-state on this MV-algebra.

Proof By Propositions 6.10 and 6.11, A/K is a good pseudo-BCK algebra that is a $\vee$-semilattice and $\hat{s}$ on A/K is a Bosbach state such that $\mathrm{Ker}(\hat{s}) = \{1/K\}$. By Proposition 3.4.7 in [208], $(A/K)/\mathrm{Ker}(\hat{s})$ is term-equivalent to an MV-algebra, that is, it is Archimedean and $\hat{s}$ is an MV-state on it.

Since $A/K = (A/K)/\mathrm{Ker}(\hat{s})$, the same is also true for A/K, and this proves the theorem. $\square$

We recall that if a bounded pseudo-BCK algebra A is good, in view of Proposition 1.24, we can define a binary operation $\oplus$ via $x \oplus y = x^\sim \to y^{\sim-} = y^- \rightsquigarrow x^{\sim-}$ that corresponds to an "MV-addition". And for any pseudo-MV algebra A we know, [96], that an MV-state is a state-morphism iff $m(a \oplus b) = m(a) \oplus_\mathbb{Ł} m(b)$ for all $a, b \in A$. Inspired by this we can characterize state-morphisms as follows.

Lemma 6.2 *Let m be a Bosbach state on a bounded pseudo-BCK algebra A. The following statements are equivalent:*

(a) *m is a state-morphism;*
(b) *$m(a^\sim \to b^{-\sim}) = \min\{m(a) + m(b), 1\}$ for all $a, b \in A$;*
(c) *$m(b^- \rightsquigarrow a^{\sim-}) = \min\{m(a) + m(b), 1\}$ for all $a, b \in A$.*

Proof

(a) $\Rightarrow$ (b)   Assume that m is a state-morphism on A. Since m is a Bosbach state, by Proposition 6.3(3) we have

$$m(a^\sim \to b^{-\sim}) = m(a^\sim \to b) = m(a^\sim) \to_{\text{Ł}} m(b)$$
$$= m(a)^- \to_{\text{Ł}} m(b) = \min\{m(a) + m(b), 1\}.$$

(b) $\Rightarrow$ (a)   Applying Proposition 6.6 and (b) we have

$$m(a \to b) = m(a^{-\sim} \to b^{-\sim}) = \min\{m(a^-) + m(b), 1\}$$
$$= \min\{1 - m(a) + m(b), 1\} = m(a) \to_{\text{Ł}} m(b).$$

Similarly, $m(a \rightsquigarrow b) = m(a) \to_{\text{Ł}} m(b)$.
Moreover, taking $a = b = 0$ in (b) we get $m(0) = 0$.
Hence m is a state-morphism.

In the same way we can prove that (a) $\Leftrightarrow$ (c). $\square$

Proposition 6.12 *Let s be a Bosbach state on a bounded pseudo-BCK algebra A. The following are equivalent:*

(a) *s is a state-morphism;*
(b) *$\mathrm{Ker}(s)$ is a normal and maximal deductive system of A.*

Proof

(a) $\Rightarrow$ (b)   We will follow the idea used in Proposition 3.4.10 in [208].
Let s be a state-morphism. We show that $[\mathrm{Ker}(s) \cup \{a\}) = A$ for any $a \notin \mathrm{Ker}(s)$. Take $a \notin \mathrm{Ker}(s)$ and an arbitrary $x \in A$. Since $[0, 1]$ is a simple MV-algebra and $s(a) \neq 1$, it follows that $[s(a)) = [0, 1]$, so $s(a) \to^n s(x) = 1$ for some $n \in \mathbb{N}$. But we have $s(a) \to^n s(x) = s(a \to^n x)$ and hence $a \to^n x \in \mathrm{Ker}(s)$.
Now, $1 = (a \to^n x) \rightsquigarrow (a \to^n x) = a \to^n ((a \to^n x) \rightsquigarrow x)$, where $a, a \to^n x \in \mathrm{Ker}(s) \cup \{a\}$, which means that $x \in [\mathrm{Ker}(s) \cup \{a\})$. Thus $[\mathrm{Ker}(s) \cup \{a\}) = A$, proving that $\mathrm{Ker}(s)$ is a maximal deductive system of A.
(b) $\Rightarrow$ (a)   Let $K = \mathrm{Ker}(s)$. According to Theorem 6.1, $A/K = (A/K)/\mathrm{Ker}(\hat{s})$ is a BCK algebra that is term equivalent to an MV-algebra. Assume F is a deductive system of A/K and let $K(F) = \{a \in A \mid a/K \in F\}$. Then $K(F)$ is a deductive system of A containing K. The maximality of K implies $K = K(F)$ and $F = \{1/K\}$. By Theorem 6.1, A/K can be assumed to be an Archimedean MV-algebra having only one maximal deductive system, $\{1/F\}$. Therefore A/K is

an MV-subalgebra of the MV-algebra of the real interval $[0, 1]$. This yields that the mapping $a \mapsto a/K$ $(a \in A)$ is the Bosbach state s that is a state-morphism.

$\square$

Lemma 6.3 *Let m be a state-morphism on a bounded pseudo-BCK algebra A and $K = \mathrm{Ker}(m)$. Then:*

$$a/K \leq b/K \quad \text{if and only if} \quad m(a) \leq m(b);$$

$$a/K = b/K \quad \text{if and only if} \quad m(a) = m(b).$$

Proof By Proposition 6.7 it follows that m is a Bosbach state on A. Applying Lemma 6.1, $a/K \leq b/K$ iff $m(b) = m(a \vee_1 b)$.

But $m(a \vee_1 b) = \max\{m(a), m(b)\}$ and hence $m(a) \leq m(b)$.

The second assertion follows from the first one. $\square$

Proposition 6.13 *Let m be a state-morphism on a bounded pseudo-BCK algebra A. Then $(m(A), \oplus, ^-, 0)$ is a subalgebra of the standard MV-algebra $([0, 1], \oplus, ^-, 0)$ and the mapping $a/\mathrm{Ker}(m) \mapsto m(a)$ is an isomorphism of $A/\mathrm{Ker}(m)$ onto $m(A)$.*

Proof Since m is a state-morphism, it is a homomorphism of $(A, \rightarrow, \rightsquigarrow, 0, 1)$ onto $(m(A), \rightarrow, \rightarrow, 0, 1)$. Hence $A/\mathrm{Ker}(m) \cong m(A)$. $\square$

Proposition 6.14 *Let A be a bounded pseudo-BCK algebra and m_1, m_2 be two state-morphisms such that $\mathrm{Ker}(m_1) = \mathrm{Ker}(m_2)$. Then $m_1 = m_2$.*

Proof The assertion holds in the case of pseudo-MV algebras (Proposition 4.5 in [96]).

By Proposition 6.7, m_1 and m_2 are two Bosbach states. The conditions yield $A/\mathrm{Ker}(m_1) = A/\mathrm{Ker}(m_2)$, and as in the proof of Proposition 6.12, we have that $A/\mathrm{Ker}(m_1)$ is in fact an MV-subalgebra of the MV-algebra of the real interval $[0, 1]$. But $\mathrm{Ker}(\hat{m}_1) = \{1/K\} = \mathrm{Ker}(\hat{m}_2)$. Hence, by Proposition 4.5 in [96], $\hat{m}_1 = \hat{m}_2$, consequently, $m_1 = m_2$. $\square$

Definition 6.3 Let A be a bounded pseudo-BCK algebra. We say that a Bosbach state s is *extremal* if for any $0 < \lambda < 1$ and for any two Bosbach states s_1, s_2 on A, $s = \lambda s_1 + (1 - \lambda) s_2$ implies $s_1 = s_2$.

Summarizing the previous characterizations of state-morphisms, we have the following result.

Theorem 6.2 *Let s be a Bosbach state on a bounded pseudo-BCK algebra A. Then the following are equivalent:*

(a) *s is an extremal Bosbach state;*
(b) *$s(x \vee_1 y) = \max\{s(x), s(y)\}$ for all $x, y \in A$;*

(c) $s(x \vee_2 y) = \max\{s(x), s(y)\}$ *for all* $x, y \in A$;
(d) s *is a state-morphism*;
(e) $\mathrm{Ker}(s)$ *is a maximal deductive system*.

Proof The equivalence of (b)–(e) was proved in Propositions 6.8 and 6.12.

(d) $\Rightarrow$ (a) Let $s = \lambda s_1 + (1 - \lambda)s_2$, where s_1, s_2 are Bosbach states and $0 < \lambda < 1$. Then $\mathrm{Ker}(s) = \mathrm{Ker}(s_1) \cap \mathrm{Ker}(s_2)$ and the maximality of $\mathrm{Ker}(s)$ gives that $\mathrm{Ker}(s_1)$ and $\mathrm{Ker}(s_2)$ are maximal and normal deductive systems. (e) yields that s_1 and s_2 are state-morphisms and Proposition 6.14 entails $s_1 = s_2 = s$.

(a) $\Rightarrow$ (d) Let s be an extremal state on A. Define $\hat{s}$ by Lemma 6.1 on $A/\mathrm{Ker}(s)$. We assert that $\hat{s}$ is an extremal MV-state on the MV-algebra $A/\mathrm{Ker}(s)$. Indeed, let $\hat{m} = \lambda\mu_1 + (1 - \lambda)\mu_2$, where $0 < \lambda < 1$ and μ_1 and μ_2 are states on $A/\mathrm{Ker}(s)$. There exist two Bosbach states s_1 and s_2 on A such that $s_i(a) := \mu_i(a/\mathrm{Ker}(s))$, $a \in A$ for $i = 1, 2$. Then $s = \lambda s_1 + (1 - \lambda)s_2$ which gives $s_1 = s_2 = s$, thus $\mu_1 = \mu_2 = \hat{s}$. Since $A/\mathrm{Ker}(s)$ is in fact an MV-algebra, we conclude from Theorem 6.1.30 in [108] that $\hat{s}$ is a state-morphism on $A/\mathrm{Ker}(s)$. Consequently, so is s on A. (The equivalence (a) $\Leftrightarrow$ (d) was proved in Theorem 6.1.30 in [108] for the case of bounded commutative BCK-algebras, hence it also holds for MV-algebras.) $\square$

We present the following characterization of the existence of Bosbach states on a bounded pseudo-BCK algebra.

Theorem 6.3 *Let A be a bounded pseudo-BCK algebra. The following statements are equivalent*:

(a) *A admits a Bosbach state*;
(b) *there exists a normal deductive system $F \neq A$ of A such that A/F is termwise equivalent to an MV-algebra*;
(c) *there exists a normal and maximal deductive system F such that A/F is termwise equivalent to an MV-algebra.*

Proof

(a) $\Rightarrow$ (b) If m is a Bosbach state, then, according to Theorem 6.1, the normal deductive system $F = \mathrm{Ker}(m)$ satisfies (b).

(b) $\Rightarrow$ (a) If A/F is an MV-algebra, then it possesses at least one MV-state, say μ. The function $m(a) := \mu(a/F)$ $(a \in A)$ is a Bosbach state on A.

(a) $\Rightarrow$ (c) If A possesses at least one state, by the Krein-Mil'man Theorem, $\partial_e \mathcal{BS}(A) = \mathcal{SM}(A) \neq \emptyset$. Then there is a state-morphism m on A and by Theorem 6.2(e), the deductive system $F = \mathrm{Ker}(m)$ is maximal and normal, so by Theorem 6.1, F satisfies (c).

(c) $\Rightarrow$ (a) The proof is the same as that of (b) $\Rightarrow$ (a). $\square$

Remark 6.2 In the case of pseudo-BL algebras and bounded Rℓ-monoids the existence of a state-morphism is equivalent to the existence of a maximal filter which is normal (see [131] and [110], respectively). This result is based on the fact that, if A is one of the above mentioned structures and H is a maximal and normal filter of A, then A/H is an MV-algebra.

In the case of pseudo-BCK algebras this result is not true, as we can see in the next examples.

Example 6.5 Consider the bounded pseudo-BCK algebra $A = \{0, a, b, 1\}$ with $\rightarrow = \rightsquigarrow$ given by the following table:

$\rightarrow$	0	a	b	1
0	1	1	1	1
a	b	1	1	1
b	b	b	1	1
1	0	a	b	1

Then $F = \{1\}$ is a unique proper deductive system of A. Moreover, F is evidently normal. But $A \cong A/\{1\}$ is not an MV-algebra and in view of Theorem 6.3, A has no Bosbach states.

Example 6.6 Consider $A = \{0, a, b, c, 1\}$ with $0 < a < b < c < 1$ and the operations $\rightarrow$, $\rightsquigarrow$ given by the following tables:

$\rightarrow$	0	a	b	c	1		$\rightsquigarrow$	0	a	b	c	1
0	1	1	1	1	1		0	1	1	1	1	1
a	b	1	1	1	1		a	b	1	1	1	1
b	b	c	1	1	1		b	b	b	1	1	1
c	0	a	b	1	1		c	0	b	b	1	1
1	0	a	b	c	1		1	0	a	b	c	1

Then $(A, \leq, \rightarrow, \rightsquigarrow, 0, 1)$ is a good pseudo-BCK lattice. But there is no Bosbach state on A. Indeed, assume that A admits a Bosbach state s such that $s(0) = 0$, $s(a) = \alpha$, $s(b) = \beta$, $s(c) = \gamma$, $s(1) = 1$. From $s(x) + s(x \rightarrow y) = s(y) + s(y \rightarrow x)$, taking $x = a$, $y = 0$, $x = b$, $y = 0$ and $x = c$, $y = 0$, respectively we get $\alpha = 1/2$, $\beta = 1/2$, $\gamma = 1$. On the other hand, we have:

$$s(a) + s(a \rightsquigarrow b) = s(a) + s(1) = 1/2 + 1 = 3/2,$$

$$s(b) + s(b \rightsquigarrow a) = s(b) + s(b) = 1/2 + 1/2 = 1,$$

so condition (B_2) does not hold.

Thus there is no Bosbach state, in particular, no state-morphism on A.

Remark 6.3 The previous Examples 6.5 and 6.6 of bounded linearly ordered stateless pseudo-BCK algebras exhibit another difference between bounded pseudo-BCK algebras and bounded pseudo-BL algebras since *every linearly ordered pseudo-BL algebra admits a Bosbach state* (see [98, 104]).

Remark 6.4 We say that a net of Bosbach states $\{s_\alpha\}$ *converges weakly* to a Bosbach state s if $s(a) = \lim_\alpha s_\alpha(a)$ for every $a \in A$. According to the definition of Bosbach states, the set of Bosbach states is a (possibly empty) compact Hausdorff topological space in the weak topology.

Extremal Bosbach states are very important because they generate all Bosbach states: by the Krein–Mil'man Theorem, Theorem 5.17 in [155], every Bosbach state is a weak limit of a net of convex combinations of extremal Bosbach states.

Let $\mathcal{BS}(A)$, $\partial_e \mathcal{BS}(A)$ and $\mathcal{SM}(A)$ denote the set of all Bosbach states, all extremal Bosbach states, and all state-morphisms on $(A, \leq, \rightarrow, \rightsquigarrow, 0, 1)$, respectively. Theorem 6.2 says

$$\partial_e \mathcal{BS}(A) = \mathcal{SM}(A)$$

and they are compact subsets of $\mathcal{BS}(A)$ in the weak topology.

Let A be a bounded pseudo-BCK algebra and $x, y \in A$. We recall that x is orthogonal to y, denoted $x \perp y$, iff $x^{-\sim} \leq y^\sim$ (see Definition 1.11). If $x \perp y$, we define a partial operation $+$ on A by $x + y := x \oplus y = y^\sim \rightarrow x^{\sim-}$.

Definition 6.4 Let A be a good pseudo-BCK algebra. A *Riečan state* on A is a function $s : A \longrightarrow [0, 1]$ such that the following conditions hold for all $x, y \in A$:

(R_1) if $x \perp y$, then $s(x + y) = s(x) + s(y)$;
(R_2) $s(1) = 1$.

The notion of a Riečan state extends the additive measures introduced by Riečan for BL-algebras ([245]).

Proposition 6.15 *If s is a Riečan state on a good pseudo-BCK algebra A, then the following properties hold for all $x, y \in A$:*

(1) $s(x^-) = s(x^\sim) = 1 - s(x)$;
(2) $s(0) = 0$;
(3) $s(x^{-\sim}) = s(x^{\sim-}) = s(x^{--}) = s(x^{\sim\sim}) = s(x)$;
(4) *if $x \leq y$, then $s(x) \leq s(y)$ and $s(y \rightarrow x^{-\sim}) = s(y \rightsquigarrow x^{\sim-}) = 1 + s(x) - s(y)$;*
(5) $s((x \vee_1 y) \rightarrow x^{-\sim}) = s((x \vee_1 y) \rightsquigarrow x^{-\sim}) = 1 - s(x \vee_1 y) + s(x)$ *and* $s((x \vee_2 y) \rightarrow x^{-\sim}) = s((x \vee_2 y) \rightsquigarrow x^{-\sim}) = 1 - s(x \vee_2 y) + s(x)$;
(6) $s((x \vee_1 y) \rightarrow y^{-\sim}) = s((x \vee_1 y) \rightsquigarrow y^{-\sim}) = 1 - s(x \vee_1 y) + s(y)$ *and* $s((x \vee_2 y) \rightarrow y^{-\sim}) = s((x \vee_2 y) \rightsquigarrow y^{-\sim}) = 1 - s(x \vee_2 y) + s(y)$.

Proof

(1) Since $x \perp x^-$ and $x + x^- = 1$, we have: $s(x) + s(x^-) = s(1) = 1$, so $s(x^-) = 1 - s(x)$. Similarly, $s(x^\sim) = 1 - s(x)$.
(2) This follows from the fact that $0 \perp 0$ and $0 + 0 = 0$.
(3) Applying the fact that $x \perp 0$ and $x + 0 = x^{\sim-}$, we get $s(x) = s(x^{\sim-})$ and similarly the other equalities.
(4) Since $x \leq y$, it follows that $x \perp y^-$ and $x + y^- = y \rightarrow x^{-\sim}$.

Hence $s(x) + s(y^-) = s(y \to x^{-\sim})$, so $s(x) - s(y) = s(y \to x^{-\sim}) - 1 \le 0$, that is, $s(x) \le s(y)$. We get $s(y \to x^{-\sim}) = 1 + s(x) - s(y)$.

Similarly, from $x \le y$ we have $y^\sim \perp x$ and $y^\sim + x = y \rightsquigarrow x^{\sim-}$ and we get

$$s\left(y \rightsquigarrow x^{\sim-}\right) = 1 + s(x) - s(y).$$

(5) This follows from (4), since $x \le x \vee_1 y$ and $x \le x \vee_2 y$.

(6) This follows from (4), since $y \le x \vee_1 y$ and $y \le x \vee_2 y$.   □

Theorem 6.4 *Any Bosbach state on a good pseudo-BCK algebra is a Riečan state.*

Proof Let A be a good pseudo-BCK algebra and s be a Bosbach state on A. Assume $x \perp y$, that is, $x^{-\sim} \le y^\sim$. We have: $s(y^\sim) + s(y^\sim \to x^{-\sim}) = s(x^{-\sim}) + s(x^{-\sim} \to y^\sim)$. Since $s(x^{-\sim} \to y^\sim) = 1$, we get: $1 - s(y) + s(y^\sim \to x^{-\sim}) = s(x) + 1$, so $s(y^\sim \to x^{-\sim}) = s(x) + s(y)$.

Thus $s(x \oplus y) = s(x) + s(y)$.

Since by hypothesis $s(1) = 1$, it follows that s is a Riečan state on A.   □

Remark 6.5 The converse is not true in general, as we can see in the next example, which shows that there exists a Riečan state on a good pseudo-BCK algebra A that is not a Bosbach state. Moreover, the good pseudo-BCK algebra A from this example has no Bosbach states.

Example 6.7 Consider $A = \{0, a, b, c, 1\}$ with $0 < a < b < c < 1$ from Example 6.6. Then $(A, \le, \to, \rightsquigarrow, 0, 1)$ is a good pseudo-BCK algebra. The function $s : A \longrightarrow [0, 1]$ defined by

$$s(0) := 0, \qquad s(a) := 1/2, \qquad s(b) := 1/2, \qquad s(c) := 1, \qquad s(1) := 1$$

is a unique Riečan state. Indeed, the elements $x, y \in A$ with $x \perp y$ are those given in the table below:

x	y	$x^{-\sim}$	$y^\sim$	$x + y$
0	0	0	1	0
0	a	0	b	b
0	b	0	b	b
0	c	0	0	1
0	1	0	0	1
a	0	b	1	b
a	a	b	b	1
a	b	b	b	1
b	0	b	1	b
b	a	b	b	1
b	b	b	b	1
c	0	1	1	1
1	0	1	1	1

On the other hand, as was shown in Example 6.6, A has no Bosbach states.

We can see that the kernel of s is the set $\{c, 1\}$, which is a deductive system but not normal.

Example 6.8 The bounded pseudo-BCK algebra A in Example 6.5 has no Bosbach states, but the function $s : A \longrightarrow [0, 1]$ such that $s(0) := 0$, $s(a) = s(b) := 1/2$ and $s(1) := 1$ is a unique Riečan state on A.

The next example is in accordance with Theorem 6.4.

Example 6.9 Consider again the good pseudo-BCK algebra A_1 from Example 1.16. We claim that the Bosbach state $s : A_1 \longrightarrow [0, 1]$ defined by $s(0) := 0$, $s(a) = s(b) = s(c) = s(d) = s(1) := 1$ is also a Riečan state on A_1. Indeed, the elements $x, y \in A_1$ with $x \perp y$ are those given in the table below:

x	y	$x^{-\sim}$	$y^{\sim}$	$x + y$
0	0	0	1	0
0	a	0	0	1
0	b	0	0	1
0	c	0	0	1
0	d	0	0	1
0	1	0	0	1
a	0	1	1	1
b	0	1	1	1
c	0	1	1	1
d	0	1	1	1
1	0	1	1	1

One can easily check that s is a Riečan state.

Theorem 6.5 *Let A be a pseudo-BCK(pDN) algebra and s be a Riečan state on A such that $s(x \vee_1 y) = s(y \vee_1 x)$ and $s(x \vee_2 y) = s(y \vee_2 x)$ for all $x, y \in A$. Then s is a Bosbach state on A.*

Proof Let s be a Riečan state on a good pseudo-BCK(pDN) algebra A. According to Proposition 6.15(2) we have $s(0) = 0$. Since $y^{-\sim} = y$ for all $y \in A$, by Proposition 6.15(6) we get $s((x \vee_1 y) \rightarrow y) = 1 - s(x \vee_1 y) + s(y)$ and $s((x \vee_2 y) \rightarrow y) = 1 - s(x \vee_2 y) + s(y)$. Applying Proposition 1.7, we have:

$$s(x \rightarrow y) = 1 - s(x \vee_1 y) + s(y) \quad \text{and} \quad s(x \rightsquigarrow y) = 1 - s(x \vee_2 y) + s(y).$$

Finally, by Proposition 6.2 it follows that s is a Bosbach state on A.   □

Corollary 6.1 *Riečan states on Wajsberg pseudo-hoops with (pDN) coincide with Bosbach states.*

Proof This follows from Theorem 6.5, taking into consideration that in any Wajsberg pseudo-hoop A we have $s(x \vee_1 y) = s(y \vee_1 x)$ and $s(x \vee_2 y) = s(y \vee_2 x)$ for all $x, y \in A$.   □

Corollary 6.2 *Riečan states on locally finite Wajsberg pseudo-hoops coincide with Bosbach states.*

Proof According to Theorem 2.1 every locally finite pseudo-hoop satisfies the (pDN) condition. The assertion follows from Corollary 6.1.   □

Theorem 6.6 *Let A be a good pseudo-BCK algebra satisfying the identities*:

$$(x \to y)^{-\sim} = (x \vee_1 y) \to y^{-\sim},$$
$$(x \rightsquigarrow y)^{-\sim} = (x \vee_2 y) \rightsquigarrow y^{-\sim}.$$

If s is a Riečan state on A such that $s(x \vee_1 y) = s(y \vee_1 x)$ and $s(x \vee_2 y) = s(y \vee_2 x)$ for all $x, y \in A$, then s is a Bosbach state on A.

Proof Let s be a Riečan state on a good pseudo-BCK algebra A. According to Proposition 6.15(2) we have $s(0) = 0$. Applying Proposition 6.15(3), (6) we get:

$$s(x \to y) = s\big((x \to y)^{-\sim}\big) = s\big((x \vee_1 y) \to y^{-\sim}\big) = 1 - s(x \vee_1 y) + s(y) \quad \text{and}$$
$$s(x \rightsquigarrow y) = s\big((x \rightsquigarrow y)^{-\sim}\big) = s\big((x \vee_2 y) \rightsquigarrow y^{-\sim}\big) = 1 - s(x \vee_2 y) + s(y).$$

Thus, by Proposition 6.2, it follows that s is a Bosbach state on A.   □

Example 6.10 Consider again the good pseudo-BCK lattice A_1 from Example 1.16. One can check that A_1 satisfies the identities from Theorem 6.6. Hence every Riečan state s on A_1 satisfying the conditions $s(x \vee_1 y) = s(y \vee_1 x)$ and $s(x \vee_2 y) = s(y \vee_2 x)$ for all $x, y \in A_1$, is a Bosbach state on A_1.

Moreover, we have:

$0 \vee_1 0 = 0 \vee_2 0 = 0$;

$0 \vee_1 x = 0 \vee_2 x = x$ and $x \vee_1 0 = x \vee_2 0 = 1$,   for all $x \in A_1 \setminus \{0, 1\}$;

$1 \vee_1 x = x \vee_1 1 = 1 \vee_2 x = x \vee_2 1 = 1$ for all $x \in A_1$.

It follows that the function $s : A_1 \longrightarrow [0, 1]$ defined by $s(0) := 0$, $s(a) = s(b) = s(c) = s(d) = s(1) := 1$ satisfies the conditions $s(x \vee_1 y) = s(y \vee_1 x)$ and $s(x \vee_2 y) = s(y \vee_2 x)$ for all $x, y \in A_1$. Then according to Theorem 6.6, s is a Riečan state on A_1, which is in accordance with Example 6.9.

Example 6.11 Let A be the good pseudo-BCK algebra from Example 6.6. One can easily check that A satisfies the two identities from Theorem 6.6. According to Example 6.7, the function $s : A \longrightarrow [0, 1]$ defined by $s(0) := 0$, $s(a) := 1/2$,

$s(b) := 1/2$, $s(c) := 1$, $s(1) := 1$ is the unique Riečan state on A. But $s(a \vee_2 b) = s(b) = 1/2$, while $s(b \vee_2 a) = s(c) = 1$, so $s(a \vee_2 b) \neq s(b \vee_2 a)$. It follows that s does not satisfy the conditions from Theorem 6.6. Moreover, we have shown in Example 6.7 that s is not a Bosbach state on A.

6.2   Bosbach States on Subinterval Algebras of a Bounded Pseudo-hoop

Since every pseudo-hoop is a pseudo-BCK(pP) algebra, the results regarding the states on bounded pseudo-BCK algebras presented in the previous section are also valid for bounded pseudo-hoops. In this section we will study the restrictions of Bosbach states on the subinterval algebras of bounded pseudo-hoops.

Proposition 6.16 *Let $(A, \odot, \rightarrow, \rightsquigarrow, 0, 1)$ be a bounded pseudo-hoop, $a \in CC(A)$ and s be a Bosbach state on A. Then the following hold:*

(1) $s((x \rightarrow y) \odot a) = s(a) + s(x \rightarrow y) - 1$;
(2) $s(a \odot (x \rightsquigarrow y)) = s(a) + s(x \rightsquigarrow y) - 1$.

Proof

(1) Applying Proposition 2.21(1) and (B_1) we get:

$$s(a) + s(x \rightarrow y) = s(a) + s\big(a \rightarrow ((x \rightarrow y) \odot a)\big)$$
$$= s\big((x \rightarrow y) \odot a\big) + s\big(((x \rightarrow y) \odot a) \rightarrow a\big)$$
$$= s\big((x \rightarrow y) \odot a\big) + 1$$

(since $(x \rightarrow y) \odot a \leq a$, it follows that $((x \rightarrow y) \odot a) \rightarrow a = 1$).
   Thus $s((x \rightarrow y) \odot a) = s(a) + s(x \rightarrow y) - 1$.
(2) The proof is similar to (1), applying Proposition 2.21(2) and (B_2).   □

Lemma 6.4 *Let $(A, \odot, \rightarrow, \rightsquigarrow, 0, 1)$ be a bounded pseudo-hoop, $a \in CC(A)$ and s be a Bosbach state on A. Then the following hold for all $n \in \mathbb{N}$:*

(1) $s(a^{n+1}) = s(a^n) + s(a) - 1$;
(2) $s(a^n) = n s(a) - n + 1$.

Proof

(1) By (B_1) we have $s(a^n) + s(a^n \rightarrow a^{n+1}) = s(a^{n+1}) + s(a^{n+1} \rightarrow a^n)$.
       Since $a^{n+1} \leq a^n$, we have $a^{n+1} \rightarrow a^n = 1$ and applying Corollary 2.3 we
   get $s(a^n) + s(a) = s(a^{n+1}) + s(1)$, so $s(a^{n+1}) = s(a^n) + s(a) - 1$.
(2) This follows from (1) by induction.   □

Proposition 6.17 *Let* $(A, \odot, \rightarrow, \rightsquigarrow, 0, 1)$ *be a bounded strongly simple pseudo-hoop,* $a \in CC(A)$ *and* s *be a Bosbach state on* A. *If* $a \neq 1$, *then* $s(a) \neq 1$.

Proof Since A is strongly simple and $a \neq 1$, by Proposition 2.11 it follows that $[a) = A$. Hence $0 \in [a)$ and according to Remark 1.13(2) there exists an $m \in \mathbb{N}$ such that $a^m = 0$. Suppose that $s(a) = 1$. Then, applying Lemma 6.4(2), from $s(a^m) = ms(a) - m + 1$ we get $0 = 1$, a contradiction. Thus $s(a) \neq 1$. $\qquad\square$

In what follows, A_0^a, A_a^1 and A_a^b $(a \leq b)$ will be the subinterval algebras on a bounded pseudo-hoop A introduced in Chap. 2.

Theorem 6.7 *Let* $(A, \odot, \rightarrow, \rightsquigarrow, 0, 1)$ *be a bounded pseudo-hoop,* $a \in JC(A)$ *and* s *be a Bosbach state on* A. *If* $s(a) \neq 1$, *then* $s_a^1 : [a, 1] \longrightarrow [0, 1]$, $s_a^1(x) := \frac{s(x) - s(a)}{1 - s(a)}$ *is a Bosbach state on* A_a^1.

Proof We verify the defining conditions of a Bosbach state:

(B_1)

$$s_a^1(x) + s_a^1\left(x \rightarrow_a y\right) = s_a^1(x) + s_a^1(x \rightarrow y)$$

$$= \frac{s(x) - s(a)}{1 - s(a)} + \frac{s(x \rightarrow y) - s(a)}{1 - s(a)}$$

$$= \frac{s(x) + s(x \rightarrow y) - 2s(a)}{1 - s(a)}$$

$$= \frac{s(y) + s(y \rightarrow x) - 2s(a)}{1 - s(a)}$$

$$= \frac{s(y) - s(a)}{1 - s(a)} + \frac{s(y \rightarrow x) - s(a)}{1 - s(a)}$$

$$= s_a^1(y) + s_a^1(y \rightarrow x)$$

$$= s_a^1(y) + s_a^1\left(y \rightarrow_a x\right).$$

(B_2) This can be proved similarly as (B_1).
(B_3) Obviously, $s_a^1(a) = 0$ and $s_a^1(1) = 1$.

Thus s_a^1 is a Bosbach state on A_a^1. $\qquad\square$

Corollary 6.3 *Let* $(A, \odot, \rightarrow, \rightsquigarrow, 0, 1)$ *be a bounded strongly simple pseudo-hoop,* $a \in CC(A) \cap JC(A)$, $a \neq 1$ *and* s *be a Bosbach state on* A.
Then $s_a^1 : [a, 1] \longrightarrow [0, 1]$, $s_a^1(x) := \frac{s(x) - s(a)}{1 - s(a)}$ *is a Bosbach state on* A_a^1.

Proof This follows from Proposition 6.17 and Theorem 6.7. $\qquad\square$

Theorem 6.8 *Let* $(A, \odot, \rightarrow, \rightsquigarrow, 0, 1)$ *be a bounded pseudo-hoop,* $a \in CC(A)$ *and* s *be a Bosbach state on* A*. If* $s(a) \neq 0$*, then* $s_0^a : [0, a] \longrightarrow [0, 1]$*,* $s_0^a(x) := \frac{s(x)}{s(a)}$ *is a Bosbach state on* A_0^a*.*

Proof We verify the defining conditions of a Bosbach state.

(B_1) Applying Proposition 6.16(1) we get

$$s_0^a(x) + s_0^a\big(x \rightarrow_0^a y\big) = s_0^a(x) + s_0^a\big((x \rightarrow y) \odot a\big)$$

$$= \frac{s(x)}{s(a)} + \frac{s((x \rightarrow y) \odot a)}{s(a)}$$

$$= \frac{s(x) + s(a) + s(x \rightarrow y) - 1}{s(a)}$$

$$= \frac{s(y) + s(a) + s(y \rightarrow x) - 1}{s(a)}$$

$$= \frac{s(y)}{s(a)} + \frac{s((y \rightarrow x) \odot a)}{s(a)}$$

$$= s_0^a(y) + s_0^a\big(y \rightarrow_0^a x\big).$$

(B_2) This can be proved in a similar way by applying Proposition 6.16(2).
(B_3) Obviously, $s_0^a(0) = 0$ and $s_0^a(a) = 1$.

Thus s_0^a is a Bosbach state on A_0^a. $\square$

Theorem 6.9 *Let* $(A, \odot, \rightarrow, \rightsquigarrow, 0, 1)$ *be a bounded pseudo-hoop,* $a, b \in CC(A) \cap JC(A)$*,* $a \leq b$ *and* s *be a Bosbach state on* A*. If* $s(a) \neq s(b)$*, then* $s_a^b : [a, b] \longrightarrow [0, 1]$*,* $s_a^b(x) := \frac{s(x) - s(a)}{s(b) - s(a)}$ *is a Bosbach state on* A_a^b*.*

Proof We verify the defining conditions of a Bosbach state in the same way as in the above theorem:

(B_1)

$$s_a^b(x) + s_a^b\big(x \rightarrow_a^b y\big) = s_a^b(x) + s_a^b\big((x \rightarrow y) \odot b\big)$$

$$= \frac{s(x) - s(a)}{s(b) - s(a)} + \frac{s((x \rightarrow y) \odot b) - s(a)}{s(b) - s(a)}$$

$$= \frac{s(x) - 2s(a) + s(b) + s(x \rightarrow y) - 1}{s(b) - s(a)}$$

$$= \frac{s(y) + s(y \rightarrow x) + s(b) - 2s(a) - 1}{s(b) - s(a)}$$

$$= \frac{s(y) - s(a)}{s(b) - s(a)} + \frac{s(b) + s(y \rightarrow x) - s(a) - 1}{s(b) - s(a)}$$

$$= \frac{s(y) - s(a)}{s(b) - s(a)} + \frac{s((y \to x) \odot b) - s(a)}{s(b) - s(a)}$$

$$= s_a^b(y) + s_a^b\big((y \to x) \odot b\big) = s_a^b(y) + s_a^b\big(y \to_a^b x\big).$$

(B_2) Similar to (B_1).

(B_3) Obviously, $s_a^b(a) = 0$ and $s_a^b(b) = 1$.

Thus s_a^b is a Bosbach state on A_a^b.   □

6.3 States on FL_w-Algebras

In this section we establish some properties of states on FL_w-algebras, pseudo-MTL algebras, bounded Rℓ-monoids and pseudo-BL algebras. We prove that in the case of a good FL_w-algebra satisfying the Glivenko property, Riečan and Bosbach states coincide. As a consequence, Riečan states on good Rℓ-monoids coincide with Bosbach states. It is proved that there is a bijection between the state-morphisms on a pseudo-BL algebra and its maximal and normal filters.

Proposition 6.18 *Let A be an FL_w-algebra and let $s : A \longrightarrow [0, 1]$ be a function such that $s(1) = 1$. Then the following are equivalent*:

(a) $1 + s(x \wedge y) = s(x \vee y) + s(d_1(x, y))$ *for all $x, y \in A$*;
(b) $1 + s(x \wedge y) = s(x) + s(x \to y)$ *for all $x, y \in A$*;
(c) $s(x) + s(x \to y) = s(y) + s(y \to x)$ *for all $x, y \in A$,*

where $+$ is the usual addition of real numbers.

Proof

(a) $\Rightarrow$ (b) If $a \leq b$, then $a \wedge b = a, a \vee b = b, a \to b = 1$ and

$$d_1(a, b) = (a \to b) \wedge (b \to a) = 1 \wedge (b \to a) = b \to a,$$

hence by the hypothesis, $1 + s(a) = s(b) + s(b \to a)$.
Letting $a = x \wedge y$ and $b = y$, it follows that

$$1 + s(x \wedge y) = s(y) + s(y \to x \wedge y) = s(y) + s(y \to x)$$

(here we applied (rl-c_{21})).

(b) $\Rightarrow$ (c) $s(x) + s(x \to y) = 1 + s(x \wedge y) = 1 + s(y \wedge x) = s(y) + s(y \to x)$.

(c) $\Rightarrow$ (a) We have that $d_1(x, y) = x \vee y \to x \wedge y$, hence, applying the hypothesis we get that:

$$s(x \vee y) + s\big(d_1(x, y)\big) = s(x \vee y) + s(x \vee y \to x \wedge y)$$

$$= s(x \wedge y) + s(x \wedge y \to x \vee y) = s(x \wedge y) + s(1)$$

$$= 1 + s(x \wedge y)$$

$(x \wedge y \leq x \vee y$ implies $x \wedge y \to x \vee y = 1)$.   □

Proposition 6.19 *Let A be an FL$_w$-algebra and let s : A $\longrightarrow$ [0, 1] be a function such that s(1) = 1. Then the following are equivalent:*

(a) $1 + s(x \wedge y) = s(x \vee y) + s(d_2(x, y))$ *for all x, y $\in$ A;*
(b) $1 + s(x \wedge y) = s(x) + s(x \rightsquigarrow y)$ *for all x, y $\in$ A;*
(c) $s(x) + s(x \rightsquigarrow y) = s(y) + s(y \rightsquigarrow x)$ *for all x, y $\in$ A.*

Proof The proof is similar to that of Proposition 6.18. $\square$

Proposition 6.20 *Let A be an FL$_w$-algebra and s : A $\longrightarrow$ [0, 1] such that s(0) = 0. Then the following are equivalent:*

(a) *s is a Bosbach state on A;*
(b) *for all x, y $\in$ A, y $\leq$ x implies s(x $\rightarrow$ y) = s(x $\rightsquigarrow$ y) = 1 $-$ s(x) + s(y);*
(c) *for all x, y $\in$ A, s(x $\rightarrow$ y) = s(x $\rightsquigarrow$ y) = 1 $-$ s(x $\vee$ y) + s(y).*

Proof

(a) $\Rightarrow$ (b) This follows from (B_1) and (B_2).
(b) $\Rightarrow$ (c) Applying (b), since $y \leq x \vee y$ and $x \vee y \rightarrow y = x \rightarrow y$, it follows that:

$$s(y) + s\big(y \rightarrow (x \vee y)\big) = s(x \vee y) + s\big((x \vee y) \rightarrow y\big) = s(x \vee y) + s(x \rightarrow y),$$

so $s(x \rightarrow y) = 1 - s(x \vee y) + s(y)$ and similarly $s(x \rightsquigarrow y) = 1 - s(x \vee y) + s(y)$.
(c) $\Rightarrow$ (a) Using (c), we get:

$$s(x) + s(x \rightarrow y) = s(x) + 1 - s(x \vee y) + s(y)$$
$$= 1 - s(y \vee x) + s(x) + s(y) = s(y) + s(y \rightarrow x).$$

Similarly,

$$s(x) + s(x \rightsquigarrow y) = s(x) + 1 - s(x \vee y) + s(y)$$
$$= 1 - s(y \vee x) + s(x) + s(y) = s(y) + s(y \rightsquigarrow x).$$

Moreover, by (c) we have: $s(1) = s(x \rightarrow x) = 1 - s(x) + s(x) = 1$.
Thus s is a Bosbach state on A. $\square$

Proposition 6.21 *Let s be a Bosbach state on an FL$_w$-algebra A. Then for all x, y $\in$ A the following properties hold:*

(rl-b$_1$) $s(d_1(x, y)) = s(d_2(x, y))$;
(rl-b$_2$) $s(x \odot y) = 1 - s(x \rightarrow y^-)$ *and* $s(y \odot x) = 1 - s(x \rightsquigarrow y^\sim)$;
(rl-b$_3$) $s(x) + s(y) = s(x \odot y) + s(y^- \rightarrow x)$;
(rl-b$_4$) $s(x) + s(y) = s(y \odot x) + s(y^\sim \rightsquigarrow x)$;
(rl-b$_5$) $s(x \rightarrow y) = s(x^{-\sim} \rightarrow y^{-\sim})$ *and* $s(x \rightsquigarrow y) = s(x^{\sim -} \rightsquigarrow y^{\sim -})$.

Proof

(rl-b$_1$) By Propositions 6.18 and 6.19,

$$s\big(d_1(x, y)\big) = 1 + s(x \wedge y) - s(x \vee y) = s\big(d_2(x, y)\big).$$

(rl-b$_2$) By Proposition 6.1(6) we have $s((x \odot y)^-) = 1 - s(x \odot y)$.
But $(x \odot y)^- = x \to y^-$, so $s(x \odot y) = 1 - s(x \to y^-)$.
Similarly, $s(y \odot x) = 1 - s((y \odot x)^\sim) = 1 - s(x \rightsquigarrow y^\sim)$.

(rl-b$_3$) Applying (B_1) and (rl-b$_2$) we have:

$$\begin{aligned}
s(x \odot y) + s\big(y^- \to x\big) &= s(x \odot y) + s(x) + s\big(x \to y^-\big) - s\big(y^-\big) \\
&= s(x \odot y) + s(x) + 1 - s(x \odot y) - 1 + s(y) \\
&= s(x) + s(y).
\end{aligned}$$

(rl-b$_4$) Applying (B_2) and (rl-b$_2$) we have:

$$\begin{aligned}
s(y \odot x) + s\big(y^\sim \rightsquigarrow x\big) &= s(y \odot x) + s(x) + s\big(x \rightsquigarrow y^\sim\big) - s\big(y^\sim\big) \\
&= s(y \odot x) + s(x) + 1 - s(y \odot x) - 1 + s(y) \\
&= s(x) + s(y).
\end{aligned}$$

(rl-b$_5$) Applying Proposition 6.20, (rl-c$_{27}$) and Proposition 6.1(5) we get:

$$\begin{aligned}
s\big(x^{-\sim} \to y^{-\sim}\big) &= 1 - s\big(x^{-\sim} \vee y^{-\sim}\big) + s\big(y^{-\sim}\big) \\
&= 1 - s\big((x^{-\sim} \vee y^{-\sim})^{-\sim}\big) + s\big(y^{-\sim}\big) \\
&= 1 - s\big((x \vee y)^{-\sim}\big) + s\big(y^{-\sim}\big) \\
&= 1 - s(x \vee y) + s(y) = s(x \to y).
\end{aligned}$$

Similarly, $s(x \rightsquigarrow y) = s(x^{\sim -} \rightsquigarrow y^{\sim -})$.   □

Corollary 6.4 *Let s be a Bosbach state on an FL_w-algebra A. Then for all $x, y \in A$ the following properties hold:*

(1) $s(x \to y) = s(x \to y^{-\sim}) = s(y^- \rightsquigarrow x^-) = s(x^{-\sim} \to y^{-\sim}) = s(x^{-\sim} \to y)$;
(2) $s(x \rightsquigarrow y) = s(x \rightsquigarrow y^{\sim -}) = s(y^\sim \to x^\sim) = s(x^{\sim -} \rightsquigarrow y^{\sim -}) = s(x^{\sim -} \rightsquigarrow y)$;
(3) $s(x^\sim \to y) = s(y^- \rightsquigarrow x^{\sim -}) = s(x^\sim \to y^{-\sim}) = s(y^- \rightsquigarrow x)$.

Proof

(1) and (2) follow from Proposition 6.3(2) and (rl-b$_5$).
(3) Replacing x with $x^\sim$ in (1) we get:

$$s\big(x^\sim \to y\big) = s\big(x^\sim \to y^{-\sim}\big) = s\big(y^- \rightsquigarrow x^{\sim -}\big) = s\big(x^\sim \to y^{-\sim}\big) = s\big(x^\sim \to y\big).$$

From (2) we have $s(x \rightsquigarrow y) = s(y^\sim \to x^\sim)$, so $s(y \rightsquigarrow x) = s(x^\sim \to y^\sim)$.

Replacing y by y^- it follows that $s(y^- \rightsquigarrow x) = s(x^\sim \rightarrow y^{-\sim})$.
Finally, we conclude that:

$$s(x^\sim \rightarrow y) = s(y^- \rightsquigarrow x^{\sim-}) = s(x^\sim \rightarrow y^{-\sim}) = s(y^- \rightsquigarrow x). \qquad \square$$

Example 6.12 Consider the good FL$_w$-algebra A_1 from Example 3.5.
The function $s : A_1 \rightarrow [0, 1]$ defined by:

$$s(0) := 0,$$

$$s(a_1) = s(a_2) = s(b_2) = s(s) = s(a) = s(b)$$

$$= s(n) = s(c) = s(d) = s(m) = s(1) := 1$$

is the unique Bosbach state on A_1.

Proposition 6.22 *Let s be a Bosbach state on an FL$_w$-algebra A. Then the following properties hold for all $x, y \in A$:*

(1) $x / \mathrm{Ker}(s) = y / \mathrm{Ker}(s)$ *iff* $s(x \wedge y) = s(x \vee y)$;
(2) *If* $s(x \wedge y) = s(x \vee y)$, *then* $s(x) = s(y) = s(x \wedge y)$.

Proof

(1) According to Lemma 6.1 and Proposition 6.18(a) we have:

$$x / \mathrm{Ker}(s) = y/\mathrm{Ker}(s) \quad \text{iff} \quad d_1(x, y) \in \mathrm{Ker}(s) \quad \text{iff}$$

$$s(d_1(x, y)) = 1 \quad \text{iff} \quad s(x \wedge y) = s(x \vee y).$$

(2) Since $s(x \wedge y) \leq s(x), s(y) \leq s(x \vee y)$ it follows by hypothesis that $s(x) = s(y) = s(x \wedge y)$. $\qquad \square$

Proposition 6.23 *Let s be a Bosbach state on a pseudo-MTL algebra A. Then the following properties hold for all $x, y \in A$:*

(mtl-b$_1$) $s(d_1(x, y)) = s(d_1(x \rightarrow y, y \rightarrow x))$ *and* $s(d_2(x, y)) = s(d_2(x \rightsquigarrow y, y \rightsquigarrow x))$;
(mtl-b$_2$) $s(d_1(x^-, y^-)) = s(d_1(x^\sim, y^\sim)) = s(d_1(x, y))$;
(mtl-b$_3$) $s(x^- \rightarrow y^-) = s(x^\sim \rightarrow y^\sim) = 1 + s(x) - s(x \vee y)$;
(mtl-b$_4$) $s(x^- \rightarrow y^-) = s(y^{-\sim} \rightarrow x^{-\sim})$ *and* $s(x^\sim \rightarrow y^\sim) = s(y^{\sim-} \rightarrow x^{\sim-})$;
(mtl-b$_5$) $s(x) + s(y) = s(x \vee y) + s(x \wedge y)$;
(mtl-b$_6$) $x \vee y = 1$ *implies* $1 + s(x \wedge y) = s(x) + s(y)$;
(mtl-b$_7$) $1 + s(d_1(x, y)) = s(x \rightarrow y) + s(y \rightarrow x)$ *and* $1 + s(d_2(x, y)) = s(x \rightsquigarrow y) + s(y \rightsquigarrow x)$;
(mtl-b$_8$) $1 + s(x) + s(y) = s(d_1(x, y)) + s((x \rightarrow y) \rightarrow y) + s((y \rightarrow x) \rightarrow x)$ *and* $1 + s(x) + s(y) = s(d_2(x, y)) + s((x \rightsquigarrow y) \rightsquigarrow y) + s((y \rightsquigarrow x) \rightsquigarrow x)$.

Proof

(*mtl-b₁*) By Proposition 6.18(a) and by the pseudo-prelinearity condition it follows that:

$$1 + s(d_1(x, y)) = 1 + s((x \to y) \wedge (y \to x))$$
$$= s((x \to y) \vee (y \to x)) + s(d_1(x \to y, y \to x))$$
$$= 1 + s(d_1(x \to y, y \to x)).$$

Thus $s(d_1(x, y)) = s(d_1(x \to y, y \to x))$.
Similarly, $s(d_2(x, y)) = s(d_2(x \rightsquigarrow y, y \rightsquigarrow x))$.

(*mtl-b₂*) Applying Proposition 6.18(a), Proposition 6.1(6) and taking into consideration (*psbck-c₄₁*) and (*psmtl-c₁*), we get:

$$s(d_1(x^-, y^-)) = 1 + s(x^- \wedge y^-) - s(x^- \vee y^-)$$
$$= 1 + s((x \vee y)^-) - s((x \wedge y)^-)$$
$$= 1 - s(x \vee y) + s(x \wedge y) = s(d_1(x, y)).$$

Similarly, $s(d_1(x^\sim, y^\sim)) = s(d_1(x, y))$.

(*mtl-b₃*) Applying Proposition 6.18(b), Proposition 6.1(6) and taking into consideration (*psbck-c₄₁*) and (*psmtl-c₁*), we get:

$$s(x \vee y) + s(x^- \to y^-) = s(x \vee y) + 1 + s(x^- \wedge y^-) - s(x^-)$$
$$= s(x \vee y) + s((x \vee y)^-) + s(x) = 1 + s(x),$$

hence $s(x^- \to y^-) = 1 + s(x) - s(x \vee y)$.
Similarly, $s(x^\sim \to y^\sim) = 1 + s(x) - s(x \vee y)$.

(*mtl-b₄*) By (*psbck-c₁₅*) and by Proposition 6.4(2) it follows that:

$$s(x^- \to y^-) = s(x^- \rightsquigarrow y^-) \leq s(y^{-\sim} \to x^{-\sim}) \leq s(x^{-\sim-} \rightsquigarrow y^{-\sim-})$$
$$= s(x^- \rightsquigarrow y^-) = s(x^- \to y^-).$$

Thus $s(x^- \to y^-) = s(y^{-\sim} \to x^{-\sim})$.
Similarly, $s(x^\sim \to y^\sim) = s(y^{\sim-} \to x^{\sim-})$.

(*mtl-b₅*) Applying Proposition 6.1(5) and taking into consideration (*psbck-c₄₁*) and (*psmtl-c₁*), we have:

$$s(x \vee y) + s(x \wedge y) = s((x \vee y)^{-\sim}) + s((x \wedge y)^{-\sim})$$
$$= s(x^{-\sim} \vee y^{-\sim}) + s(x^{-\sim} \wedge y^{-\sim}).$$

Applying (*mtl-b₃*) we get:

$$s(x^{-\sim} \vee y^{-\sim}) = 1 + s(x^{-\sim}) - s(x^{-\sim-} \to y^{-\sim-}) = 1 + s(x) - s(x^- \to y^-).$$

By Proposition 6.18(b) and (*mtl-b$_4$*) we obtain:

$$s\left(x^{-\sim} \wedge y^{-\sim}\right) = s\left(y^{-\sim}\right) + s\left(y^{-\sim} \to x^{-\sim}\right) - 1 = s(y) + s\left(x^- \to y^-\right) - 1.$$

Hence $s(x^{-\sim} \vee y^{-\sim}) + s(x^{-\sim} \wedge y^{-\sim}) = s(x) + s(y)$.

We conclude that $s(x) + s(y) = s(x \vee y) + s(x \wedge y)$.

(*mtl-b$_6$*) This follows from (*mtl-b$_5$*).

(*mtl-b$_7$*) This follows by (*mtl-b$_6$*) and the pseudo-prelinearity condition.

(*mtl-b$_8$*) Since $y \le x \to y$ and $x \le y \to x$, applying Proposition 6.1(1) we get:

$$1 + s(y) = s(x \to y) + s\left((x \to y) \to y\right),$$
$$1 + s(x) = s(y \to x) + s\left((y \to x) \to x\right).$$

Adding these two equalities and applying (*mtl-b$_7$*) we have:

$$1 + s(x) + s(y) = -1 + s(x \to y) + s(y \to x)$$
$$+ s\left((x \to y) \to y\right) + s\left((y \to x) \to x\right)$$
$$= s\left(d_1(x, y)\right) + s\left((x \to y) \to y\right) + s\left((y \to x) \to x\right).$$

Similarly, $1 + s(x) + s(y) = s(d_2(x, y)) + s((x \rightsquigarrow y) \rightsquigarrow y) + s((y \rightsquigarrow x) \rightsquigarrow x)$. $\qquad\square$

Proposition 6.24 *If s is a Bosbach state on a pseudo-MTL algebra A, then for all $x, y \in A$ the following are equivalent*:

(a) $x / \operatorname{Ker}(s) = y / \operatorname{Ker}(s)$;
(b) $s(x \wedge y) = s(x \vee y)$;
(c) $s(x) = s(y) = s(x \wedge y)$.

Proof

(a) $\Leftrightarrow$ (b) We have:

$$x / \operatorname{Ker}(s) = y / \operatorname{Ker}(s) \quad \text{iff} \quad d_1(x, y) \in \operatorname{Ker}(s) \quad \text{iff} \quad s\left(d_1(x, y)\right) = 1.$$

But, according to Proposition 6.18(a) we have $1 + s(x \wedge y) = s(x \vee y) + s(d_1(x, y))$, so $x / \operatorname{Ker}(s) = y / \operatorname{Ker}(s)$ iff $s(x \wedge y) = s(x \vee y)$.

(b) $\Rightarrow$ (c) Since $s(x \wedge y) \le s(x), s(y) \le s(x \vee y)$, applying (b) we get $s(x) = s(y) = s(x \wedge y)$.

(c) $\Rightarrow$ (b) According to (*mtl-b$_5$*) we have $s(x) + s(y) = s(x \vee y) + s(x \wedge y)$ and applying (c) we get $s(x \wedge y) = s(x \vee y)$. $\qquad\square$

Proposition 6.25 *Let s be a Bosbach state on a bounded Rℓ-monoid A. Then the following properties hold for all $x, y \in A$:*

(*divrl-b$_1$*) $s(d_1(x, y)) \le s(d_1(x \to y, y \to x))$ *and* $s(d_2(x, y)) \le s(d_2(x \rightsquigarrow y, y \rightsquigarrow x))$;

(*divrl-b$_2$*) $s(d_1(x^-, y^-)) \geq s(d_1(x, y))$ *and* $s(d_1(x^\sim, y^\sim)) \geq s(d_1(x, y))$;
(*divrl-b$_3$*) $s(d_2(x^-, y^-)) \geq s(d_2(x, y))$ *and* $s(d_2(x^\sim, y^\sim)) \geq s(d_2(x, y))$;
(*divrl-b$_4$*) *if* $x \leq y$, *then* $s(y) - s(x) = s(y \odot x^-) = s(x^\sim \odot y)$.

Proof

(*divrl-b$_1$*) By Proposition 6.18(a) it follows that:

$$1 + s\big(d_1(x, y)\big) = 1 + s\big((x \to y) \wedge (y \to x)\big)$$
$$= s\big((x \to y) \vee (y \to x)\big) + s\big(d_1(x \to y, y \to x)\big)$$
$$\leq s(1) + s\big(d_1(x \to y, y \to x)\big) = 1 + s\big(d_1(x \to y, y \to x)\big).$$

Thus $s(d_1(x, y)) \leq (d_1(x \to y, y \to x))$.
Similarly, $s(d_2(x, y)) \leq s(d_2(x \rightsquigarrow y, y \rightsquigarrow x))$.
(*divrl-b$_2$*) Applying Proposition 6.18(a), Proposition 6.1(6) and taking into consideration (*psbck-c$_{41}$*) and (*psbck-c$_{42}$*), we get:

$$s\big(d_1\big(x^-, y^-\big)\big) = 1 + s\big(x^- \wedge y^-\big) - s\big(x^- \vee y^-\big)$$
$$\geq 1 + s\big((x \vee y)^-\big) - s\big((x \wedge y)^-\big)$$
$$= 1 - s(x \vee y) + s(x \wedge y) = s\big(d_1(x, y)\big).$$

Similarly, $s(d_1(x^\sim, y^\sim)) \geq s(d_1(x, y))$.
(*divrl-b$_3$*) This follows in a similar manner as (*divrl-b$_2$*).
(*divrl-b$_4$*) Applying (*rl-b$_2$*), (*psbck-c$_{19}$*) and Corollary 6.4 we get:

$$s\big(y \odot x^-\big) = 1 - s\big(x^- \rightsquigarrow y^-\big) = 1 - s\big(y^{-\sim} \to x^{-\sim}\big) = 1 - s(y \to x).$$

By (*B$_1$*) and $x \leq y$ it follows that $s(y \to x) = 1 + s(x) - s(y)$.
Thus $s(y \odot x^-) = s(y) - s(x)$ and similarly $s(x^\sim \odot y) = s(y) - s(x)$.　　　$\square$

Theorem 6.10 *Let A be a good FL$_w$-algebra satisfying the Glivenko property. Then every Riečan state on A is a Bosbach state.*

Proof Let s be a Riečan state on A, so $s(0) = 0$ and $s(1) = 1$.
Consider $y \leq x$. Then applying Proposition 6.15(3), (4) we have:

$$s(x \to y) = s\big((x \to y)^{-\sim}\big) = 1 - s(x) + s\big(y^{-\sim}\big) = 1 - s(x) + s(y).$$

Similarly, $s(x \rightsquigarrow y) = 1 - s(x) + s(y)$.
Applying Proposition 6.20 it follows that s is a Bosbach state on A.　　　$\square$

Remark 6.6 Since any good Rℓ-monoid satisfies the Glivenko property (Remark 4.16), it follows that Riečan states on good Rℓ-monoids coincide with Bosbach states. This result was proved in [111], showing that every extremal Riečan state on a good Rℓ-monoid is an extremal Bosbach state and applying the Krein-Mil'man Theorem.

As we already mentioned (Remark 6.2), in the case of pseudo-BL algebras and bounded Rℓ-monoids the existence of a state-morphism is equivalent to the existence of a maximal filter which is normal. In what follows we will prove this fact for the case of pseudo-BL algebras (see [131]).

Proposition 6.26 *Let A be a pseudo-BL algebra and H a normal and maximal filter of A. Then there is a unique state-morphism s on A such that* Ker$(s) = H$.

Proof Since by Proposition 4.26 A/H is a locally finite MV-algebra, according to [38] we can suppose that A/H is an MV-subalgebra of $[0, 1]_L$. Then the mapping s : $A \longrightarrow A/H$ defines a state-morphism on A such that Ker$(s) = H$. The uniqueness of s follows from Proposition 6.14.   □

Corollary 6.5 *The function s ↦ Ker(s) establishes a bijection between the state-morphisms on a pseudo-BL algebra A and the normal and maximal filters of A.*

Example 6.13 (Example 2.5 in [131]) Consider $A = \{0, a, b, 1\}$ with $0 < a < b < 1$ and define the operations $\odot$, $\rightarrow$ by the following tables:

$\odot$	0	a	b	1
0	0	0	0	0
a	0	0	a	a
b	0	a	b	b
1	0	a	b	1

$\rightarrow$	0	a	b	1
0	1	1	1	1
a	a	1	1	1
b	0	a	1	1
1	0	a	b	1

Then $(A, \wedge, \vee, \odot, \rightarrow, 0, 1)$ is a BL-algebra denoted by BL$_4$A ([260]).

The function $s : A \rightarrow [0, 1]$ defined by $s(0) := 0$, $s(a) := \frac{1}{2}$, $s(b) := 1$, $s(1) := 1$ is the unique Bosbach state on BL$_4$A.

On the other hand we can see that $F = \{b, 1\}$ is the unique maximal filter of A. (F is also normal, since in a BL-algebra every filter is normal.)

Example 6.14 (Example 4.13 in [109]) Consider $A = \{0, a, b, c, 1\}$ with $0 < a, b < c < 1$ and a, b incomparable (see Fig. 1.2) and let $\odot$, $\rightarrow$ be the operations given by the following tables:

$\odot$	0	a	b	c	1
0	0	0	0	0	0
a	0	a	0	a	a
b	0	0	b	b	b
c	0	a	b	c	c
1	0	a	b	c	1

$\rightarrow$	0	a	b	c	1
0	1	1	1	1	1
a	b	1	b	1	1
b	a	a	1	1	1
c	0	a	b	1	1
1	0	a	b	c	1

Then $(A, \wedge, \vee, \odot, \rightarrow, 0, 1)$ is a proper bounded commutative Rℓ-monoid (since $(a \rightarrow b) \vee (b \rightarrow a) = c \neq 1$, it follows that A is not a BL-algebra).

Moreover, $F_a = \{a, c, 1\}$ and $F_b = \{b, c, 1\}$ are the unique maximal (and normal) filters of A. The only extremal states on A are s_a and s_b where Ker$(s_a) = F_a$ and Ker$(s_b) = F_b$. Every Bosbach state on A is a convex combination of s_a and s_b.

Finally, given a Riečan state s on a good FL_w-algebra A, we will construct a Riečan state on $A/\mathrm{Ker}(s)$.

Theorem 6.11 *Let A be a good FL_w-algebra. If s is a Riečan state on A, then the function $\hat{s} : A/\mathrm{Ker}(s) \to [0, 1]$ defined by $\hat{s}(x/\mathrm{Ker}(s)) := s(x)$ is a Riečan state on $A/\mathrm{Ker}(s)$.*

Proof First, we prove that $\hat{s}$ is well-defined.

Indeed, if $x/\mathrm{Ker}(s) = y/\mathrm{Ker}(s)$, then by Proposition 6.22 it follows that $s(x \wedge y) = s(x \vee y)$. Then by Proposition 6.18(a) we have $s(d_1(x, y)) = 1$.

It follows that $d_1(x, y) \in \mathrm{Ker}(s)$ and similarly, $d_2(x, y) \in \mathrm{Ker}(s)$.

Thus $x \equiv_{\mathrm{Ker}(s)} y$.

Moreover, if $x \equiv_{\mathrm{Ker}(s)} y$, then $s(x) = s(y)$.

Indeed, $x \equiv_{\mathrm{Ker}(s)} y$ is equivalent to $s(x \to y) = s(y \to x) = 1$ and by Proposition 6.18(c) it follows that $s(x) = s(y)$.

Applying the method used in [110] we prove now that $\hat{s}$ is a Riečan state on $A/\mathrm{Ker}(s)$.

First we recall that if $\hat{x} \leq \hat{y}$, then there is an element $x_1 \in \hat{x}$ such that $x_1 \leq y$.

Indeed, it suffices to take $x_1 = x \wedge y$.

Assume that $\hat{x} \perp \hat{y}$, that is, $\hat{y}^{-\sim} \leq \hat{x}^{-}$, hence $\hat{x}^{-\sim} \leq \hat{y}^{-\sim} = \hat{y}^{\sim-\sim} = \hat{y}^{\sim}$, so $\hat{x}^{-\sim} \leq \hat{y}^{\sim}$. Consider $x_1 \in \hat{x}^{-\sim}$ such that $x_1 \leq y^{\sim}$. Hence $x_1^{-\sim} \leq y^{\sim}$, so $x_1 \perp y$.

Therefore:

$$\hat{s}(\hat{x} + \hat{y}) = \hat{s}\big((\hat{y}^{\sim} \odot \hat{x}^{\sim})^{-}\big) = \hat{s}\big(((y^{\sim} \odot x^{\sim})^{-})\big) = s\big((y^{\sim} \odot x^{\sim})^{-}\big) = s(x \oplus y)$$

$$= s\big(x^{-\sim} \oplus y^{-\sim}\big) = \hat{s}\big(\hat{x}_1 + \hat{y}^{-\sim}\big) = s\big(x_1 + y^{-\sim}\big) = s(x_1) + s\big(y^{-\sim}\big)$$

$$= s(x) + s(y) = \hat{s}(\hat{x}) + \hat{s}(\hat{y}).$$

(We took into consideration that $x_1 \equiv_{\mathrm{Ker}(s)} x$ implies $s(x_1) = s(x)$.) $\square$

6.4 On the Existence of States on Residuated Structures

In this section we will investigate the existence of states on some classes of FL_w-algebras, pseudo-MTL algebras and pseudo-BL algebras.

Theorem 6.12 *Any perfect pseudo-BL algebra admits a unique state-morphism.*

Proof According to Proposition 6.26, if H is a normal and maximal filter of a pseudo-BL algebra A, then there is a unique state-morphism on A such that $\mathrm{Ker}(s) = H$. On the other hand, a perfect pseudo-BL algebra has a unique maximal filter, namely $\mathrm{Rad}(A)$. By Theorem 5.6, $\mathrm{Rad}(A)$ is a normal filter of A. Thus, if A is a perfect pseudo-BL algebra, then $\mathrm{Rad}(A)$ is the unique normal and maximal filter of A. Hence a perfect pseudo-BL algebra admits a unique state-morphism. $\square$

Theorem 6.13 *Any perfect* FL_w*-algebra admits a Bosbach state.*

Proof Let A be a perfect FL_w-algebra, so $A = \mathrm{Rad}(A) \cup \mathrm{Rad}(A)^*$. Consider the map $s : A \longrightarrow [0, 1]$ defined by

$$s(x) := \begin{cases} 1 & \text{if } x \in \mathrm{Rad}(A) \\ 0 & \text{if } x \in \mathrm{Rad}(A)^*. \end{cases}$$

We will show that s is a Bosbach state on A. Obviously, $s(1) = 1$ and $s(0) = 0$. In order to prove the conditions (B_1) and (B_2) we consider the following cases:

(1) $x, y \in \mathrm{Rad}(A)$.

Obviously, $s(x) = s(y) = 1$. Since $\mathrm{Rad}(A)$ is a filter of A and $x \le y \to x$, $y \le x \to y$, it follows that $x \to y, y \to x \in \mathrm{Rad}(A)$. Hence $s(x \to y) = s(y \to x) = 1$.

Similarly $s(x \rightsquigarrow y) = s(y \rightsquigarrow x) = 1$, so the conditions (B_1) and (B_2) are verified.

(2) $x, y \in \mathrm{Rad}(A)^*$.

In this case $s(x) = s(y) = 0$ and we will prove that $x \to y, y \to x \in \mathrm{Rad}(A)$. Indeed, suppose that $x \to y \in \mathrm{Rad}(A)^*$. Since $x \le x^{-\sim}$, it follows that $x^{-\sim} \to y \le x \to y$, so $x^{-\sim} \to y \in \mathrm{Rad}(A)^*$. But, $x^- \le x^{-\sim} \to y$, hence $x^- \in \mathrm{Rad}(A)^*$, that is, $x \in \mathrm{Rad}(A)$, which is a contradiction. It follows that $x \to y \in \mathrm{Rad}(A)$ and similarly, $y \to x \in \mathrm{Rad}(A)$. Hence $s(x \to y) = s(y \to x) = 1$. In the same way we can prove that $s(x \rightsquigarrow y) = s(y \rightsquigarrow x) = 1$, so (B_1) and (B_2) are verified.

(3) $x \in \mathrm{Rad}(A)$, $y \in \mathrm{Rad}(A)^*$.

Obviously, $s(x) = 1$ and $s(y) = 0$. Because $x \le y \to x$ we get $y \to x \in \mathrm{Rad}(A)$.

We show that $x \to y \in \mathrm{Rad}(A)^*$. Indeed, suppose that $x \to y \in \mathrm{Rad}(A)$.

Because $y \le y^{-\sim}$ we have $x \to y \le x \to y^{-\sim}$, so $x \to y^{-\sim} \in \mathrm{Rad}(A)$. This means that $(x \odot y^\sim)^- \in \mathrm{Rad}(A)$, that is, $x \odot y^\sim \in \mathrm{Rad}(A)^*$. On the other hand, since $\mathrm{Rad}(A)$ is a filter of A and $x, y^\sim \in \mathrm{Rad}(A)$ we have $x \odot y^\sim \in \mathrm{Rad}(A)$, a contradiction. We conclude that $x \to y \in \mathrm{Rad}(A)^*$, so $s(x \to y) = 0$ and $s(y \to x) = 1$. Similarly, $s(x \rightsquigarrow y) = 0$ and $s(y \rightsquigarrow x) = 1$.

Thus conditions (B_1) and (B_2) are verified.

(4) $x \in \mathrm{Rad}(A)^*$ and $y \in \mathrm{Rad}(A)$.

Obviously, $s(x) = 0$ and $s(y) = 1$.

Since $y \le x \to y$, it follows that $x \to y \in \mathrm{Rad}(A)$.

We show that $y \to x \in \mathrm{Rad}(A)^*$. Indeed, suppose that $y \to x \in \mathrm{Rad}(A)$.

From $x \le x^{-\sim}$ we get $y \to x \le y \to x^{-\sim}$, so $y \to x^{-\sim} \in \mathrm{Rad}(A)$. Hence $(y \odot x^\sim)^- \in \mathrm{Rad}(A)$, that is, $y \odot x^\sim \in \mathrm{Rad}(A)^*$. But $y, x^\sim \in \mathrm{Rad}(A)$, so $y \odot x^\sim \in \mathrm{Rad}(A)$, a contradiction. It follows that $y \to x \in \mathrm{Rad}(A)^*$, so $s(y \to x) = 0$ and $s(x \to y) = 1$. Similarly, $s(y \rightsquigarrow x) = 0$ and $s(x \rightsquigarrow y) = 1$.

Hence (B_1) and (B_2) are verified. $\qquad\square$

Example 6.15 The pseudo-MTL algebra in Example 4.4 is a perfect pseudo-MTL algebra with $\mathrm{Rad}(A) = \{a, b, c, 1\}$ and $\mathrm{Rad}(A)^* = \{0\}$ and it admits the Bosbach state $s : A \to [0, 1]$, $s(0) := 0$, $s(a) = s(b) = s(c) = s(1) := 1$.

Remark 6.7 It was proved in [98] that every linearly ordered pseudo-BL algebra admits a Bosbach state. We say that a pseudo-BL algebra A is *representable* if it can be represented as a subdirect product of linearly ordered pseudo-BL algebras. It follows that every representable pseudo-BL algebra admits a Bosbach state.

6.5   Local States on Perfect Pseudo-MTL Algebras

In this section we introduce the notion of a local state on a perfect pseudo-MTL algebra A and we prove an extension theorem for this type of state. More precisely we prove that, if A is relatively free of zero divisors, then any local state on A can be extended to a Riečan state on A.

Definition 6.5 If A is a perfect pseudo-MTL algebra, then a *local state* (or *local additive measure*) on A is a function $s : \mathrm{Rad}(A)^* \longrightarrow [0, \infty)$ satisfying the conditions:

(ls_1) $s(x \oplus y) = s(x) + s(y)$ for all $x, y \in \mathrm{Rad}(A)^*$;
(ls_2) $s(0) = 0$.

According to Proposition 5.10 it follows that the function s is well defined, i.e. $x \oplus y \in \mathrm{Rad}(A)^*$ for all $x, y \in \mathrm{Rad}(A)^*$.

Example 6.16

(1) The function $s : \mathrm{Rad}(A)^* \longrightarrow [0, \infty)$, $s(x) := 0$ for all $x \in \mathrm{Rad}(A)^*$ is a local state on the perfect pseudo-MTL algebra A.
(2) If S is a Riečan state on the perfect pseudo-MTL algebra A, then $s := S_{|\mathrm{Rad}(A)^*}$ is a local state on A.

If s is a local state on the perfect pseudo-MTL algebra A, then we define the function $s^* : \mathrm{Rad}(A) \longrightarrow [0, \infty)$ by $s^*(x) := 1 - s(x^- \oplus x^\sim)$ for all $x \in \mathrm{Rad}(A)$.

Proposition 6.27 *If s is a local state on the perfect pseudo-MTL algebra A, then the following hold for all $x, y \in \mathrm{Rad}(A)^*$:*

(1) $s(x^{-\sim}) = s(x)$;
(2) $s(x) + s(y^{--}) = s((y^- \odot x^\sim)^-)$ and $s(x) + s(y^{\sim\sim}) = s((x^- \odot y^\sim)^\sim)$;
(3) $s(y^{--}) + s((x^- \odot y^\sim)^\sim) = s(y^{\sim\sim}) + s((y^- \odot x^\sim)^-)$;
(4) $s(x) \leq s((x^- \odot x^\sim)^-)$, $s(x) \leq s((x^- \odot x^\sim)^\sim)$.

Proof

(1) Since $x \perp 0$ and $x \oplus 0 = x^{-\sim}$, it follows that $s(x^{-\sim}) = s(x)$.
(2) Since $y \in \mathrm{Rad}(A)^*$, we have $y^{--}, y^{\sim\sim} \in \mathrm{Rad}(A)^*$, so according to Proposition 5.13 we have $x \perp y^{--}$ and $y^{\sim\sim} \perp x$. It follows that:

$$x \oplus y^{--} = y^{--\sim} \to x^{\sim-} = (y^{--\sim} \odot x^{\sim})^- = (y^- \odot x^{\sim})^-,$$
$$y^{\sim\sim} \oplus x = y^{\sim\sim-} \leadsto x^{\sim-} = (x^- \odot y^{\sim\sim-})^\sim = (x^- \odot y^\sim)^\sim.$$

Thus $s(x) + s(y^{--}) = s((y^- \odot x^\sim)^-)$ and $s(x) + s(y^{\sim\sim}) = s(x^- \odot y^\sim)^\sim$.
(3) This follows from (2).
(4) This follows by setting $y = x$ in (2) and taking into consideration that $s(x^{--}), s(x^{\sim\sim}) \geq 0$. $\qquad\square$

Proposition 6.28 *If s is a local state on the perfect pseudo-MTL algebra A, then the following hold for all $x, y \in \mathrm{Rad}(A)$:*

(1) $s^*(1) = 1$;
(2) $s^*(x^{-\sim}) = s^*(x)$;
(3) $s^*(x \oplus y) = 1 - [s(y^- \odot x^-) + s(y^\sim \odot x^\sim)]$;
(4) $1 + s^*(x) \leq s^*(x^{--}) + s^*(x^{\sim\sim})$;
(5) $s^*(x \oplus y) = s^*(x) + s^*(y)$ iff $s(y^- \odot x^-) = s(y^\sim \odot x^\sim) = 0$ *and* $s(x^-) + s(x^\sim) + s(y^-) + s(y^\sim) = 1$;
(6) $\min\{s(x^-), s(x^\sim)\} \leq \frac{1}{2}$.

Proof

(1) $s^*(1) = 1 - s(1^- \oplus 1^\sim) = 1 - s(0 \oplus 0) = 1 - s(0) = 1$.
(2) This follows immediately from the definition of s^* and ($psbck\text{-}c_{18}$).
(3) Since $x \in \mathrm{Rad}(A)$, $x \leq x \oplus y$ and $\mathrm{Rad}(A)$ is a filter of A, we have $x \oplus y \in \mathrm{Rad}(A)$. It follows that $(x \oplus y)^-, (x \oplus y)^\sim \in \mathrm{Rad}(A)^*$, hence $(x \oplus y)^- \perp (x \oplus y)^\sim$.

By the definition of s^* we get

$$s^*(x \oplus y) = 1 - s((x \oplus y)^- \oplus (x \oplus y)^\sim) = 1 - [s((x \oplus y)^-) + s((x \oplus y)^\sim)].$$

Taking into consideration the identities:

$$s((x \oplus y)^-) = s((y^- \odot x^-)^{\sim-}) = s(y^- \odot x^-) \quad \text{and}$$
$$s((x \oplus y)^\sim) = s((y^\sim \odot x^\sim)^{-\sim}) = s(y^\sim \odot x^\sim)$$

we get $s^*(x \oplus y) = 1 - [s(y^- \odot x^-) + s(y^\sim \odot x^\sim)]$.
(4) We have

$$s^*(x^{--}) = 1 - s(x^{---} \oplus x^{--\sim}) = 1 - s(x^{---}) - s(x^-)$$

and similarly

$$s^*\left(x^{\sim\sim}\right) = 1 - s\left(x^{\sim\sim-} \oplus x^{\sim\sim\sim}\right) = 1 - s\left(x^{\sim}\right) - s\left(x^{\sim\sim\sim}\right)$$

(here we applied the fact that $x, y \in \mathrm{Rad}(A)$ implies $x^{---}, x^{--\sim}, x^{\sim\sim-},$ $x^{\sim\sim\sim} \in \mathrm{Rad}(A)^*$, so $x^{---} \perp x^{--\sim}$ and $x^{\sim\sim-} \perp x^{\sim\sim\sim}$).

Thus $s^*(x^{--}) + s^*(x^{\sim\sim}) = 1 - [s(x^{---}) + s(x^{\sim\sim\sim})] + s^*(x)$, so $1 + s^*(x) \le s^*(x^{--}) + s^*(x^{\sim\sim})$.

(5) Applying (3) we get:

$$s^*(x \oplus y) = s^*(x) + s^*(y) \quad \text{iff}$$

$$1 + s\left(y^- \odot x^-\right) + s\left(y^\sim \odot x^\sim\right) = s\left(x^- \oplus x^\sim\right) + s\left(y^- \oplus y^\sim\right) \quad \text{iff}$$

$$1 + s\left(\left(y^- \odot x^-\right) \oplus \left(y^\sim \odot x^\sim\right)\right) = s\left(\left(x^- \oplus x^\sim\right) \oplus \left(y^- \oplus y^\sim\right)\right).$$

Since $1 + s((y^- \odot x^-) \oplus (y^\sim \odot x^\sim)) \ge 1$ and $s((x^- \oplus x^\sim) \oplus (y^- \oplus y^\sim)) \le 1$, we get that:

$$s^*(x \oplus y) = s^*(x) + s^*(y) \quad \text{iff}$$

$$s\left(\left(y^- \odot x^-\right) \oplus \left(y^\sim \odot x^\sim\right)\right) = 0 \quad \text{and}$$

$$s\left(\left(x^- \oplus x^\sim\right) \oplus \left(y^- \oplus y^\sim\right)\right) = 1 \quad \text{iff}$$

$$s\left(y^- \odot x^-\right) = s\left(y^\sim \odot x^\sim\right) = 0 \quad \text{and}$$

$$s\left(x^-\right) + s\left(x^\sim\right) + s\left(y^-\right) + s\left(x^\sim\right) = 1.$$

(6) We have $s^*(x) = 1 - s(x^- \oplus x^\sim) = 1 - [s(x^-) + s(x^\sim)]$, so $s(x^-) + s(x^\sim) \le 1$. Thus $\min\{s(x^-), s(x^\sim)\} \le \frac{1}{2}$. $\qquad\square$

Theorem 6.14 (Extension theorem) *Let A be a perfect pseudo-MTL algebra which is relatively free of zero divisors. Then every local state on A can be extended to a Riečan state on A.*

Proof Let $s : \mathrm{Rad}(A)^* \longrightarrow [0, \infty)$ be a local state on A and the function $S : A \longrightarrow [0, 1]$ defined by

$$S(x) := \begin{cases} 1 & \text{if } x \in \mathrm{Rad}(A) \\ s(x) \cap [0, 1] & \text{if } x \in \mathrm{Rad}(A)^*. \end{cases}$$

We prove that S is a Riečan state on A. More precisely, since it is obvious that $S(1) = 1$, we must prove that for all $x, y \in A$ such that $x \perp y$, we have $S(x + y) = S(x) + S(y)$.

Consider the following cases:

(1) $x, y \in \mathrm{Rad}(A)$.
    According to Lemma 5.4, $x \not\perp y$.

(2) $x \in \text{Rad}(A)$ and $y \in \text{Rad}(A)^*$.

Applying Lemma 5.3 we get $x \perp y$ iff $y = 0$. We have $S(x + y) = S(x + 0) = S(x \oplus 0) = S(x^{-\sim}) = 1$ (since $x^{-\sim} \in \text{Rad}(A)$). On the other hand, $S(x) + S(y) = S(x) + S(0) = 1 + s(0) = 1$. Thus $S(x + y) = S(x) + S(y)$.

(3) $x \in \text{Rad}(A)^*$ and $y \in \text{Rad}(A)$.

Applying Lemma 5.3 we get $x \perp y$ iff $x = 0$. We have $S(x + y) = S(0 + y) = S(0 \oplus y) = S(y^{-\sim}) = 1$ and $S(x) + S(y) = S(0) + S(y) = s(0) + 1 = 1$.

Thus $S(x + y) = S(x) + S(y)$.

(4) $x, y \in \text{Rad}(A)^*$.

By Proposition 5.13 it follows that $x \perp y$ and by the definition of a local additive measure we have:

$$S(x + y) = S(x \oplus y) = s(x \oplus y) \cap [0, 1] = \big(s(x) + s(y)\big) \cap [0, 1] = S(x) + S(y).$$

Thus S is a Riečan state on A.   □

We call S the *extension* of the local state s.

Example 6.17 If s is the local state from Example 6.16(1), then the function $S : A \longrightarrow [0, 1]$ defined by $S(x) := 1$ for all $x \in \text{Rad}(A)$ and $S(x) := 0$ for all $x \in \text{Rad}(A)^*$ is an extension of s.

Theorem 6.15 *Let A be a perfect pseudo-MTL algebra relatively free of zero divisors. The extension S of a local state s on A is a Bosbach state on A if and only if $s(x) = 0$ for all $x \in \text{Rad}(A)^*$.*

Proof According to Theorem 6.13, the map $s : A \rightarrow [0, 1]$ defined by

$$s(x) := \begin{cases} 1 & \text{if } x \in \text{Rad}(A) \\ 0 & \text{if } x \in \text{Rad}(A)^* \end{cases}$$

is a Bosbach state on A.

Conversely, let A be a free of zero divisors perfect pseudo-MTL algebra, s be a local additive measure on A and S the extension of s. We will investigate the conditions for S to be a Bosbach state on A. Obviously, $S(1) = 1$ and $S(0) = 0$. In order to check the conditions (B_1) and (B_2) from the definition of a Bosbach state (Definition 6.1), we consider the following cases:

(1) $x, y \in \text{Rad}(A)$.

Obviously, $S(x) = S(y) = 1$. Since $\text{Rad}(A)$ is a filter of A and $x \leq y \rightarrow x$, $y \leq x \rightarrow y$, it follows that $x \rightarrow y, y \rightarrow x \in \text{Rad}(A)$.

Hence $S(x \rightarrow y) = S(y \rightarrow x) = 1$.

Similarly $S(x \rightsquigarrow y) = S(y \rightsquigarrow x) = 1$, so conditions (B_1) and (B_2) are verified.

(2) $x, y \in \text{Rad}(A)^*$.

We will prove that $x \rightarrow y, y \rightarrow x \in \mathrm{Rad}(A)$. Indeed, suppose that $x \rightarrow y \in \mathrm{Rad}(A)^*$. Since $x \leq x^{-\sim}$, it follows that $x^{-\sim} \rightarrow y \leq x \rightarrow y$, so $x^{-\sim} \rightarrow y \in \mathrm{Rad}(A)^*$. But $x^- \leq x^{-\sim} \rightarrow y$, hence $x^- \in \mathrm{Rad}(A)^*$, that is, $x \in \mathrm{Rad}(A)$, which is a contradiction. It follows that $x \rightarrow y \in \mathrm{Rad}(A)$ and similarly, $y \rightarrow x \in \mathrm{Rad}(A)$. Hence $S(x \rightarrow y) = S(y \rightarrow x) = 1$. In the same way we can prove that $S(x \rightsquigarrow y) = S(y \rightsquigarrow x) = 1$, so (B_1) and (B_2) are verified iff $s(x) = s(y)$ for all $x, y \in \mathrm{Rad}(A)^*$. Since $s(0) = 0$, it follows that (B_1) and (B_2) are verified iff $s(x) = 0$ for all $x \in \mathrm{Rad}(A)^*$.

(3) $x \in \mathrm{Rad}(A)$, $y \in \mathrm{Rad}(A)^*$.

Obviously, $S(x) = 1$ and $S(y) = s(y)$. As $x \leq y \rightarrow x$, we get $y \rightarrow x \in \mathrm{Rad}(A)$.

We show that $x \rightarrow y \in \mathrm{Rad}(A)^*$. Indeed, assume that $x \rightarrow y \in \mathrm{Rad}(A)$.

Since $y \leq y^{-\sim}$ we have $x \rightarrow y \leq x \rightarrow y^{-\sim}$, so $x \rightarrow y^{-\sim} \in \mathrm{Rad}(A)$. This means that $(x \odot y^\sim)^- \in \mathrm{Rad}(A)$, that is, $x \odot y^\sim \in \mathrm{Rad}(A)^*$. On the other hand, since $\mathrm{Rad}(A)$ is a filter of A and $x, y^\sim \in \mathrm{Rad}(A)$ we have $x \odot y^\sim \in \mathrm{Rad}(A)$, a contradiction. We conclude that $x \rightarrow y \in \mathrm{Rad}(A)^*$, so $S(x \rightarrow y) = s(x \rightarrow y)$ and $S(y \rightarrow x) = 1$. Similarly, $S(x \rightsquigarrow y) = s(x \rightsquigarrow y)$ and $S(y \rightsquigarrow x) = 1$. Thus conditions (B_1) and (B_2) are verified iff $s(x \rightarrow y) = s(y)$ and $s(x \rightsquigarrow y) = s(y)$. For $y = 0$, from the first identity we get $s(x^-) = s(0) = 0$ and replacing x with $y^\sim$ we have $s(y^{-\sim}) = 0$. By Proposition 6.27(1) it follows that $s(y) = 0$.

(4) $x \in \mathrm{Rad}(A)^*$, $y \in \mathrm{Rad}(A)$.

Obviously, $S(x) = s(x)$ and $S(y) = 1$. Since $y \leq x \rightarrow y$, we get $x \rightarrow y \in \mathrm{Rad}(A)$.

We show that $y \rightarrow x \in \mathrm{Rad}(A)^*$. Indeed, assume that $y \rightarrow x \in \mathrm{Rad}(A)$.

From $x \leq x^{-\sim}$ we have $y \rightarrow x \leq y \rightarrow x^{-\sim}$, so $y \rightarrow x^{-\sim} \in \mathrm{Rad}(A)$. This means that $(y \odot x^\sim)^- \in \mathrm{Rad}(A)$, that is, $y \odot x^\sim \in \mathrm{Rad}(A)^*$. But $y, x^\sim \in \mathrm{Rad}(A)$, so $y \odot x^\sim \in \mathrm{Rad}(A)$, a contradiction. Hence $y \rightarrow x \in \mathrm{Rad}(A)^*$, so $S(y \rightarrow x) = s(y \rightarrow x)$ and $S(x \rightarrow y) = 1$. Similarly, $S(y \rightsquigarrow x) = s(y \rightsquigarrow x)$ and $S(x \rightsquigarrow y) = 1$. It follows that conditions (B_1) and (B_2) are verified iff $s(x) = s(y \rightarrow x)$ and $s(x) = s(y \rightsquigarrow x)$. Taking $x = 0$ in the first identity, we have $s(y^-) = 0$ and replacing y with $x^\sim$ it follows that $s(x^{-\sim}) = 0$. Applying Proposition 6.27(1) we get $s(x) = 0$. $\qquad\square$

Chapter 7
Measures on Pseudo-BCK Algebras

In this chapter we generalize the measures on BCK-algebras introduced by Dvurečenskij and Pulmannova in [94] and [108] to pseudo-BCK algebras that are not necessarily bounded. In particular, we show that if A is a downwards-directed pseudo-BCK algebra and m a measure on it, then the quotient over the kernel of m can be embedded into the negative cone of an Abelian, Archimedean ℓ-group as its subalgebra. This result will enable us to characterize nonzero measure-morphisms as measures whose kernel is a maximal deductive system.

7.1 Measures on Pseudo-BCK Algebras

Consider the bounded BCK(P) algebra $\mathcal{A}_{\text{Ł}} = ([0, 1], \leq, \rightarrow_{\text{Ł}}, 0, 1)$, where $\rightarrow_{\text{Ł}}$ is the Łukasiewicz implication $x \rightarrow_{\text{Ł}} y = \min\{1 - x + y, 1\}$ (i.e. the standard MV-algebra $([0, 1], \odot_{\text{Ł}}, ^{-}, 1)$).

Definition 7.1 Let $(A, \leq, \rightarrow, \rightsquigarrow, 1)$ be a pseudo-BCK algebra. If $m : A \longrightarrow [0, \infty)$ is such that for all $x, y \in A$:

(1) $m(x \rightarrow y) = m(x \rightsquigarrow y) = m(y) - m(x)$ whenever $y \leq x$, then m is said to be a *measure*;
(2) if $0 \in A$ and m is a measure with $m(0) = 1$, then m is said to be a *state-measure*;
(3) if $m(x \rightarrow y) = m(x \rightsquigarrow y) = \max\{0, m(y) - m(x)\}$, then m is said to be a *measure-morphism*;
(4) if $0 \in A$, $m(0) = 1$ and m is a measure-morphism, then m is said to be a *state-measure-morphism*.

Of course, the function vanishing on A is always a (trivial) measure.

We note that our definition of a measure (a state-measure) defines a map that maps a pseudo-BCK algebra that is in the "negative cone" to the positive cone of the reals $\mathbb{R}$. For a relationship with the previous type of Bosbach state see the second part of the present section and Remark 7.4.

L.C. Ciungu, *Non-commutative Multiple-Valued Logic Algebras*,
Springer Monographs in Mathematics, DOI 10.1007/978-3-319-01589-7_7,
© Springer International Publishing Switzerland 2014

Example 7.1 Let $(G, \vee, \wedge, \vee, +, -, 0)$ be an ℓ-group with negative cone G^- (see Example 1.2). Assume that m is a positive-valued function on G^- that preserves addition in G^-. Then m is a measure on the bounded pseudo-BCK algebra G^-, and conversely if m is a measure on G^-, then m is additive on G^- and positive-valued.

We recall that not every negative cone, even of an Abelian ℓ-group, admits a nontrivial measure. For an example see Example 9.6 in [155].

Proposition 7.1 *Let m be a measure on a pseudo-BCK algebra A. For all $x, y \in A$, we have:*

(1) $m(1) = 0$;
(2) $m(x) \geq m(y)$ *whenever* $x \leq y$;
(3) $m(x \vee_1 y) = m(y \vee_1 x)$ *and* $m(x \vee_2 y) = m(y \vee_2 x)$;
(4) $m(x \vee_1 y) = m(x \vee_2 y)$;
(5) $m(x \to y) = m(x \rightsquigarrow y)$.

Proof

(1) Since $1 \leq 1$ we get $m(1) = m(1 \to 1) = m(1) - m(1) = 0$.
(2) Since $x \leq y$ it follows that $m(y \to x) = m(x) - m(y)$, so $m(x) - m(y) \geq 0$.
(3) First, let $x \leq y$. Then by Proposition 1.5(4), we have $m(x \vee_1 y) = m(y)$. Using the property of measures, we have:

$$m\big((y \vee_1 x) \to x\big) = m(x) - m(y \vee_1 x) = m(x) - m\big((y \to x) \rightsquigarrow x\big)$$
$$= m(x) - m(x) + m(y \to x)$$
$$= m(x) - m(x) + m(x) - m(y)$$
$$= m(x) - m(y),$$

giving $m(y \vee_1 x) = m(y)$. Hence $m(x \vee_1 y) = m(y \vee_1 x) = m(y)$, whenever $x \leq y$.

Now let $x, y \in A$ be arbitrary. Using the first part of the present proof and (2), we have:

$$m(x \vee_1 y) = m\big(x \vee_1 (x \vee_1 y)\big) = m\big((x \vee_1 y) \vee_1 x\big) \leq m(y \vee_1 x)$$
$$= m\big(y \vee_1 (y \vee_1 x)\big) = m\big((y \vee_1 x) \vee_1 y\big) \leq m(x \vee_1 y).$$

Thus $m(x \vee_1 y) = m(y \vee_1 x)$ for all $x, y \in A$.

In a similar way we prove that $m(x \vee_2 y) = m(y \vee_2 x)$.
(4) First, again let $x \leq y$. Then $m(x \vee_1 y) = m(y)$ and $m(y) = m(x \vee_2 y) = m(y \vee_2 x)$. This gives $m(x \vee_1 y) = m(x \vee_2 y) = m(y)$.

Now let $x, y \in A$ be arbitrary. Using (3), we have:

$$m(x \vee_1 y) = m\big(x \vee_1 (x \vee_1 y)\big) = m\big(x \vee_2 (x \vee_1 y)\big) = m\big((x \vee_1 y) \vee_2 x\big)$$
$$\leq m(y \vee_2 x) = m(x \vee_2 y) = m\big(x \vee_2 (x \vee_2 y)\big) = m\big((x \vee_2 y) \vee_2 x\big)$$
$$= m\big((x \vee_2 y) \vee_1 x\big) \leq m(y \vee_1 x) = m(x \vee_1 y).$$

It follows that $m(x \vee_1 y) = m(x \vee_2 y)$.

(5) According to Proposition 1.7 and (4),

$$m(x \to y) = m((x \vee_1 y) \to y) = m(y) - m(x \vee_1 y)$$
$$= m(y) - m(x \vee_2 y) = m((x \vee_2 y) \rightsquigarrow y) = m(x \rightsquigarrow y). \qquad \square$$

Lemma 7.1 *Let m be a state-measure on a bounded pseudo-BCK algebra A. Then for all $x \in A$, we have $m(x^-) = m(x^\sim) = 1 - m(x)$.*

Proof Since $0 \leq x$, we have:

$$m(x^-) = m(x \to 0) = m(0) - m(x) = 1 - m(x).$$
$$m(x^\sim) = m(x \rightsquigarrow 0) = m(0) - m(x) = 1 - m(x). \qquad \square$$

Proposition 7.2 *Let A be a pseudo-BCK algebra. Then:*

(1) $y \leq x$ *implies* $m((x \to y) \rightsquigarrow y) = m((x \rightsquigarrow y) \to y) = m(x)$ *whenever m is a measure on A;*
(2) *if m is a measure on A, then* $\mathrm{Ker}_0(m) = \{x \in A \mid m(x) = 0\}$ *is a normal deductive system of A;*
(3) *any measure-morphism on A is a measure on A.*

Proof

(1) From $y \leq x \to y$ we get

$$m((x \to y) \rightsquigarrow y) = m(y) - m(x \to y) = m(y) - (m(y) - m(x)) = m(x).$$

Similarly, $m((y \rightsquigarrow x) \to x) = m(x)$.
(2) According to Proposition 7.1(1), $1 \in \mathrm{Ker}_0(m)$.
    Assume that $x, x \to y \in \mathrm{Ker}_0(m)$.
    Since $x \leq x \vee_1 y$, by Proposition 7.1(2), we have $0 = m(x) \geq m(x \vee_1 y)$, so $m(x \vee_1 y) = 0$.
    In addition, $0 = m(x \to y) = m((x \vee_1 y) \to y) = m(y) - m(x \vee_1 y) = m(y)$, so $y \in \mathrm{Ker}_0(m)$. (Here we applied the fact that $y \leq x \vee_1 y$ and Proposition 1.7.)
    Thus $\mathrm{Ker}_0(m)$ is a deductive system of A.
    The normality of $\mathrm{Ker}_0(m)$ follows from Proposition 7.1(5).
(3) We have $m(1) = m(1 \to 1) = \max\{0, m(1) - m(1)\} = 0$, so if $y \leq x$ then $0 = m(1) = m(y \to x) = \max\{0, m(x) - m(y)\}$ and $m(x) \leq m(y)$, thus $m(x \to y) = \max\{0, m(y) - m(x)\} = m(y) - m(x)$.
    Similarly, $m(x \rightsquigarrow y) = m(y) - m(x)$. $\qquad \square$

Example 7.2 Consider the bounded pseudo-BCK lattice A_1 from Example 1.16. The function $m : A_1 \longrightarrow [0, \infty)$ defined by: $m(0) := 1$, $m(a) = m(b) = m(c) = m(d) = m(1) := 0$ is the unique measure on A_1. Moreover, m is even a state-measure on A_1.

Proposition 7.3 *Let A be a bounded pseudo-BCK algebra. If M is a Bosbach state, then $m := 1 - M$ is a state-measure.*

Proof Let $y \leq x$, that is, $y \to x = y \rightsquigarrow x = 1$.

By Proposition 6.1(1), $M(x \to y) = M(x \rightsquigarrow y) = 1 - M(x) + M(y)$.

It follows that:

$$m(x \to y) = m(x \rightsquigarrow y) = 1 - M(x \to y) = M(x) - M(y)$$
$$= 1 - M(y) - \big(1 - M(x)\big) = m(y) - m(x).$$

Since by (B_3) $m(0) = 1 - M(0) = 1$, we conclude that m is a state-measure. $\square$

Proposition 7.4 *Let A be a bounded pseudo-BCK algebra. If m is a state-measure on A, then $M := 1 - m$ is a Bosbach state on A.*

Proof We have: $y \leq x \vee_1 y$ and using the definition of the measure, we get

$$m\big((x \vee_1 y) \to y\big) = m\big((x \vee_1 y) \rightsquigarrow y\big) = m(y) - m(x \vee_1 y).$$

Using Proposition 1.7, we have: $x \vee_1 y \to y = x \to y$, so we get $m(x \to y) = m(y) - m(x \vee_1 y)$.

Similarly, $m(y \to x) = m(x) - m(y \vee_1 x)$.

In the same way we get:

$$m(x \rightsquigarrow y) = m(y) - m(x \vee_2 y) \quad \text{and} \quad m(y \rightsquigarrow x) = m(x) - m(y \vee_2 x).$$

According to Proposition 7.1(3) we have $m(x \vee_1 y) = m(y \vee_1 x)$, so $m(x) + m(x \to y) = m(y) + m(y \to x)$.

Similarly, $m(x) + m(x \rightsquigarrow y) = m(y) + m(y \rightsquigarrow x)$.

Therefore:

$$M(x) + M(x \to y) = M(y) + M(y \to x) \quad \text{and}$$
$$M(x) + M(x \rightsquigarrow y) = M(y) + M(y \rightsquigarrow x).$$

Furthermore, $M(0) = 0$ by the hypothesis and $M(1) = 1$ by Proposition 7.1(1). Thus M is a Bosbach state. $\square$

If A is a bounded pseudo-BCK algebra, in a similar way as for Bosbach states, we can define extremal state-measures, as well as the weak-topology. Denote the set of state-measures by $\mathcal{SM}_1(A)$, the set of state-measure-morphisms by $\mathcal{SMM}_1(A)$, and the set of extremal state-measures by $\partial_e \mathcal{SM}_1(A)$, respectively.

Theorem 7.1 *Let A be a bounded pseudo-BCK algebra. Define a map $\Psi :$ $\mathcal{SM}_1(A) \to \mathcal{BS}(A)$ by $\Psi(m) := 1 - m$, $m \in \mathcal{SM}_1(A)$. Then Ψ is an affine-homeomorphism such that m is a state-measure-morphism if and only if $\Psi(m)$ is a state-morphism. In particular, m is an extremal state-measure if and only if m is a state-measure-morphism.*

Proof Propositions 7.3–7.4 show that Ψ is a bijection preserving convex combinations and weak topologies.

If, say, s is a state-morphism on A, i.e. $s(x \to y) = \min\{1 - m(x) + m(y), 1\}$, then it is straightforward to show that $m = 1 - s$ is a state-measure-morphism on A, i.e. $m(x \to y) = \max\{m(y) - m(x), 0\}$ (likewise for the second arrow $\rightsquigarrow$).

In view of Theorem 6.2, we see that a state-measure is extremal iff it is a state-measure-morphism.   □

Remark 7.1 As a corollary of Theorem 7.1 and Remark 6.4, we have that if A is a bounded pseudo-BCK algebra, then

$$\partial_e \mathcal{SM}_1(A) = \mathcal{SMM}_1(A).$$

Theorem 7.2 *Let m be a measure on a pseudo-BCK algebra A.*

Then $A / \mathrm{Ker}_0(m)$ is a pseudo-BCK algebra and the mapping $\hat{m} : A / \mathrm{Ker}_0(m) \longrightarrow [0, +\infty)$ defined by $\hat{m}(\bar{x}) := m(x)$, $\bar{x} := x / \mathrm{Ker}_0(m) \in A / \mathrm{Ker}_0(m)$, is a measure on $A / \mathrm{Ker}_0(m)$, and $A / \mathrm{Ker}_0(m)$ is sup-commutative.

Proof By Proposition 7.1(5) we have $m(x \to y) = m(x \rightsquigarrow y)$.

According to Proposition 7.2(2), $\mathrm{Ker}_0(m)$ is a normal deductive system of A.

Consider $\bar{x} = \bar{y}$. Then $x \to y$, $y \to x \in \mathrm{Ker}_0(m)$ and $m(x \to y) = m((x \vee_1 y) \to y) = 0 = m(y) - m(x \vee_1 y)$, so $m(y) = m(x \vee_1 y)$.

Similarly, $m(x) = m(y \vee_1 x)$. But $m(y) = m(x \vee_1 y) = m(y \vee_1 x) = m(x)$.

Hence $\hat{m}$ is a well-defined function on $A / \mathrm{Ker}_0(m)$.

To show that $\hat{m}$ is a measure, assume $\bar{y} \leq \bar{x}$. By Proposition 1.5(4), $\bar{y} \vee_1 \bar{x} = \bar{x}$.

Then $\hat{m}(\bar{x} \to \bar{y}) = m(x \to y) = m((x \vee_1 y) \to y) = m(y) - m(x \vee_1 y)$.

But $m(x \vee_1 y) = m(y \vee_1 x) = \hat{m}(\bar{y} \vee_1 \bar{x}) = \hat{m}(\bar{x}) = m(x)$.

Therefore $\hat{m}(\bar{x} \to \bar{y}) = \hat{m}(\bar{y}) - \hat{m}(\bar{x})$. Similarly, $\hat{m}(\bar{x} \rightsquigarrow \bar{y}) = \hat{m}(\bar{y}) - \hat{m}(\bar{x})$.

In the same way as in the proof of Proposition 6.11 we can show that $A / \mathrm{Ker}_0(M)$ is both $\vee_1$-commutative and $\vee_2$-commutative.   □

In view of Theorem 6.1 and Theorem 7.1 we know that if m is a state-measure on a bounded pseudo-BCK algebra A, then $A / \mathrm{Ker}(m)$ is in fact an MV-algebra, so that according to Mundici's famous representation theorem, [41], $A / \mathrm{Ker}(m)$ is an interval in an ℓ-group with strong unit. In the following result we generalize this ℓ-group representation of the quotient for measures on unbounded pseudo-BCK algebras that are downwards-directed.

Theorem 7.3 *Let m be a measure on an unbounded pseudo-BCK algebra A that is a downwards-directed set. Then the arrows $\to$ and $\rightsquigarrow$ on $A / \mathrm{Ker}_0(m)$ coincide. Moreover, there is a unique (up to isomorphism) Archimedean ℓ-group G such that $A / \mathrm{Ker}_0(m)$ is a subalgebra of the pseudo-BCK algebra G^- and $A / \mathrm{Ker}_0(m)$ generates the ℓ-group G.*

Proof We note that if a is an arbitrary element of A, then $([a, 1], \leq, \to, \rightsquigarrow, a, 1)$ is a pseudo-BCK algebra.

We define $K_0 := \mathrm{Ker}_0(m)$. Given $x, y \in A$, choose an element $a \in A$ such that $a \leq x, y$. If $m(a) = 0$, then $a / K_0 = (x \to y) / K_0 = (x \rightsquigarrow y) / K_0 = 1 / K_0$.

Assume $m(a) > 0$ and define $m_a(z) := m(z)/m(a)$ for any $z \in [a, 1]$. Then m_a is a state-measure on $[a, 1]$ and in view of Theorem 7.1, $s_a := 1 - m_a$ is a Bosbach state on $[a, 1]$.

Theorem 6.1 entails that $[a, 1]/\mathrm{Ker}(s_a)$ can be converted into an Archimedean MV-algebra. In particular, $(x \to y)/\mathrm{Ker}(s_a) = (x \leadsto y)/\mathrm{Ker}_0(s_a)$. This yields $s_a((x \to y) \to (x \leadsto y)) = 1$ and $m((x \to y) \to (x \leadsto y)) = 0$. In a similar way, $m((x \leadsto y) \to (x \to y)) = 0$. This proves $\to/\mathrm{Ker}_0(m) = \leadsto/\mathrm{Ker}_0(m)$.

In addition, we can prove that, for all $x, y \in A$,

$$\big((x \to y) \vee (y \to x)\big)/\mathrm{Ker}_0(m) = 1/\mathrm{Ker}_0(m) = \big((x \leadsto y) \vee (y \leadsto x)\big)/\mathrm{Ker}_0(m).$$

It is clear that if $m = 0$, then $\mathrm{Ker}_0(m) = A$ and $A/\mathrm{Ker}_0(m) = \{1/\mathrm{Ker}_0(m)\}$, so that the trivial ℓ-group $G = \{0_G\}$, where 0_G is a neutral element of G, satisfies our conditions.

Therefore, let $m \neq 0$. By Lemma 4.1.8 in [208], $A/\mathrm{Ker}_0(m)$ is a distributive lattice. As in Proposition 1.10 we can show that $A/\mathrm{Ker}_0(m)$ satisfies the (RCP) condition, and therefore, $A/\mathrm{Ker}_0(m)$ is a Łukasiewicz BCK algebra, see [112]. Therefore, [103, 112], there is a unique (up to isomorphism of ℓ-groups) ℓ-group G such that $A/\mathrm{Ker}_0(m)$ can be embedded into the pseudo-BCK algebra of the negative cone G^-. Moreover, $A/\mathrm{Ker}_0(m)$ generates G.

Since the arrows in $A/\mathrm{Ker}_0(m)$ coincide, we see that G is Abelian, and since every interval $[a/K_0, 1/K_0]$ is an Archimedean MV-algebra, so is G. $\square$

Remark 7.2 We note that if m is a measure-morphism on A, then:

(1) $m(u \to^n x) = \max\{0, m(x) - nm(u)\}$ for any $n \geq 0$;
(2) $m(x_1 \to (\cdots \to (x_n \to a) \cdots)) = \max\{0, m(a) - m(x_1) - \cdots - m(x_n)\}$.

Proposition 7.5 *Let m be a measure-morphism on a pseudo-BCK algebra A such that $m \neq 0$. Then $\mathrm{Ker}_0(m)$ is a normal and maximal deductive system of A.*

Proof Since m is a measure-morphism, then by Proposition 7.2(2), $\mathrm{Ker}_0(m)$ is a normal deductive system.

Choose $a \in A$ such that $m(a) \neq 0$. Let F be the deductive system generated by $\mathrm{Ker}_0(m)$ and by the element a. Let $z \in A$ be an arbitrary element of A. There is an integer $n \geq 1$ such that $(n - 1)m(a) \leq m(z) < nm(a)$. By Remark 7.2(1), we have that $m(a \to^n z) = \max\{0, m(z) - nm(a)\} = 0$, so $z \in F$ and $A \subseteq F$ proving that $\mathrm{Ker}_0(m)$ is a maximal deductive system. $\square$

If $m \neq 0$ is a measure on a bounded pseudo-BCK algebra A, then passing to a state-measure $s_m(a) := m(a)/m(1)$, $a \in A$, and using Theorem 7.1, we see that m is a measure-morphism iff $\mathrm{Ker}_0(m)$ is a maximal deductive system.

The same result is true for unbounded pseudo-BCK algebras that are downwards-directed:

Theorem 7.4 *Let $m \neq 0$ be a measure on an unbounded pseudo-BCK algebra A that is downwards-directed. Then m is a measure-morphism if and only if $\mathrm{Ker}_0(m)$ is a maximal deductive system.*

Proof By Proposition 7.5, $\text{Ker}_0(m)$ is a maximal deductive system of A.

Suppose now $\text{Ker}_0(m)$ is a maximal deductive system of A. In view of Theorem 7.3, $A/\text{Ker}_0(m)$ can be embedded as a subalgebra into the pseudo-BCK algebra G^-, where G^- is the negative cone of an Abelian and Archimedean ℓ-group G that is generated by $A/\text{Ker}_0(m)$. Let $\hat{m}(a/\text{Ker}_0(m)) := m(a)$ $(a \in A)$. Then $\text{Ker}_0(\hat{m}) = \{1/\text{Ker}_0(m)\}$ and $0_G := 1/\text{Ker}_0(m)$ is the neutral element of G.

Fix an element $a \in A$ with $m(a) > 0$. Since $\text{Ker}_0(m)$ is maximal in A, $\text{Ker}_0(\hat{m}) = \{1/\text{Ker}_0(m)\}$ is maximal in $A/\text{Ker}_0(m)$ and consequently, $\{1/\text{Ker}_0(m)\}$ is a maximal deductive system of the pseudo-BCK algebra G^- because $A/\text{Ker}_0(m)$ generates G. Therefore the ℓ-ideal $L := \{0_g\} = \{1/\text{Ker}_0(m)\}$ is a maximal ℓ-ideal of G.

We recall that every maximal ℓ-ideal, L, of an ℓ-group is prime ($a, b \in G^+$ with $a \wedge b = 0$ implies $a \in L$ or $b \in L$), whence G/L is a linearly ordered ℓ-group (see e.g. Proposition 9.9 in [76]).

Since $G = G/L$, G is Archimedean and linearly ordered, and by the Hölder theorem, Theorem 24.16 in [76], G is an ℓ-subgroup of the ℓ-group of real numbers, $\mathbb{R}$. Let s be the unique extension of $\hat{m}$ onto G, then s is additive on G and $s(g) \geq 0$ for any $g \in G^-$. Since G is an ℓ-subgroup of $\mathbb{R}$, s is a unique additive function on G that is positive on the negative cone (see Example 7.1) with the property $s(a/\text{Ker}_0(m)) = m(a) > 0$ for our fixed element $a \in A$. Because $A/\text{Ker}_0(m)$ can be embedded into $\mathbb{R}^-$, we see that s is a measure-morphism on G^-.

Consequently, m is a measure-morphism on A. $\square$

Proposition 7.6 *Let m_1 and m_2 be two measure-morphisms on a downwards-directed pseudo-BCK algebra A such that there is an element $a \in A$ with $m_1(a) = m_2(a) > 0$. If $\text{Ker}_0(m_1) = \text{Ker}_0(m_2)$, then $m_1 = m_2$.*

In addition, let $a \in A$ be fixed. If m is a measure-morphism on A such that $m(a) > 0$, then m cannot be expressed as a convex combination of two measures m_1 and m_2 such that $m_1(a) = m_2(a) = m(a)$.

Proof

(1) By Theorem 7.4, $A/\text{Ker}_0(m_1) = A/\text{Ker}_0(m_2)$ is a pseudo-BCK subalgebra of $\mathbb{R}^-$. The condition $m_1(a) = m_2(a) > 0$ entails $\hat{m}_1 = \hat{m}_2$, so $m_1 = m_2$.
(2) Let $m = \lambda m_1 + (1 - \lambda)m_2$ where m_1 and m_2 are measures on A such that $m_1(a) = m_2(a) = m(a)$ and $0 < \lambda < 1$. Then $\text{Ker}_0(m) \subseteq \text{Ker}_0(m_1) \cap \text{Ker}_0(m_2)$. The maximality of $\text{Ker}_0(m)$ entails that both $\text{Ker}_0(m_1)$ and $\text{Ker}_2(m)$ are maximal ideals and by Theorem 7.4, we see that m_1 and m_2 are measure-morphisms on A. The condition $m_1(a) = m_2(a) = m(a)$ yields by (1) that $m = m_1 = m_2$. $\square$

Proposition 7.7 *Let m be a state-measure on a good pseudo-BCK algebra A. Then $M := 1 - m$ is a Riečan state on A.*

Proof Let x, y be such that $x \perp y$, that is, $y^{-\sim} \leq x^-$ and using the fact that m is a measure, we obtain $m(x^- \to y^{-\sim}) = m(x^- \rightsquigarrow y^{-\sim}) = m(y^{-\sim}) - m(x^-)$.

Now, because A is good we get $m(x^- \to y^{\sim-}) = m(y) - 1 + m(x)$, which implies $M(x \oplus y) = M(x) + M(y)$. Since $M(1) = 1 - m(1) = 1$, we conclude that M is a Riečan state on A. $\square$

Proposition 7.8 *Let A be a pseudo-BCK(pDN) algebra and s be a Riečan state on A. Then $S := 1 - s$ is a state-measure.*

Proof Let s be a Riečan state on A.

Consider $y \leq x$. Changing x to y in Proposition 6.15(5) we get:

$$s\big((y \vee_1 x) \to y^{-\sim}\big) = s\big((y \vee_1 x) \rightsquigarrow y^{-\sim}\big) = 1 - s(y \vee_1 x) + s(y).$$

But according to Proposition 1.5(4) we have $y \vee_1 x = x$, so

$$s\big(x \to y^{-\sim}\big) = s\big(x \rightsquigarrow y^{-\sim}\big) = 1 - s(x) + s(y).$$

Taking into consideration the (pDN) condition we get

$$s(x \to y) = s(x \rightsquigarrow y) = 1 - s(x) + s(y).$$

It follows that $S(x \to y) = S(x \rightsquigarrow y) = S(y) - S(x)$.

Moreover, we have $S(0) = 1$, so S is a state-measure on A. $\square$

Remark 7.3 We can also define a measure as a map $m : A \longrightarrow (-\infty, 0]$ such that

$$m(x \to y) = m(x \rightsquigarrow y) = m(x) - m(y) \quad \text{whenever } y \leq x.$$

Properties (2) of Proposition 7.1 and (1) of Proposition 7.2 become:

(2′) $m(x) \leq m(y)$ whenever $x \leq y$ and m is a measure on A;
(1′) $y \leq x$ implies $m((x \to y) \rightsquigarrow y) = m((x \rightsquigarrow y) \to y) = -m(x)$ whenever m is a measure on A.

If $m(0) = 0$ then m is a state on A.

Proposition 7.3 will be modified so that $m = 1 + M$.

Consider again the bounded pseudo-BCK lattice A_1 from Example 1.16.

The function $m : A_1 \longrightarrow (-\infty, 0]$ defined by: $m(0) := -1$, $m(a) = m(b) = m(c) = m(d) = m(1) := 0$ is the unique measure on A_1.

Remark 7.4 If a pseudo-BCK algebra is defined on the negative cone, as in Examples 1.2 and 1.4, we map the negative cone to the positive cone in $\mathbb{R}$. According to the second definition, we map the negative cone to negative numbers.

7.2 Pseudo-BCK Algebras with Strong Unit

In this section we will study state-measures on pseudo-BCK algebras with strong unit. We apply the results of the previous section to show how to characterize state-measure-morphisms as extremal state-measures, or as those with the maximal deductive system. In particular, we show that for unital pseudo-BCK algebras that are downwards-directed, the quotient over the kernel can be embedded into the negative cone of an Abelian, Archimedean ℓ-group with strong unit.

According to [108], we say that an element u of a pseudo-BCK algebra A is a *strong unit* if, for the deductive system $F(u) = [u]$ of A that is generated by u, we have $F(u) = A$. For example, if $(A, \leq, \to, \leadsto, 0, 1)$ is a bounded pseudo-BCK algebra, then $u = 0$ is a strong element. If G is an ℓ-group with strong unit $u \geq 0$, then the negative cone G^- is an unbounded pseudo-BCK algebra with strong unit $-u$.

Remark 7.5 We note that a deductive system F of a pseudo-BCK algebra with a strong unit u is a proper subset of A if and only if $u \notin F$.

By a *unital pseudo-BCK algebra* we mean a pair (A, u) where A is a pseudo-BCK algebra with a fixed strong unit u. We say that a measure m on (A, u) is a *state-measure* if $m(u) = 1$. If, in addition, m is a measure-morphism such that $m(u) = 1$, we also call it a *state-measure-morphism*. We denote by $\mathcal{SM}(A, u)$ and $\mathcal{SMM}(A, u)$ the set of all state-measures and state-measure-morphisms on (A, u), respectively. The set $\mathcal{SM}(A, u)$ is convex, i.e. if $m_1, m_2 \in \mathcal{SM}(A, u)$ and $\lambda \in [0, 1]$, then $m := \lambda m_1 + (1 - \lambda)m_2 \in \mathcal{SM}(A, u)$; it could be empty. A state-measure m is *extremal* if $m = \lambda m_1 + (1 - \lambda)m_2$ for $\lambda \in (0, 1)$ yields $m = m_1 = m_2$. We denote by $\partial_e \mathcal{SM}(A, u)$ the set of all extremal state-measures on (A, u).

Example 7.3 Let G be an ℓ-group with strong unit $u \geq 0$. Then a mapping m on G^- is a state-measure on $(G^-, -u)$ if and only if (i) $m : G^- \to [0, \infty)$, (ii) $m(g + h) = m(g) + m(h)$ for $g, h \in G^-$, and (iii) $m(-u) = 1$. A state-measure m is extremal if and only if $m(g \wedge h) = \max\{m(g), m(h)\}$, $g, h \in G^-$ (see Proposition 4.7 in [96]). In addition, $(-u) \to^n g = (g + nu) \wedge 0$ for any $n \geq 1$.

Let $\Omega \neq \emptyset$ be a compact Hausdorff topological space and let $\mathcal{B}(\Omega)$ be the Borel σ-algebra of Ω generated by all open subsets of Ω. Any element of $\mathcal{B}(\Omega)$ is said to be a *Borel set*, and any σ-additive (signed) measure is said to be a *Borel measure*.

Let $\mathcal{P}(\Omega)$ denote all probability measures, that is, all positive regular Borel measures $\mu \in \mathcal{M}(\Omega)$ such that $\mu(\Omega) = 1$.

We recall that a Borel measure μ is called *regular* if

$$\inf\{\mu(O) : Y \subseteq O, O \text{ open}\} = \mu(Y) = \sup\{\mu(C) : C \subseteq Y, C \text{ closed}\}$$

for any $Y \in \mathcal{B}(\Omega)$.

Example 7.4 Let $\Omega \neq \emptyset$ be a compact Hausdorff topological space and let $C(\Omega)$ be the set of all continuous functions on Ω. Then $C(\Omega)$ is an ℓ-group with respect to the pointwise ordering and usual addition of functions and the element $u = 1$, the constant function equal to 1, is a strong unit. According to the Riesz Representation Theorem, see e.g. p. 87 in [155], a mapping $m : A \longrightarrow [0, \infty)$ is a state-measure on

$(C(\Omega)^-, -1)$ if and only if there is a unique regular Borel probability measure μ on $\mathcal{B}(\Omega)$ such that

$$m(f) := -\int_\Omega f(x)\,d\mu(x), \quad f \in C(\Omega)^-,$$

and vice-versa, given a regular Borel probability measure μ, the above integral always defines a state-measure.

If Ω is a separable space, then a state-measure is extremal if and only if it is a state-measure-morphism if and only if $\mu = \delta_x$ for some point $x \in \Omega$, where $\delta_x(M) = 1$ iff $x \in M$, and $\delta_x(M) = 0$ otherwise, so $m(f) = f(x)$.

Definition 7.2 We say that a net of state-measures $\{m_\alpha\}$ *converges weakly* to a state-measure m if $m(a) = \lim_\alpha m_\alpha(a)$ for every $a \in A$.

Proposition 7.9 *The state spaces $\mathcal{SM}(A, u)$ and $\mathcal{SMM}(A, u)$ are compact Hausdorff topological spaces.*

Proof If $\mathcal{SM}(A, u)$ is void, the statement is evident. Thus suppose that (A, u) admits at least one state-measure. For any state-measure m and any $x \in A$ we have by Proposition 1.7: $m(u \to x) = m((u \vee_1 x) \to x) = m(x) - m(u \vee_1 x)$. But $u \le u \vee_1 x$, hence $m(u \vee_1 x) \le m(u) = 1$, so $m(u \to x) \ge m(x) - 1$ and $m(x) \le m(u \to x) + 1$. Therefore

$$m(x) \le m(u \to x) + 1 \le m\big(u \to^2 x\big) + 2 \le \cdots \le m\big(u \to^{n-1} x\big) + n - 1.$$

Since u is strong, given $x \in A$, let n_x denote an integer $n_x \ge 1$ such that $u \to^{n_x} x = 1$. Then $u \le u \to^{n_x-1} x$ and $m(u \to^{n_x-1} x) \le m(u) = 1$. Consequently, $m(x) \le m(u \to^{n_x-1} x) + n_x - 1 \le n_x$. Hence $\mathcal{SM}(A, u) \subseteq \prod_{x \in A}[0, n_x]$. By Tychonoff's Theorem, the product of closed intervals is compact. The set of state-measures $\mathcal{SMM}(A, u)$ can be expressed as an intersection of closed subsets of $[0, \infty)^A$, namely of the following sets (for $x, y \in A$):

$$M_{x,y} = \big\{m \in [0, \infty)^A \mid m(x \to y) = m(x \rightsquigarrow y) = m(y) - m(x)\big\}, \quad x \le y,$$

$$M_x = \big\{m \in [0, \infty)^A \mid m(x) \ge 0\big\}, \big\{m \in [0, \infty)^A \mid m(u) = 1\big\}.$$

Therefore $\mathcal{SM}(A, u)$ is a closed subset of the given product of intervals, and hence, it is compact.

Similarly, the set of state-measure-morphisms $\mathcal{SMM}(A, u)$ is a subset of $\prod_{x \in A}[0, n_x]$ and it can be expressed as an intersection of closed subsets of $[0, \infty)^A$, namely of the following sets (for $x, y \in A$):

$$M_{x,y} = \big\{m \in [0, \infty)^A \mid m(x \to y) = m(x \rightsquigarrow y) = \max\big\{0, m(y) - m(x)\big\}\big\},$$

$$M_x = \big\{m \in [0, \infty)^A \mid m(x) \ge 0\big\}, \big\{m \in [0, \infty)^A \mid m(u) = 1\big\}.$$

Therefore $\mathcal{SMM}(A, u)$ is a closed subset of the given product of intervals, and hence, it is compact. $\qquad\square$

Proposition 7.10 *Let u be a strong unit of a pseudo-BCK algebra A and m be a measure on A. Then m vanishes on A if and only if $m(u) = 0$.*

Proof Assume $m(u) = 0$. Then $m(u \to x) = m((u \vee_1 x) \to x) = m(x) - m(u \vee_1 x)$. But $u \leq u \vee_1 x$, hence $0 \leq m(u \vee_1 x) \leq m(u) = 0$, so $m(x) = m(u \to x)$ and

$$m(x) = m(u \to x) = m\left(u \to^2 x\right) = \cdots = \left(m \to^n x\right) = m(1) = 0$$

when $u \to^n x = 1$ for some integer $n \geq 1$.

If now $m(u) > 0$, then m does not vanish trivially on A. $\square$

Lemma 7.2 *Let m_1, m_2 be state-measure-morphisms on a unital pseudo-BCK algebra (A, u). If $\mathrm{Ker}_0(m_1) = \mathrm{Ker}_0(m_2)$, then $m_1 = m_2$.*

In addition, any state-measure-morphism cannot be expressed as a convex combination of other state-measure-morphisms.

Proof The sets $m_1(A) = \{m_1(a) \mid a \in A\}$ and $m_2(A) = \{m_2(a) \mid a \in A\}$ of real numbers can be endowed with a total operation $*_\mathbb{R}$ such that $(m_1(A), *_\mathbb{R}, 0)$ and $(m_2(A), *_\mathbb{R}, 0)$ are subalgebras of the BCK algebra $([0, \infty), *_\mathbb{R}, 0)$ in the sense of Chap. 5 in [108], where $s *_\mathbb{R} t = \max\{0, s - t\}$, $s, t \in [0, \infty)$. The number 1 is a strong unit in all such algebras.

If we let $\hat{m}_1$ and $\hat{m}_2$ be the state-measure-morphisms on the quotient pseudo-BCK algebras $A/\mathrm{Ker}_0(m_1)$ and $A/\mathrm{Ker}_0(m_2)$ defined by $\hat{m}_i(a/\mathrm{Ker}_0(m_i)) = m_i(a)$, we have again $\hat{m}_i(A/\mathrm{Ker}_0(m_i)) = m_i(A)$ for $i = 1, 2$.

Define a mapping $\phi : m_1(A) \to m_2(A)$ by $\phi(m_1(a)) = m_2(a)$ $(a \in A)$. It is possible to show that this is a BCK-algebra injective homomorphism. By Lemma 6.1.22 in [108], this means that $m_1(A) = m_2(A)$, that is, $m_1(a) = m_2(a)$ for all $a \in A$.

Suppose now that $m = \lambda m_1 + (1 - \lambda)m_2$, where m, m_1, m_2 are state-measure-morphisms and $\lambda \in (0, 1)$. Then $\mathrm{Ker}_0(m) \subseteq \mathrm{Ker}_0(m_1) \cap \mathrm{Ker}_0(m_2)$. By Proposition 7.5, all kernels $\mathrm{Ker}_0(m)$, $\mathrm{Ker}_0(m_1)$, $\mathrm{Ker}_0(m_2)$ are maximal deductive systems, so $\mathrm{Ker}_0(m) = \mathrm{Ker}_0(m_1) = \mathrm{Ker}_0(m_2)$ and by the first part of the present proof, $m = m_1 = m_2$. $\square$

Proposition 7.11 *Let u be a strong unit of a pseudo-BCK algebra A and let J be a deductive system of A and $J_0 := J \cap [u, 1]$. Then J_0 is a deductive system of the pseudo-BCK algebra $([u, 1], \leq, \to, \rightsquigarrow, u, 1)$. If $F(J_0)$ is the deductive system of A generated by J_0, then*

$$F(J_0) = F. \tag{$*$}$$

Moreover, J_0 is maximal in $[u, 1]$ if and only if J is maximal in A.

Proof Suppose that J is a deductive system of A. Then $J_0 := J \cap [u, 1]$ is evidently a deductive system of $[u, 1]$. It is clear that $F(J_0) \subseteq F$.

On the other hand, take $x \in J$. Since u is a strong unit, by Lemma 1.9, there is an integer $n \geq 1$ such that $u \to^n x = 1 = u \to (\cdots \to (u \to x) \cdots)$.

Set $x_n = u \vee_1 x$ and $x_{n-i} = u \vee_1 (u \to^i x)$ for $i = 1, \ldots, n - 1$.

An easy calculation shows that $x_i \in J_0$ for any $i = 1, \ldots, n$.

Moreover, $u \to (u \to (\cdots \to (u \to x) \cdots)) = x_1 \to (x_2 \to (\cdots \to (x_n \to x) \cdots)) = 1$ which by Lemma 1.9 proves $x \in F(J_0)$.

Now let J be a maximal deductive system of A. Assume that F is a deductive system of $[u, 1]$ containing J_0 with $F \neq [u, 1]$, and let $\hat{F}(F)$ be the deductive system of A generated by F.

Then $F \subseteq \hat{F}(F) \cap [u, 1]$.

If now $x \in \hat{F}(F) \cap [u, 1]$, there are $f_1, \ldots, f_n \in F$ such that $f_1 \to (\cdots \to (f_n \to x) \cdots) = 1$ giving $x \in F$. Hence $F = \hat{F}(F) \cap [u, 1]$.

We assert that $\hat{F}(F)$ is a deductive system of A containing J, and $\hat{F}(F) \neq A$. If not, then $u \in \hat{F}(F)$ and therefore by Lemma 1.9, there are $x_1, \ldots, x_n \in F$ such that $x_1 \to (\cdots \to (x_n \to u) \cdots) = 1$. If we set $z_n = x_n \vee_1 u$ and $z_{n-i} = x_{n-i} \vee_1 (x_i \to (\cdots \to (x_n \to u) \cdots))$, for $i = 1, \ldots, n - 1$, then each z_i belongs to F and $z_1 \to (\cdots \to (z_n \to u) \cdots) = 1$ which implies $u \in F$, which is a contradiction.

The maximality of J entails $J = \hat{F}(F)$. Since $J_0 \subseteq F = \hat{F}(F) \cap [u, 1] = J \cap [0, 1] = J_0$. That is, J_0 is a maximal deductive system of $[u, 1]$ as was claimed.

Assume now that J_0 is a maximal deductive system of $[u, 1]$ and let $G \neq A$ be a deductive system of A containing J. Then $G_0 := G \cap [u, 1]$ is a deductive system of $[u, 1]$ containing J_0, and by $(\ast)$, we get $G = F(G_0)$. We assert $u \notin G_0$. Suppose the converse. Then $u \in G$ and for any $x \in A$, there is an integer $n \geq 1$ such that $u \to^n x = u \to (\cdots (u \to x) \cdots) = 1$ proving $x \in G$, so $A \subseteq G$, which is absurd.

The maximality of J_0 entails $J_0 = G_0$ and in view of $(\ast)$, we have $J = F(J_0) = F(G_0) = G$, thus J is a maximal deductive system of A. $\square$

Proposition 7.12 *Let m be a state-measure on a unital pseudo-BCK algebra (A, u), and let m_u be the restriction of m onto the interval $[u, 1]$. Then m_u is a state-measure-morphism on $([u, 1], \leq, \to, \rightsquigarrow, u, 1)$.*

Consider the following conditions:

(a) *m is a state-measure-morphism on (A, u);*
(b) *m_u is a state-morphism on $[u, 1]$;*
(c) *$\mathrm{Ker}_0(m)$ is a maximal deductive system of (A, u);*
(d) *$\mathrm{Ker}_0(m_u)$ is a maximal deductive system of $[u, 1]$.*

Then (b), (c), (d) are mutually equivalent and (a) implies each of the conditions (b), (c) and (d).

Proof Let m_u be the restriction of m to $[u, 1]$. Then m_u is a state-measure on $[u, 1]$ and $\mathrm{Ker}_0(m_u) = \mathrm{Ker}_0(m) \cap [u, 1]$. Due to Proposition 7.11 we have

$$F\big(\mathrm{Ker}_0(m_u)\big) = \mathrm{Ker}_0(m).$$

(a) $\Rightarrow$ (b) This is evident.
(b) $\Leftrightarrow$ (d) This follows from Theorem 6.2(d)–(e).
(b) $\Leftrightarrow$ (c) We have $\mathrm{Ker}_0(m_u) = \mathrm{Ker}_0(m) \cap [u, 1]$.

Then by Proposition 7.11, we have the equivalence in question. $\square$

Theorem 7.5 *Let (A, u) be an unbounded unital pseudo-BCK algebra that is downwards-directed and let m be a state-measure on (A, u). Then there is a unique (up to isomorphism) Abelian and Archimedean ℓ-group G with strong unit $u_G > 0$ such that the unbounded unital pseudo-BCK algebra $(A/\operatorname{Ker}_0(m), u/\operatorname{Ker}_0(m))$ is isomorphic to the unbounded unital pseudo-BCK algebra $(G^-, -u_G)$.*

Proof Let m_u, be the restriction of m to the interval $[u, 1]$. By Theorems 7.1 and 6.1, the quotient $[u, 1]/\operatorname{Ker}_0(m_u)$ can be converted into an MV-algebra, and in view of $\operatorname{Ker}_0(m_u) = \operatorname{Ker}_0(m) \cap [u, 1]$ we have that $[u, 1]/\operatorname{Ker}_0(m_u)$ is isomorphic to $[u/\operatorname{Ker}_0(m), 1/\operatorname{Ker}_0(m)] = [u, 1]/\operatorname{Ker}_0(m)$, so that both can be viewed as isomorphic MV-algebras. Let G be the ℓ-group guaranteed by Theorem 7.3 that is generated by $A/\operatorname{Ker}_0(m)$.

Therefore $u_G := -(u/\operatorname{Ker}_0(m))$ is a strong unit for G and $0_G := 1/\operatorname{Ker}_0(m)$ is the neutral element of G.

By Mundici's famous theorem [41], the unital ℓ-group (G, u_G) is the same for $[u, 1]/\operatorname{Ker}_0(m_u)$ and $[u, 1]/\operatorname{Ker}_0(m)$. If now $g \in G^-$, then $g = g_1 + \cdots + g_n$, where $g_1, \ldots, g_n \in [u, 1]/\operatorname{Ker}_0(m)$. The set of elements $g \in G^-$ such that $g \in A/\operatorname{Ker}_0(m)$ is a pseudo-BCK algebra containing $A/\operatorname{Ker}_0(m)$. Furthermore, because $A/\operatorname{Ker}_0(m)$ generates G, this implies that the pseudo-BCK algebra $(G^-, -u_G)$ is isomorphic to the unital pseudo-BCK algebra $(A/\operatorname{Ker}_0(m), u/\operatorname{Ker}_0(m))$. $\qquad\square$

Theorem 7.6 *Let m be a state-measure on a unital pseudo-BCK algebra (A, u) that is downwards-directed and let m_u be the restriction of m to the pseudo-BCK algebra $[u, 1]$. The following statements are equivalent:*

(a) *m is a state-measure-morphism on (A, u);*
(b) *m_u is a state-morphism on $[u, 1]$;*
(c) *$\operatorname{Ker}_0(m)$ is a maximal deductive system of (A, u);*
(d) *$\operatorname{Ker}_0(m_u)$ is a maximal deductive system of $[u, 1]$;*
(e) *m is an extremal state-measure on (A, u);*
(f) *m_u is an extremal state-measure on $[u, 1]$.*

Proof By Theorem 7.12, (b), (c), (d) are mutually equivalent and (a) implies each of the conditions (b), (c), (d). Theorem 7.4 entails that (c) implies (a). From Theorem 6.2 we see that (b) and (f) are equivalent. Proposition 7.6 gives (a) implies (e).

(e) $\Rightarrow$ (a) Let m be an extremal state-measure on (A, u). Define $\hat{m}(a/\operatorname{Ker}_0(m)) := m(a)$ $(a \in A)$. We assert that $\hat{m}$ is extremal on the unital pseudo-BCK algebra $(A/\operatorname{Ker}_0(m), u/\operatorname{Ker}_0(m))$. Indeed, if $\hat{m} = \lambda \mu_1 + (1 - \lambda)\mu_2, 0 < \lambda < 1$, where μ_1 and μ_2 are two state-measures on $(A/\operatorname{Ker}_0(m), u/\operatorname{Ker}_0(m))$, then there are two state-measures m_1, m_2 on (A, u) such that $\hat{m}_1 = \mu_1$ and $\hat{m}_2 = \mu_2$. Hence $m = \lambda m_1 + (1 - \lambda)m_2$ yielding $m_1 = m_2$ and $\mu_1 = \mu_2$.

By Theorem 7.5, $A/\operatorname{Ker}_0(m)$ is isomorphic to the pseudo-BCK algebra (G^-, u_G), where G^- is the negative cone of an Abelian and Archimedean ℓ-group G that is generated by $A/\operatorname{Ker}_0(m)$ and the element $u_G := -(u/\operatorname{Ker}_0(m))$ is a strong unit for G.

Similarly as in the proof of Theorem 7.4, $\hat{m}$ can be extended to a state-measure, s, on $(G^-, u/\operatorname{Ker}_0(m))$ so that s can be extended to an additive function denoted again by s on the whole unital ℓ-group $(G, -(u/\operatorname{Ker}_0(m)))$ that is positive on G^- and $s(u/\operatorname{Ker}_0(m)) = 1$.

Moreover, s is extremal on $(G, -(u/\operatorname{Ker}_0(m)))$ which by Theorem 12.18 in [155] is possible if and only if $\operatorname{Ker}_0(\hat{m}) = \{1/\operatorname{Ker}_0(m)\}$ is a maximal deductive system of the unital pseudo-BCK algebra $(A/\operatorname{Ker}_0(m), u/\operatorname{Ker}_0(m))$.

Since the mapping $a \mapsto a/\operatorname{Ker}_0(m)$ is surjective, this implies that $\operatorname{Ker}_0(m)$ is a maximal deductive system of (A, u). By the equivalence of (c) and (a) we have that m is a measure-morphism. $\square$

As a direct consequence of Theorem 7.6 and the Krein-Mil'man Theorem we have:

Corollary 7.1 *Let* (A, u) *be a unital pseudo-BCK algebra that is downwards-directed. Then*

$$\partial_e \mathcal{SM}(A, u) = \mathcal{SMM}(A, u)$$

and every state-measure on (A, u) *is a weak limit of a net of convex combinations of state-measure-morphisms.*

7.3 Coherence, de Finetti Maps and Borel States

In this section, we will generalize to pseudo-BCK algebras the relation between de Finetti maps and Bosbach states, following the results proved by Kühr and Mundici in [211] who showed that de Finetti's coherence principle, which has its origin in Dutch bookmaking, has a strong relationship to MV-states on MV-algebras. We then generalize this to state-measures on unital pseudo-BCK algebras that are downwards-directed.

We recall the following definitions and notation used in [211]. Let A be a nonempty set and let $\mathcal{W}$ be a fixed system of maps from $[0, 1]^A$. We endow $\mathcal{W}$ with the weak topology induced from the product topology on $[0, 1]^A$. By conv $\mathcal{W}$ and cl $\mathcal{W}$ we denote the convex hull and the closure of $\mathcal{W}$, respectively. In addition, if $\mathcal{W}$ is convex, $\partial_e \mathcal{W}$ will denote the set of all extremal points of $\mathcal{W}$. We note that the weak topology of Bosbach states is in fact the relativized product topology on $[0, 1]^A$.

Definition 7.3 ([211]) Let $A' = \{a_1, a_2, \ldots, a_n\}$ be a finite subset of A. Then a map $\beta : A' \longrightarrow [0, 1]$ is said to be *coherent* over A' if

$$\text{for all } \sigma_1, \sigma_2, \ldots, \sigma_n \in \mathbb{R}, \quad \text{there is a } V \in \mathcal{W} \quad \text{s.t.} \quad \sum_{i=1}^{n} \sigma_i \big(\beta(a_i) - V(a_i) \big) \geq 0.$$

By a *de Finetti map* on A we mean a function $\beta : A \longrightarrow [0, 1]$ which is coherent over every finite subset of A. We denote by $\mathcal{F}_{\mathcal{W}}$ the set of all de Finetti maps on A.

An interpretation of Definition 7.3 is as follows [211]: Two players, the book-maker and the bettor, wager money on the possible occurrence of elementary events $a_1, \ldots, a_n \in A$. The bookmaker sets a betting odd $\beta(a_i) \in [0, 1]$, and the bettor chooses stakes $\sigma_i \in \mathbb{R}$. The bettor pays the bookmaker $\sigma_i \beta(a_i)$, and will receive $\sigma_i V(a_i)$ from the bookmaker's possible world V. As scholars, we can assume that σ_i may be positive as well as negative. If the orientation of money transfer is given via bettor-to-bookmaker, then the inequality in Definition 7.3 means that the book-maker's book should be *coherent* in the sense that the bettor cannot choose stakes $\sigma_1, \ldots, \sigma_n$ which ensure that he will win money for every $V \in \mathcal{W}$.

Remark 7.6 Let A be a bounded pseudo-BCK algebra and denote by $\mathcal{BS}(A)$ the set of Bosbach states on A and by $\mathcal{W}$ the set of state-morphisms on A. Note that, according to Theorem 6.2, $\mathcal{W}$ coincides with the set of extremal Bosbach states, and by the Krein-Mil'man Theorem,

$$\mathcal{BS}(A) = \mathrm{cl}\,\mathrm{conv}\,\partial_e \mathcal{BS}(A) = \mathrm{cl}\,\mathrm{conv}\,\mathcal{SM}(A).$$

Theorem 7.7 *Let A be a bounded pseudo-BCK algebra and let $\mathcal{W} = \mathcal{SM}(A) \neq \emptyset$. Then*

$$\mathcal{F}_{\mathcal{W}} = \mathcal{BS}(A).$$

Proof According to Theorem 6.2 and Remark 6.4, $\mathcal{W}$ is closed. Now we can apply Proposition 3.1 in [211] since we have $\mathcal{W} \subseteq \mathcal{BS}$, $\partial_e \mathcal{BS} = \mathcal{W} (\subseteq \mathcal{W})$, $\mathcal{W}$ closed. So we get $\mathcal{BS} = \mathcal{F}_{\mathcal{W}}$.   □

Theorem 7.7 has an important consequence, namely that every Bosbach state (if it exists) on a bounded pseudo-BCK algebra is a de Finetti map coming from the set of $[0, 1]$-valued functions on A, generated by the set of state-morphisms. Moreover, applying Remark 7.6 we have that this de Finetti map is exactly the weak limit of a net of convex combinations of state-morphisms.

There is also another relationship concerning the representability of Bosbach states via integrals.

Let A be a bounded pseudo-BCK algebra and let $\mathcal{W} = \mathcal{SM}(A)$. Every element $a \in A$ determines a (continuous) function $f_a : \mathcal{W} \to [0, 1]$ via

$$f_a(V) := V(a), \quad V \in \mathcal{W}.$$

Definition 7.4 We say that a mapping $s : A \longrightarrow [0, 1]$ is a *Borel state* (of $\mathcal{W}$) if there is a regular Borel probability measure μ defined on the Borel σ-algebra of the topological space $\mathcal{W}$ generated by all open subsets of $\mathcal{W}$ such that

$$s(a) = \int_{\mathcal{W}} f_a(V) d\mu(V).$$

Let $\mathcal{B}_{\mathcal{W}}$ be the set of all Borel states of $\mathcal{W}$.

Theorem 7.8 *Let A be a bounded pseudo-BCK algebra. For any Bosbach state s on A, there is a Borel probability measure μ on $\mathcal{B}(\mathcal{W})$ such that*

$$s(a) = \int_{\mathcal{W}} f_a(V) d\mu(V).$$

Proof Since $\mathcal{W} = \mathcal{SM}(A)$ is closed (see Remark 6.4), by Theorem 4.2 in [211] we have $\mathcal{W} \subseteq \mathcal{B}_{\mathcal{W}}$, $\partial_e \mathcal{B}_{\mathcal{W}} \subseteq \mathcal{W}$ and $\mathcal{F}_{\mathcal{W}} = \mathcal{B}_{\mathcal{W}}$. Therefore by Theorem 7.7, $\mathcal{BS}(A) = \mathcal{B}_{\mathcal{W}}$, i.e. every Bosbach state is a Borel state on A. $\square$

We note that if we set $\Omega = \mathcal{BS}(A)$, then for any $a \in A$, the function $\tilde{a}$: $\mathcal{BS}(A) \longrightarrow [0, 1]$ defined by $\tilde{a}(s) := s(a)$, $s \in \mathcal{SB}(A)$, is continuous. Therefore, we can strengthen Theorem 7.8 as follows.

Theorem 7.9 *Let A be a bounded pseudo-BCK algebra. For any Bosbach state s on A, there is a unique Borel probability measure μ on $\mathcal{B}(\mathcal{BS}(A))$ such that*

$$s(a) = \int_{\mathcal{SM}(A)} \tilde{a}(x) d\mu(x).$$

Proof Suppose that the set of all Bosbach states on A is nonempty. By the Krein-Mil'man Theorem (Remark 7.6), the set of extremal Bosbach states is also nonempty and it coincides with the set of state-morphisms. Define $F_0 := \bigcap\{\mathrm{Ker}(s) \mid s \in \mathcal{SM}(A)\}$. In view of Theorem 6.1, Lemma 6.2 and Proposition 6.12, F_0 is a normal ideal, and similarly as in Theorem 6.1, we can show that A/F_0 is an Archimedean MV-algebra, and for any Bosbach state s on A, the mapping $\hat{s}(a/F_0) = s(a)$ $(a \in A)$ is an MV-state (= Bosbach state) on A/F_0; we set $\bar{a} := a/F_0$ $(a \in A)$. Moreover, the state spaces $\mathcal{BS}(A)$ and $\mathcal{BS}(A/F_0)$ are affinely homeomorphic compact nonempty Hausdorff topological spaces under the mapping $s \in \mathcal{BS}(A) \mapsto \hat{s} \in \mathcal{SB}(A/F_0)$ (i.e. they are homeomorphic in the weak topologies of states preserving convex combinations of states). In addition, the compact subsets of extremal Bosbach space are also homeomorphic under this mapping. By [202], on the Borel σ-algebra $\mathcal{B}(\mathcal{BS}(A))$, there is a unique Borel probability measure μ such that

$$s(a) = \hat{s}(a/F_0) = \int_{\mathcal{SM}(A/F_0)} \tilde{\bar{a}} d\mu.$$

This integral can be rewritten identifying the compact spaces and Borel σ-algebras in the form

$$s(a) = \int_{\mathcal{SM}(A)} \tilde{a}(x) d\mu(x).$$ $\square$

It is interesting to note that de Finetti was a great propagator of probabilities as finitely additive measures. The main result of [202] and Theorem 7.9 state that whenever s is a Bosbach state, it generates a σ-additive probability such that s is in fact an integral over this Borel probability measure. Thus Theorem 7.9 joins de

Finetti's "finitely additive probabilities" with σ-additive measures on an appropriate Borel σ-algebra.

We now generalize Theorem 7.7 and Theorem 7.8 to unbounded pseudo-BCK algebras that are downwards-directed.

Theorem 7.10 *Let (A, u) be a pseudo-BCK algebra that is downwards-directed and let $\mathcal{W} = \mathcal{SMM}(A, u) \neq \emptyset$. Then*

$$\mathcal{F}_{\mathcal{W}} = \mathcal{SM}(A, u).$$

Proof This follows from Theorem 7.6 and using the same steps as those in Theorem 7.7. $\square$

Theorem 7.11 *Let (A, u) be a pseudo-BCK algebra that is downwards-directed. For any state-measure m on (A, u), where $\mathcal{W} = \mathcal{SMM}(A, u) \neq \emptyset$, there is a Borel probability measure μ on $\mathcal{B}(\mathcal{W})$ such that*

$$m(a) = \int_{\mathcal{W}} f_a(V) d\mu(V).$$

Proof This follows from Theorem 7.6 and it follows steps analogous to those in Theorem 7.8. $\square$

Chapter 8
Generalized States on Residuated Structures

In the case of states on multiple-valued logic algebras, the domain of a state varied, while the codomain remained the real interval [0, 1] with its additive structure. On the other hand, in the theory of probability models ([123, 253]), the probability is seen as a new kind of semantics. Instead of the validity of sentences (events) one studies the probability of their achievement. For multiple-valued logics it is more profitable to study the truth degree of sentences instead of their validity. In many cases, the evaluation of the truth degree of sentences is made in an abstract structure (MV-algebra, BL-algebra, etc.), and not in the standard algebra [0, 1] (see for example [158]). This point of view suggests that we define a probability with values in an abstract algebra (in our case, an FL_w-algebra). This chapter begins by showing that in the definition of a Bosbach state the MV-algebra structure was used for the codomain of the state. We found several equivalent conditions which define a Bosbach state, and these conditions are expressed in terms of FL_w-algebra operations. If we replace [0, 1], the standard MV-algebra, with an arbitrary FL_w-algebra or FL_{ew}-algebra, the equivalence of these conditions is no longer preserved. In fact, the group of equivalent conditions splits into two parts. The conditions as a whole are not equivalent, but within each subgroup the equivalence is preserved. Each of the two groups of equivalent conditions leads to a notion of a generalized Bosbach state. In this way the notions of a generalized Bosbach state of type I and of type II appear.

We distinguish two kinds of generalized states; namely we define generalized states of type I and II, we study their properties and we prove that every strong type II state is an order-preserving type I state. We prove that any perfect FL_w-algebra admits a strong type I and type II state. Some conditions are given for a generalized state of type I on a linearly ordered bounded $R\ell$-monoid to be a state operator.

We introduce the notion of a generalized state-morphism and we prove that any generalized state morphism is an order-preserving type I state and, under certain conditions, an order-preserving type I state is a generalized state-morphism. The notion of a strong perfect FL_w-algebra is introduced and it is proved that any strong perfect FL_w-algebra admits a generalized state-morphism. The notion of a generalized Riečan state is also given, and the main results are proved based on the notion of

L.C. Ciungu, *Non-commutative Multiple-Valued Logic Algebras*,
Springer Monographs in Mathematics, DOI 10.1007/978-3-319-01589-7_8,
© Springer International Publishing Switzerland 2014

the Glivenko property defined in the non-commutative case. The main results consist of proving that any order-preserving type I state is a generalized Riečan state and in certain circumstances the two states coincide. We introduce the notion of a generalized local state on a perfect pseudo-MTL algebra A and we prove that, if A is relatively free of zero divisors, then every generalized local state can be extended to a generalized Riečan state. The notions of extension property and Horn-Tarski property are introduced for a pair (A, L) of FL_w-algebras with L complete and it is proved that under some conditions, if the pair $(Reg(A), L)$ has the extension property or Horn-Tarski property, then the pair (A, L) has these properties too. Finally, we outline how the generalized states give an approach to the theory of probabilistic models for non-commutative fuzzy logics associated to a pseudo t-norm.

8.1 Generalized Bosbach States on FL_w-Algebras

Starting from the observation that in the definition of a Bosbach state the standard MV-algebra structure for its codomain was used, for the case of FL_w-algebras the notion of a state was generalized as a function with values in an FL_w-algebra ([74, 75]). Properties of generalized states are useful for the development of an algebraic theory of probabilistic models for non-commutative fuzzy logics.

In this section we define the generalized Bosbach state of types I and II and we study their properties. The notion of a strong generalized Bosbach state is introduced and it is proved that any perfect FL_w-algebra admits a strong type I and type II state. Some conditions are given for a generalized state of type I on a linearly ordered bounded $R\ell$-monoid to be a state operator.

Let A be an FL_w-algebra and $s : A \longrightarrow [0, 1]$ be a function such that $s(0) = 0$ and $s(1) = 1$. It was proved in Propositions 6.18 and 6.19 that the following are equivalent for all $x, y \in A$:

(P_1^1) $1 + s(x \wedge y) = s(x \vee y) + s(d_1(x, y))$;
(P_2^1) $1 + s(x \wedge y) = s(x) + s(x \rightarrow y)$;
(P_3^1) $s(x) + s(x \rightarrow y) = s(y) + s(y \rightarrow x)$

and the following are also equivalent for all $x, y \in A$:

(P_1^2) $1 + s(x \wedge y) = s(x \vee y) + s(d_2(x, y))$;
(P_2^2) $1 + s(x \wedge y) = s(x) + s(x \rightsquigarrow y)$;
(P_3^2) $s(x) + s(x \rightsquigarrow y) = s(y) + s(y \rightsquigarrow x)$.

One can easily check that:

(I) The condition (P_1^1) is equivalent to each of the following conditions:

   (I_1) $s(d_1(x, y)) = s(x \vee y) \rightarrow_{\text{Ł}} s(x \wedge y)$;
   (I_2) $s(x \vee y) = s(d_1(x, y)) \rightarrow_{\text{Ł}} s(x \wedge y)$.

(II) The condition (P_2^1) is equivalent to each of the following conditions:

(II_1) $s(x \to y) = s(x) \to_Ł s(x \wedge y);$

(II_2) $s(x) = s(x \to y) \to_Ł s(a \wedge b).$

(III) The condition (P_3^1) is equivalent to the following condition:

(III_1) $s(x \to y) \to_Ł s(y) = s(y \to x) \to_Ł s(x).$

Similarly, we can prove that:

(I') The condition (P_1^2) is equivalent to each of the following conditions:

(I_1') $s(d_2(x, y)) = s(x \vee y) \to_Ł s(x \wedge y);$

(I_2') $s(x \vee y) = s(d_2(x, y)) \to_Ł s(x \wedge y).$

(II') The condition (P_2^2) is equivalent to each of the following conditions:

(II_1') $s(x \rightsquigarrow y) = s(x) \to_Ł s(x \wedge y);$

(II_2') $s(x) = s(x \rightsquigarrow y) \to_Ł s(x \wedge y).$

(III') The condition (P_3^2) is equivalent to the following condition:

(III_1') $s(x \rightsquigarrow y) \to_Ł s(y) = s(y \rightsquigarrow x) \to_Ł s(x).$

The above equalities suggest an extension of the definition of a Bosbach state replacing the standard MV-algebra $([0, 1], \min, \max, \odot_Ł, \to_Ł, 0, 1)$ with an arbitrary FL_w-algebra.

In the sequel, $(A, \wedge, \vee, \odot, \to, \rightsquigarrow, 0, 1)$ and $(L, \wedge, \vee, \odot, \to, \rightsquigarrow, 0, 1)$ are FL_w-algebras and $s : A \longrightarrow L$ is an arbitrary function.

(We use the same notation for the operations in both structures, but the reader should be aware that they are different.)

By d_{1A}, d_{2A} and d_{1L}, d_{2L} we denote the distance functions in the FL_w-algebras A and L, respectively (see Definition 3.2).

Proposition 8.1 *If $s(0) = 0$ and $s(1) = 1$, then the following are equivalent:*

(i) *for all $a, b \in A$, $s(d_{1A}(a, b)) = s(a \vee b) \to s(a \wedge b)$ and $s(d_{2A}(a, b)) = s(a \vee b) \rightsquigarrow s(a \wedge b)$;*

(ii) *for all $a, b \in A$ with $b \leq a$, $s(a \to b) = s(a) \to s(b)$ and $s(a \rightsquigarrow b) = s(a) \rightsquigarrow s(b)$;*

(iii) *for all $a, b \in A$, $s(a \to b) = s(a) \to s(a \wedge b)$ and $s(a \rightsquigarrow b) = s(a) \rightsquigarrow s(a \wedge b)$;*

(iv) *for all $a, b \in A$, $s(a \to b) = s(a \vee b) \to s(b)$ and $s(a \rightsquigarrow b) = s(a \vee b) \rightsquigarrow s(b).$*

Proof Let $a, b \in A$.

(i) $\Rightarrow$ (ii) Assume $b \leq a$. Then we have:

$$d_{1A}(a, b) = a \to b, \quad \text{so}$$

$$s(a \to b) = s(d_{1A}(a, b)) = s(a \vee b) \to s(a \wedge b) = s(a) \to s(b)$$

and

$$d_{2A}(a, b) = a \rightsquigarrow b, \quad \text{so}$$

$$s(a \rightsquigarrow b) = s\big(d_{2A}(a, b)\big) = s(a \vee b) \rightsquigarrow s(a \wedge b) = s(a) \rightsquigarrow s(b).$$

(ii) $\Rightarrow$ (i) Since $a \wedge b \le a \vee b$, we have:

$$s\big(d_{1A}(a, b)\big) = s(a \vee b \rightarrow a \wedge b) = s(a \vee b) \rightarrow s(a \wedge b) \quad \text{and}$$

$$s\big(d_{2A}(a, b)\big) = s(a \vee b \rightsquigarrow a \wedge b) = s(a \vee b) \rightsquigarrow s(a \wedge b).$$

(ii) $\Rightarrow$ (iii) Applying $(rl\text{-}c_3)$ we have $a \rightarrow b = a \rightarrow a \wedge b$ and $a \rightsquigarrow b = a \rightsquigarrow a \wedge b$. Since $a \wedge b \le a$, we get:

$$s(a \rightarrow b) = s(a \rightarrow a \wedge b) = s(a) \rightarrow s(a \wedge b) \quad \text{and}$$

$$s(a \rightsquigarrow b) = s(a \rightsquigarrow a \wedge b) = s(a) \rightsquigarrow s(a \wedge b).$$

(iii) $\Rightarrow$ (ii) Assume $b \le a$, so $a \wedge b = b$. It follows that:

$$s(a \rightarrow b) = s(a) \rightarrow s(a \wedge b) = s(a) \rightarrow s(b) \quad \text{and}$$

$$s(a \rightsquigarrow b) = s(a) \rightsquigarrow s(a \wedge b) = s(a) \rightsquigarrow s(b).$$

(ii) $\Rightarrow$ (iv) Applying $(rl\text{-}c_4)$ we have $a \vee b \rightarrow b = a \rightarrow b$ and $a \vee b \rightsquigarrow b = a \rightsquigarrow b$. Since $b \le a \vee b$, we get:

$$s(a \rightarrow b) = s(a \vee b \rightarrow b) = s(a \vee b) \rightarrow s(b) \quad \text{and}$$

$$s(a \rightsquigarrow b) = s(a \vee b \rightsquigarrow b) = s(a \vee b) \rightsquigarrow s(b).$$

(iv) $\Rightarrow$ (ii) Let $b \le a$, so $a \vee b = a$. It follows that:

$$s(a \rightarrow b) = s(a \vee b) \rightarrow s(b) = s(a) \rightarrow s(b) \quad \text{and}$$

$$s(a \rightsquigarrow b) = s(a \vee b) \rightsquigarrow s(b) = s(a) \rightsquigarrow s(b). \qquad \square$$

Proposition 8.2 *If $s(0) = 0$ and $s(1) = 1$, then the following are equivalent*:

(i) *for all $a, b \in A$, $s(a \vee b) = s(d_{1A}(a, b)) \rightarrow s(a \wedge b) = s(d_{2A}(a, b)) \rightsquigarrow s(a \wedge b)$*;

(ii) *for all $a, b \in A$ $s(a) = s(a \rightarrow b) \rightarrow s(a \wedge b) = s(a \rightsquigarrow b) \rightsquigarrow s(a \wedge b)$*;

(iii) *for all $a, b \in A$ with $b \le a$, $s(a) = s(a \rightarrow b) \rightarrow s(b) = s(a \rightsquigarrow b) \rightsquigarrow s(b)$*;

(iv) *for all $a, b \in A$, $s(a \vee b) = s(a \rightarrow b) \rightarrow s(b) = s(a \rightsquigarrow b) \rightsquigarrow s(b)$*;

(v) *for all $a, b \in A$, $s(a \rightarrow b) \rightarrow s(b) = s(b \rightarrow a) \rightarrow s(a)$ and $s(a \rightsquigarrow b) \rightsquigarrow s(b) = s(b \rightsquigarrow a) \rightsquigarrow s(a)$*.

Proof Let $a, b \in A$.

(i) $\Rightarrow$ (iii) Assume $b \le a$, thus:

$$a \vee b = a, \qquad a \wedge b = b, \qquad d_{1A}(a,b) = a \to b, \qquad d_{2A}(a,b) = a \rightsquigarrow b.$$

Hence $s(a) = s(a \to b) \to s(b)$ and $s(a) = s(a \rightsquigarrow b) \rightsquigarrow s(b)$.

(iii) $\Rightarrow$ (i) Taking into consideration that $d_{1A}(a,b) = a \vee b \to a \wedge b$, $d_{2A}(a,b) = a \vee b \rightsquigarrow a \wedge b$ and $a \wedge b \le a \vee b$, we have:

$$s(a \vee b) = s(a \vee b \to a \wedge b) \to s(a \wedge b) = s\big(d_{1A}(a,b)\big) \to s(a \wedge b) \quad \text{and}$$

$$s(a \vee b) = s(a \vee b \rightsquigarrow a \wedge b) \rightsquigarrow s(a \wedge b) = s\big(d_{2A}(a,b)\big) \rightsquigarrow s(a \wedge b).$$

(ii) $\Rightarrow$ (iii) Assume $b \le a$, so $a \wedge b = b$.

Hence $s(a) = s(a \to b) \to s(b) = s(a \rightsquigarrow b) \rightsquigarrow s(b)$.

(iii) $\Rightarrow$ (ii) Applying the properties $a \wedge b \le a$, $a \to a \wedge b = a \to b$, $a \rightsquigarrow a \wedge b = a \rightsquigarrow b$ we get:

$$s(a) = s(a \to a \wedge b) \to s(a \wedge b) = s(a \to b) \to s(a \wedge b) \quad \text{and}$$

$$s(a) = s(a \rightsquigarrow a \wedge b) \rightsquigarrow s(a \wedge b) = s(a \rightsquigarrow b) \rightsquigarrow s(a \wedge b).$$

(iii) $\Rightarrow$ (iv) Since $b \le a \vee b$, $a \vee b \to b = a \to b$ and $a \vee b \rightsquigarrow b = a \rightsquigarrow b$, we have:

$$s(a \vee b) = s(a \vee b \to b) \to s(b) = s(a \to b) \to s(b) \quad \text{and}$$

$$s(a \vee b) = s(a \vee b \rightsquigarrow b) \rightsquigarrow s(b) = s(a \rightsquigarrow b) \rightsquigarrow s(b).$$

(iv) $\Rightarrow$ (v) By (iv) we have:

$$s(a \to b) \to s(b) = s(a \vee b) = s(b \vee a) = s(b \to a) \to s(a) \quad \text{and}$$

$$s(a \rightsquigarrow b) \rightsquigarrow s(b) = s(a \vee b) = s(b \vee a) = s(b \rightsquigarrow a) \rightsquigarrow s(a).$$

(v) $\Rightarrow$ (iii) Assume $b \le a$. Thus

$$s(a) = 1 \to s(a) = s(1) \to s(a) = s(b \to a) \to s(a) = s(a \to b) \to s(b) \quad \text{and}$$

$$s(a) = 1 \rightsquigarrow s(a) = s(1) \rightsquigarrow s(a) = s(b \rightsquigarrow a) \rightsquigarrow s(a) = s(a \rightsquigarrow b) \rightsquigarrow s(b). \qquad \square$$

Definition 8.1 Let $(A, \wedge, \vee, \odot, \to, \rightsquigarrow, 0, 1)$ and $(L, \wedge, \vee, \odot, \to, \rightsquigarrow, 0, 1)$ be FL_w-algebras and $s : A \longrightarrow L$ an arbitrary function such that $s(0) = 0$ and $s(1) = 1$. Then:

1. s is called a *generalized Bosbach state of type I* (or briefly, a *state of type I* or a *type I state*) if it satisfies the equivalent conditions from Proposition 8.1.
2. s is called a *generalized Bosbach state of type II* (or briefly, a *state of type II* or a *type II state*) if it satisfies the equivalent conditions from Proposition 8.2.
3. A generalized type I, II state s is called a *strong type I, II state* if $s(a \to b) = s(a \rightsquigarrow b)$ for all $a, b \in A$.

Example 8.1 Any FL_w-algebra morphism $s : A \longrightarrow L$ is an order-preserving type I state.

Example 8.2 ([114]) If A is a bounded Rℓ-monoid, then a *state operator* on A is a function $\sigma : A \longrightarrow A$ such that for any $x, y \in A$ the following conditions are satisfied:

(1) $\sigma(0) = 0$;
(2) $\sigma(x \to y) = \sigma(x) \to \sigma(x \wedge y)$ and $\sigma(x \rightsquigarrow y) = \sigma(x) \rightsquigarrow \sigma(x \wedge y)$;
(3) $\sigma(x \odot y) = \sigma(x) \odot \sigma(x \rightsquigarrow x \odot y) = \sigma(y \to x \odot y) \odot \sigma(y)$;
(4) $\sigma(\sigma(x) \odot \sigma(y)) = \sigma(x) \odot \sigma(y)$;
(5) $\sigma(\sigma(x) \to \sigma(y)) = \sigma(x) \to \sigma(y)$ and $\sigma(\sigma(x) \rightsquigarrow \sigma(y)) = \sigma(x) \rightsquigarrow \sigma(y)$;
(6) $\sigma(\sigma(x) \vee \sigma(y)) = \sigma(x) \vee \sigma(y)$.

Taking $x = y = 1$ in (2), it follows that $\sigma(1) = 1$.

Since $\sigma(0) = 0$, $\sigma(1) = 1$ and condition (2) of the above definition is condition (iii) in Proposition 8.1, it follows that any state operator σ on a bounded Rℓ-monoid is a type I state. Moreover, it was proved in [114] that $x \leq y$ implies $\sigma(x) \leq \sigma(y)$, thus σ is an order-preserving type I state.

Example 8.3 Let A be the FL$_w$-algebra from Example 4.4, L be the FL$_w$-algebra from Example 4.1 and $s : A \longrightarrow L$, $s(0) := 0$, $s(a) = s(b) = s(c) = s(1) := 1$. One can easily check that s is a type I state and a type II state.

Example 8.4 Let A be the FL$_w$-algebra from Example 4.4 and consider the functions:

$$s_1 : A \longrightarrow A, \quad s_1(0) := 0, s_1(a) := a, s_1(b) := b, s_1(c) := c, s_1(1) := 1,$$

$$s_2 : A \longrightarrow A, \quad s_2(0) := 0, s_2(a) = s_2(b) = s_2(c) = s_2(1) := 1.$$

One can easily check that s_1 is a type I state, while s_2 is a strong type I state and a strong type II state.

Remark 8.1 Not all type I states are order-preserving, and not all order-preserving type I states are type II states. Indeed, consider the following example of an FL$_{ew}$-algebra $A = \{0, a, b, c, d, 1\}$, with the following partial order relation and operations ([186]):

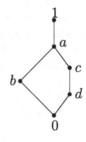

$\rightarrow$	0	a	b	c	d	1
0	1	1	1	1	1	1
a	0	1	b	c	c	1
b	c	1	1	c	c	1
c	b	1	b	1	a	1
d	b	1	b	1	1	1
1	0	a	b	c	d	1

$\odot$	0	a	b	c	d	1
0	0	0	0	0	0	0
a	0	a	b	d	d	a
b	0	b	b	0	0	b
c	0	d	0	d	d	c
d	0	d	0	d	d	d
1	0	a	b	c	d	1

Consider the maps $s_1, \ldots, s_6 : A \rightarrow A$ presented in the table below:

x	0	a	b	c	d	1
$s_1(x)$	0	a	0	1	a	1
$s_2(x)$	0	a	b	c	d	1
$s_3(x)$	0	1	0	1	1	1
$s_4(x)$	0	1	b	c	c	1
$s_5(x)$	0	1	c	b	b	1
$s_6(x)$	0	1	1	0	0	1

The type I states from A to A are s_i, with $i \in \overline{1,6}$. Out of these, the only order-preserving ones are s_2, s_3, s_4, s_5 and s_6. Indeed, s_1 is not order-preserving, as $c \leq a$ and $s_1(c) = 1 > s_1(a) = a$.

The type II states from A to A are s_3, s_4, s_5 and s_6.

The next proposition generalizes to FL_w-algebras the following result from [263]: if A and L are FL_{ew}-algebras, then every type II state $s : A \longrightarrow L$ is an order-preserving type I state.

Proposition 8.3 *Every strong type II state is an order-preserving type I state.*

Proof Since $b \leq (a \rightarrow b) \rightsquigarrow b$ and $b \leq (a \rightsquigarrow b) \rightarrow b$, applying (*psbck-$c_{11}$*) and Proposition 8.2(iii), (iv) we have:

$$s\big((a \rightarrow b) \rightsquigarrow b\big) = s\big(\big((a \rightarrow b) \rightsquigarrow b\big) \rightarrow b\big) \rightarrow s(b)$$
$$= s(a \rightarrow b) \rightarrow s(b) = s(a \vee b) \quad \text{and}$$
$$s\big((a \rightsquigarrow b) \rightarrow b\big) = s\big(\big((a \rightsquigarrow b) \rightarrow b\big) \rightsquigarrow b\big) \rightsquigarrow s(b)$$
$$= s(a \rightsquigarrow b) \rightsquigarrow s(b) = s(a \vee b).$$

From $b \leq a \rightarrow b$ and $b \leq a \rightsquigarrow b$ we get:

$$s(a \rightarrow b) = s\big((a \rightarrow b) \rightsquigarrow b\big) \rightsquigarrow s(b) = s(a \vee b) \rightsquigarrow s(b) \quad \text{and}$$
$$s(a \rightsquigarrow b) = s\big((a \rightsquigarrow b) \rightarrow b\big) \rightarrow s(b) = s(a \vee b) \rightarrow s(b).$$

Since s is strong, we have:

$$s(a \rightarrow b) = s(a \rightsquigarrow b) = s(a \vee b) \rightarrow s(b) = s(a \vee b) \rightsquigarrow s(b).$$

Taking into consideration Proposition 8.1(iv), it follows that s is a type I state. By Proposition 8.2(iii), s is order-preserving. □

Remark 8.2 In the case when L is the standard MV-algebra $[0, 1]$, order-preserving type I states $s : A \longrightarrow [0, 1]$ coincide with Bosbach states on A, as the identities (iii) from Proposition 8.1 are equivalent to the identities (II_1) and (II'_1), and type II states $s : A \longrightarrow [0, 1]$ coincide with Bosbach states on A, as the identities (v) from Proposition 8.2 are equivalent to the identities (III_1) and (III'_1).

Remark 8.3 Let A, B, L be FL_w-algebras, $s : B \longrightarrow L$ be a function and $f : A \longrightarrow L$ be an FL_w-algebra morphism. Then by Proposition 8.1(iii), if s is a type I state, then $s \circ f : A \longrightarrow L$ is a type I state, and if, moreover, s is order-preserving and f is order-preserving, then $s \circ f$ is order-preserving. By Proposition 8.2(ii), if s is a type II state, then $s \circ f$ is a type II state.

In the sequel we will use the notation $s(a)^-$ instead of $(s(a))^-$ and $s(a)^\sim$ instead of $(s(a))^\sim$.

Proposition 8.4 *If s is a type I state, then for all $a, b \in A$ the following hold:*

(1) $s(a^-) = s(a)^-$ and $s(a^\sim) = s(a)^\sim$;
(2) $s(a^{--}) = s(a)^{--}$, $s(a^{\sim\sim}) = s(a)^{\sim\sim}$, $s(a^{-\sim}) = s(a)^{-\sim}$ and $s(a^{\sim-}) = s(a)^{\sim-}$;
(3) $s(a \vee b) \rightarrow s(a) = s(b) \rightarrow s(a \wedge b)$ and $s(a \vee b) \rightsquigarrow s(a) = s(b) \rightsquigarrow s(a \wedge b)$;
(4) $s(a^{-\sim} \rightarrow a) = s(a)^{-\sim} \rightarrow s(a)$ and $s(a^{\sim-} \rightsquigarrow a) = s(a)^{\sim-} \rightarrow s(a)$;
(5) $s((a \circ_1 b) \circ_2 b) = s(a \circ_1 b) \circ_2 s(b)$, where $\circ_1, \circ_2 \in \{\rightarrow, \rightsquigarrow\}$;
(6) $s((a \circ_1 b) \circ_2 b) = (s(a \vee b) \circ_1 s(b)) \circ_2 s(b)$, where $\circ_1, \circ_2 \in \{\rightarrow, \rightsquigarrow\}$;
(7) $s(a \vee b) \rightarrow s(a) \wedge s(b) = s(a) \vee s(b) \rightarrow s(a \wedge b)$ and $s(a \vee b) \rightsquigarrow s(a) \wedge s(b) = s(a) \vee s(b) \rightsquigarrow s(a \wedge b)$;
(8) $s(a \rightarrow a \odot b) \odot s(a) \leq s(a \odot b)$ and $s(a) \odot s(a \rightsquigarrow a \odot b) \leq s(a \odot b)$.

Proof

(1) $s(a^-) = s(a \rightarrow 0) = s(a) \rightarrow s(0) = s(a) \rightarrow 0 = s(a)^-$ and similarly $s(a^\sim) = s(a)^\sim$.
(2) This follows from (1).
(3) This follows from Proposition 8.1(iii), (iv).
(4) By $(psbck\text{-}c_{14})$, Proposition 8.1(ii) and (2).
(5) By $(psbck\text{-}c_6)$ and Proposition 8.1(ii).
(6) By $(psbck\text{-}c_6)$ and Proposition 8.1(ii), (iv).
(7) From $(rl\text{-}c_4)$ we have

$$x \vee y \rightarrow z = (x \rightarrow z) \wedge (y \rightarrow z), \qquad x \vee y \rightsquigarrow z = (x \rightsquigarrow z) \wedge (y \rightsquigarrow z)$$

and from $(rl\text{-}c_3)$ we get

$$z \rightarrow x \wedge y = (z \rightarrow x) \wedge (z \rightarrow y) \quad \text{and} \quad z \rightsquigarrow x \wedge y = (z \rightsquigarrow x) \wedge (z \rightsquigarrow y),$$

for all $x, y, z \in A$.

Applying the above formulas and (3) we get:

$$s(a \vee b) \to s(a) \wedge s(b) = \big(s(a \vee b) \to s(a)\big) \wedge \big(s(a \vee b) \to s(b)\big)$$
$$= \big(s(a) \to s(a \wedge b)\big) \wedge \big(s(b) \to s(a \wedge b)\big)$$
$$= s(a) \vee s(b) \to s(a \wedge b).$$

Similarly,

$$s(a \vee b) \rightsquigarrow s(a) \wedge s(b) = \big(s(a \vee b) \rightsquigarrow s(a)\big) \wedge \big(s(a \vee b) \rightsquigarrow s(b)\big)$$
$$= \big(s(a) \rightsquigarrow s(a \wedge b)\big) \wedge \big(s(b) \rightsquigarrow s(a \wedge b)\big)$$
$$= s(a) \vee s(b) \rightsquigarrow s(a \wedge b).$$

(8) By Proposition 8.1(ii) we have

$$s(a \to a \odot b) = s(a) \to s(a \odot b) \quad \text{and} \quad s(a \rightsquigarrow a \odot b) = s(a) \rightsquigarrow s(a \odot b).$$

Applying ($psbck$-c_{25}) we get:

$$s(a \to a \odot b) \odot s(a) = \big(s(a) \to s(a \odot b)\big) \odot s(a) \leq s(a \odot b) \quad \text{and}$$
$$s(a) \odot s(a \rightsquigarrow a \odot b) = s(a) \odot \big(s(a) \rightsquigarrow s(a \odot b)\big) \leq s(a \odot b). \qquad \square$$

Proposition 8.5 *If s is an order-preserving type I state, then for all $a, b \in A$ the following hold*:

(1) $s(a) \odot s(b) \leq s(a \odot b)$;
(2) $s(a) \odot s(b)^- \leq s(a \odot b^-)$ and $s(a) \odot s(b)^\sim \leq s(a \odot b^\sim)$;
(3) $s(a \to b) \leq s(a) \to s(b)$ and $s(a \rightsquigarrow b) \leq s(a) \rightsquigarrow s(b)$;
(4) $s(a \to b) \odot s(b \to a) \leq d_{1L}(s(a), s(b))$ and $s(a \rightsquigarrow b) \odot s(b \rightsquigarrow a) \leq d_{2L}(s(a), s(b))$;
(5) $s(d_{1A}(a, b)) \leq d_{1L}(s(a), s(b))$ and $s(d_{2A}(a, b)) \leq d_{2L}(s(a), s(b))$.

Proof

(1) Obviously, $b \leq a \rightsquigarrow a \odot b$, hence $s(b) \leq s(a \rightsquigarrow a \odot b)$.
    Applying Proposition 8.4(8) we get $s(a) \odot s(b) \leq s(a) \odot s(a \rightsquigarrow a \odot b) \leq s(a \odot b)$.
(2) This follows from Proposition 8.4(1) and (1).
(3) By Proposition 8.1(iii) we get $s(a \to b) = s(a) \to s(a \wedge b)$.
    Since $a \wedge b \leq b$, we have $s(a \wedge b) \leq s(b)$ and by ($psbck$-c_{10}) it follows that $s(a) \to s(a \wedge b) \leq s(a) \to s(b)$. Thus $s(a \to b) \leq s(a) \to s(b)$.
    Similarly, $s(a \rightsquigarrow b) \leq s(a) \rightsquigarrow s(b)$.
(4) Applying (3) and ($psbck$-c_{24}) we get:

$$s(a \to b) \odot s(b \to a) \leq \big(s(a) \to s(b)\big) \odot \big(s(b) \to s(a)\big)$$
$$\leq \big(s(a) \to s(b)\big) \wedge \big(s(b) \to s(a)\big) = d_{1L}\big(s(a), s(b)\big).$$

Similarly, $s(a \rightsquigarrow b) \odot s(b \rightsquigarrow a) \leq d_{2L}(s(a), s(b))$.

(5) By (3) we have:

$$s\big(d_{1A}(a,b)\big) = s\big((a \rightarrow b) \wedge (b \rightarrow a)\big) \leq s(a \rightarrow b) \wedge s(b \rightarrow a)$$
$$\leq \big(s(a) \rightarrow s(b)\big) \wedge \big(s(b) \rightarrow s(a)\big) = d_{1L}\big(s(a), s(b)\big)$$

(from $x \wedge y \leq x, y$, we have $s(x \wedge y) \leq s(x), s(y)$, hence $s(x \wedge y) \leq s(x) \wedge s(y)$).

Similarly, $s(d_{2A}(a,b)) \leq d_{2L}(s(a), s(b))$. $\square$

Proposition 8.6 *Let s be a type II state. Then for all $a, b \in A$ the following hold*:

(1) $a \leq b$ *implies* $s(a) \leq s(b)$;
(2) $s(a) = s(a^-)^-$ *and* $s(a) = s(a^\sim)^\sim$;
(3) $s(a^{-\sim}) = s(a^{\sim-}) = s(a) = s(a)^{-\sim} = s(a)^{\sim-}$;
(4) $s(a \rightarrow b) = s((a \rightarrow b) \rightarrow b) \rightarrow s(b)$ *and* $s(a \rightsquigarrow b) = s((a \rightsquigarrow b) \rightsquigarrow b) \rightsquigarrow s(b)$;
(5) $s(a \rightarrow b) = s((a \rightarrow b) \rightsquigarrow b) \rightsquigarrow s(b)$ *and* $s(a \rightsquigarrow b) = s((a \rightsquigarrow b) \rightarrow b) \rightarrow s(b)$;
(6) $s(a^-) = s(a)^\sim$ *and* $s(a^\sim) = s(a)^-$.

Proof

(1) By $(psbck\text{-}c_6)$ and Proposition 8.2(iii) we have

$$s(a) \leq s(b \rightarrow a) \rightarrow s(a) = s(b).$$

(2) Since $0 \leq a$, by Proposition 8.2(iii) we get

$$s(a) = s(a \rightarrow 0) \rightarrow 0 = s\big(a^-\big) \rightarrow 0 = s\big(a^-\big)^-.$$

Similarly, $s(a) = s(a \rightsquigarrow 0) \rightsquigarrow 0 = s(a^\sim) \rightsquigarrow 0 = s(a^\sim)^\sim$.

(3) By (2) and $(psbck\text{-}c_{18})$ we get:

$$s\big(a^{-\sim}\big) = s\big(a^{-\sim-}\big)^- = s\big(a^-\big)^- = s(a) \quad \text{and}$$
$$s\big(a^{\sim-}\big) = s\big(a^{\sim-\sim}\big)^\sim = s\big(a^\sim\big)^\sim = s(a).$$

We also have:

$$s(a)^{-\sim} = s\big(a^-\big)^{-\sim-} = s\big(a^-\big)^- = s(a) \quad \text{and}$$
$$s(a)^{\sim-} = s\big(a^\sim\big)^{\sim-\sim} = s\big(a^\sim\big)^\sim = s(a).$$

(4) This follows from $(psbck\text{-}c_6)$ and Proposition 8.2(iii).
(5) Similar to (4).
(6) Applying (5) and (3) we have

$$s\big(a^-\big) = s(a \rightarrow 0) = s\big((a \rightarrow 0) \rightsquigarrow 0\big) \rightsquigarrow s(0) = s\big(a^{-\sim}\big)^\sim = s(a)^\sim \quad \text{and}$$
$$s\big(a^\sim\big) = s(a \rightsquigarrow 0) = s\big((a \rightsquigarrow 0) \rightarrow 0\big) \rightarrow s(0) = s\big(a^{\sim-}\big)^- = s(a)^-. \quad \square$$

Proposition 8.7 *If A and L are bounded $R\ell$-monoids and $s : A \longrightarrow L$ is an order-preserving type I state, then for all $a, b \in A$ the following hold:*

(1) $s(a \odot b) = s(a \rightarrow a \odot b) \odot s(a) = s(a) \odot s(a \rightsquigarrow a \odot b)$;
(2) $s(a \wedge b) = s(a \rightarrow b) \odot s(a) = s(a) \odot s(a \rightsquigarrow b)$.

Proof

(1) One can easily check that $a \odot b \leq a \wedge b \leq d_{1A}(a, b) \leq a \rightarrow b$ and $a \odot b \leq a \wedge b \leq d_{2A}(a, b) \leq a \rightsquigarrow b$. It follows that:

$$s(a \rightarrow a \odot b) \odot s(a) = \big(s(a) \rightarrow s(a \odot b)\big) \odot s(a)$$
$$= s(a) \wedge s(a \odot b) = s(a \odot b) \quad \text{and}$$
$$s(a) \odot s(a \rightsquigarrow a \odot b) = s(a) \odot \big(s(a) \rightsquigarrow s(a \odot b)\big)$$
$$= s(a) \wedge s(a \odot b) = s(a \odot b).$$

(2) Since $a \rightarrow a \wedge b = a \rightarrow b$ and $a \rightsquigarrow a \wedge b = a \rightsquigarrow b$, we have:

$$s(a \rightarrow b) \odot s(a) = s(a \rightarrow a \wedge b) \odot s(a) = \big(s(a) \rightarrow s(a \wedge b)\big) \odot s(a)$$
$$= s(a) \wedge s(a \wedge b) = s(a \wedge b) \quad \text{and}$$
$$s(a) \odot s(a \rightsquigarrow b) = s(a) \odot s(a \rightsquigarrow a \wedge b) = s(a) \odot \big(s(a) \rightsquigarrow s(a \wedge b)\big)$$
$$= s(a) \wedge s(a \wedge b) = s(a \wedge b). \qquad \square$$

Theorem 8.1 *If A is a linearly ordered bounded $R\ell$-monoid and $s : A \longrightarrow A$ is an order-preserving type I state such that $s^2(x) = s(x) \leq x$ for all $x \in A$, then s is a state operator on A.*

Proof Applying the hypothesis and the definition of a type I state, we will verify axioms (1)–(6) from the definition of a state operator.

(1) $s(0) = 0$:
   This follows from the definition of a type I state.
(2) $s(a \rightarrow b) = s(a) \rightarrow s(a \wedge b)$ and $s(a \rightsquigarrow b) = s(a) \rightsquigarrow s(a \wedge b)$:
   This is condition (iii) from Proposition 8.1.
(3) $s(a \odot b) = s(a) \odot s(a \rightsquigarrow a \odot b) = s(b \rightarrow a \odot b) \odot s(b)$:
   From Proposition 8.7(1) we have $s(a \odot b) = s(a) \odot s(a \rightsquigarrow a \odot b)$.
   Similarly as in the proof of Proposition 8.7(1), we have:

$$s(b \rightarrow a \odot b) \odot s(b) = \big(s(b) \rightarrow s(a \odot b)\big) \odot s(b) = s(b) \wedge s(a \odot b) = s(a \odot b).$$

(4) $s(s(a) \odot s(b)) = s(a) \odot s(b)$:
   Since $s(x) \leq x$ for all $x \in A$ we have $s(s(a) \odot s(b)) \leq s(a) \odot s(b)$.
   On the other hand, from Proposition 8.5(1), replacing a with $s(a)$ and b with $s(b)$ we get $s^2(a) \odot s^2(b) \leq s(s(a) \odot s(b))$, that is, $s(a) \odot s(b) \leq s(s(a) \odot s(b))$.
   Thus $s(s(a) \odot s(b)) = s(a) \odot s(b)$.

(5) $s(s(a) \to s(b)) = s(a) \to s(b)$ and $s(s(a) \rightsquigarrow s(b)) = s(a) \rightsquigarrow s(b)$:

Since A is linearly ordered we consider the cases:

(a) $b \le a$, so $s(b) \le s(a)$. According to condition (ii) from Proposition 8.1 we get $s(s(a) \to s(b)) = s^2(a) \to s^2(b) = s(a) \to s(b)$.

(b) $a \le b$, so $s(a) \le s(b)$. It follows that $s(a) \to s(b) = 1$, thus

$$s(s(a) \to s(b)) = s(a) \to s(b) = s(1) = 1.$$

Similarly, $s(s(a) \rightsquigarrow s(b)) = s(a) \rightsquigarrow s(b)$.

(6) $s(s(a) \vee s(b)) = s(a) \vee s(b)$.

Assume $a \le b$, hence $s(a) \le s(b)$, so $s(a) \vee s(b) = s(b)$. It follows that

$$s(s(a) \vee s(b)) = s^2(b) = s(b) = s(a) \vee s(b).$$

Similarly, if $b \le a$, then $s(b) \le s(a)$, so $s(a) \vee s(b) = s(a)$. Hence

$$s(s(a) \vee s(b)) = s^2(a) = s(a) = s(a) \vee s(b).$$

We conclude that s is a state operator on A. $\square$

Example 8.5 If A is a bounded Rℓ-monoid, then the identity $s := id_A$ is a type I state satisfying the condition $s^2(x) = s(x) = x$ for all $x \in A$. Thus s is a state operator on A (in this case in Theorem 8.1, A does not need to be linearly ordered).

Remark 8.4

(1) It was proved in [114] that any state operator σ on a linearly ordered bounded Rℓ-monoid satisfies the condition $\sigma^2 = \sigma$.

(2) If A is an arbitrary bounded Rℓ-monoid and $\sigma : A \longrightarrow A$ is an Rℓ-endomorphism satisfying the condition $\sigma^2 = \sigma$, then it is called a *state-morphism operator*. If σ is a state operator satisfying the condition $\sigma(x \odot y) = \sigma(x) \odot \sigma(y)$, then it is called a *weak state-morphism operator* and it was proved that any weak state-morphism operator is a state operator. We can see that condition $s^2(x) = s(x) \le x$ for all $x \in A$ from Theorem 8.1 can be replaced with conditions $s(x \odot y) = s(x) \odot s(y)$ for all $x, y \in A$ and $s^2 = s$. Indeed, from axiom (4) of the definition of a state operator we have: $s(s(a) \odot s(b)) = s^2(a) \odot s^2(b) = s(a) \odot s(b)$.

Proposition 8.8 *Let A and L be FL_w-algebras and let $s : A \longrightarrow L$ be a type I and type II state. Then for all $a, b \in A$ the following hold:*

(1) $s((a \to b) \to b) = s((b \to a) \to a)$;

(2) $s((a \rightsquigarrow b) \rightsquigarrow b) = s((b \rightsquigarrow a) \rightsquigarrow a)$.

Proof According to Proposition 8.4(5) and Proposition 8.2(v) we have:

(1) $s((a \to b) \to b) = s(a \to b) \to s(b) = s(b \to a) \to s(a) = s((b \to a) \to a)$.

(2) $s((a \rightsquigarrow b) \rightsquigarrow b) = s(a \rightsquigarrow b) \rightsquigarrow s(b) = s(b \rightsquigarrow a) \rightsquigarrow s(a) = s((b \rightsquigarrow a) \rightsquigarrow a)$. $\square$

Proposition 8.9 *Let $s : A \longrightarrow L$ be an order-preserving type I state or strong type II state. Then $\mathrm{Ker}(s)$ is a proper and normal filter of A.*

Proof One can easily check that $1 \in \mathrm{Ker}(s)$ and $0 \notin \mathrm{Ker}(s)$.

Consider $a, b \in A$ such that $a, a \to b \in \mathrm{Ker}(s)$, that is, $s(a) = 1$ and $s(a \to b) = 1$.

Assume that s is an order-preserving type I state.

Applying Proposition 8.1(iii) we get:

$$1 = s(a \to b) = s(a) \to s(a \wedge b) = 1 \to s(a \wedge b) = s(a \wedge b) \leq s(b).$$

Thus $s(b) = 1$, that is, $b \in \mathrm{Ker}(s)$, so $\mathrm{Ker}(s)$ is a proper filter of A.

Assume $a \to b \in \mathrm{Ker}(s)$, that is, $s(a \to b) = 1$.

From Proposition 8.1(iii) we get $s(a) \to s(a \wedge b) = 1$, so $s(a) \leq s(a \wedge b)$.

Hence $s(a \rightsquigarrow b) = s(a) \rightsquigarrow s(a \wedge b) = 1$. Thus $a \rightsquigarrow b \in \mathrm{Ker}(s)$.

Similarly, $a \rightsquigarrow b \in \mathrm{Ker}(s)$ implies $a \to b \in \mathrm{Ker}(s)$.

We conclude that $\mathrm{Ker}(s)$ is a normal filter.

Let s be a strong type II state (not necessarily order-preserving).

By Proposition 8.2(v) we have:

$$s(b) = 1 \to s(b) = s(a \to b) \to s(b) = s(b \to a) \to s(a) = s(b \to a) \to 1 = 1.$$

Thus $b \in \mathrm{Ker}(s)$, so $\mathrm{Ker}(s)$ is a proper filter of A.

Since $s(a \to b) = s(a \rightsquigarrow b)$, it follows that $\mathrm{Ker}(s)$ is a normal filter. $\qquad\square$

Proposition 8.10 *Let $s : A \longrightarrow L$ be an order-preserving type I state or strong type II state. Then in the quotient FL_w-algebra $(A/\mathrm{Ker}(s), \wedge, \vee, \odot, \to, \rightsquigarrow, 0_{\mathrm{Ker}(s)}, 1_{\mathrm{Ker}(s)})$ we have:*

(1) $a/\mathrm{Ker}(s) \leq b/\mathrm{Ker}(s)$ *iff* $s(a \to b) = 1$ *iff* $s(a \rightsquigarrow b) = 1$ *iff* $s(a) = s(a \wedge b)$ *iff* $s(b) = s(a \vee b)$;

(2) $a/\mathrm{Ker}(s) = b/\mathrm{Ker}(s)$ *iff* $s(a) = s(b) = s(a \wedge b) = s(a \vee b)$.

Proof We have

$$a/\mathrm{Ker}(s) \leq b/\mathrm{Ker}(s) \quad \text{iff}$$

$$(a \to b)/\mathrm{Ker}(s) = a/\mathrm{Ker}(s) \to b/\mathrm{Ker}(s) = 1/\mathrm{Ker}(s) = \mathrm{Ker}(s) \quad \text{iff}$$

$$s(a \to b) = 1.$$

Similarly,

$$a/\mathrm{Ker}(s) \leq b/\mathrm{Ker}(s) \quad \text{iff}$$

$$(a \rightsquigarrow b)/\mathrm{Ker}(s) = a/\mathrm{Ker}(s) \rightsquigarrow b/\mathrm{Ker}(s) = 1/\mathrm{Ker}(s) = \mathrm{Ker}(s) \quad \text{iff}$$

$$s(a \rightsquigarrow b) = 1.$$

(1) Assume that A is an order-preserving type I state.
According to Proposition 8.1(iii) we get:

$$s(a \to b) = 1 \quad \text{iff} \quad 1 = s(a) \to s(a \wedge b) \quad \text{iff} \quad s(a) \le s(a \wedge b).$$

Since $s(a \wedge b) \le s(a)$, it follows that $s(a \to b) = 1$ iff $s(a) = s(a \wedge b)$.
Similarly, $s(a \rightsquigarrow b) = 1$ iff $s(a) = s(a \wedge b)$.
On the other hand, by Proposition 8.1(iv) we have:

$$s(a \to b) = 1 \quad \text{iff} \quad 1 = s(a \vee b) \to s(b) \quad \text{iff} \quad s(a \vee b) \le s(b).$$

But $s(b) \le s(a \vee b)$, hence $s(a \to b) = 1$ iff $s(b) = s(a \vee b)$.
Similarly, $s(a \rightsquigarrow b) = 1$ iff $s(b) = s(a \vee b)$.
Suppose that s is a strong type II state.
By Proposition 8.2(ii) we have $s(a \to b) = 1$ iff $s(a) = 1 \to s(a \wedge b) = s(a \wedge b)$.
By Proposition 8.2(iv) we have $s(a \to b) = 1$ iff $s(a \vee b) = 1 \to s(b) = s(b)$.
(2) This follows from (1).   □

Theorem 8.2 *If L satisfies the* (pDN) *condition and $s : A \longrightarrow L$ is an order-preserving type I state, then $A/\operatorname{Ker}(s)$ satisfies the* (pDN) *condition.*

Proof According to Proposition 8.4(2) and taking into consideration that L satisfies the (pDN) condition, we have:

$$s\left(a \vee a^{-\sim}\right) = s\left(a^{-\sim}\right) = s(a)^{-\sim} = s(a) \quad \text{and}$$
$$s\left(a \vee a^{\sim-}\right) = s\left(a^{\sim-}\right) = s(a)^{\sim-} = s(a).$$

Applying Proposition 8.10 we get $a^{-\sim}/\operatorname{Ker}(s) = a^{\sim-}/\operatorname{Ker}(s) = a/\operatorname{Ker}(s)$.
Thus $A/\operatorname{Ker}(s)$ satisfies the (pDN) condition.   □

Theorem 8.3 *If $s : A \longrightarrow L$ is a strong type II state, then $A/\operatorname{Ker}(s)$ satisfies the* (pDN) *condition.*

Proof Applying Proposition 8.6(3) we get:

$$s\left(a \vee a^{-\sim}\right) = s\left(a^{-\sim}\right) = s(a) \quad \text{and} \quad s\left(a \vee a^{\sim-}\right) = s\left(a^{\sim-}\right) = s(a),$$

hence by Proposition 8.10 it follows that $a^{-\sim}/\operatorname{Ker}(s) = a^{\sim-}/\operatorname{Ker}(s) = a/\operatorname{Ker}(s)$.
We conclude that $A/\operatorname{Ker}(s)$ satisfies the (pDN) condition.   □

Theorem 8.4 *Any perfect FL_w-algebra admits a strong type I and type II state.*

Proof Let A be a perfect FL$_w$-algebra, so $A = \text{Rad}(A) \cup \text{Rad}(A)^*$. Consider the map $s : A \longrightarrow L$ defined by

$$s(x) := \begin{cases} 1 & \text{if } x \in \text{Rad}(A) \\ 0 & \text{if } x \in \text{Rad}(A)^*. \end{cases}$$

Obviously, $s(1) = 1$ and $s(0) = 0$.

We consider the following cases:

(1) $a, b \in \text{Rad}(A)$.

Obviously, $s(a) = s(b) = 1$. Since $\text{Rad}(A)$ is a filter of A and $b \le a \to b$, it follows that $a \to b \in \text{Rad}(A)$. Hence $s(a \to b) = 1$. Similarly, $s(a \rightsquigarrow b) = 1$.

In the same way, from $a \le a \vee b$ it follows that $a \vee b \in \text{Rad}(A)$, so $s(a \vee b) = 1$.

Thus condition (iv) from Proposition 8.1 and condition (iv) from Proposition 8.2 are verified.

(2) $a, b \in \text{Rad}(A)^*$.

In this case, $s(a) = s(b) = 0$ and we will prove that $a \to b, a \rightsquigarrow b \in \text{Rad}(A)$. Indeed, suppose that $a \to b \in \text{Rad}(A)^*$. Since $a \le a^{-\sim}$, it follows that $a^{-\sim} \to b \le a \to b$, so $a^{-\sim} \to b \in \text{Rad}(A)^*$. But $a^- \le a^{-\sim} \to b$, hence $a^- \in \text{Rad}(A)^*$, that is, $a \in \text{Rad}(A)$, which is a contradiction. It follows that $a \to b \in \text{Rad}(A)$ and similarly, $a \rightsquigarrow b \in \text{Rad}(A)$. Hence $s(a \to b) = s(a \rightsquigarrow b) = 1$.

Since $a \wedge b \le a$, we have $a \wedge b \in \text{Rad}(A)^*$, so $s(a \wedge b) = 0$.

We can see that condition (iii) from Proposition 8.1 and condition (ii) from Proposition 8.2 are verified.

(3) $a \in \text{Rad}(A)$ and $b \in \text{Rad}(A)^*$.

Obviously, $s(a) = 1$ and $s(b) = 0$.

We show that $a \to b \in \text{Rad}(A)^*$. Indeed, suppose that $a \to b \in \text{Rad}(A)$.

Because $b \le b^{-\sim}$, we have $a \to b \le a \to b^{-\sim}$, so $a \to b^{-\sim} \in \text{Rad}(A)$. This means that $(a \odot b^\sim)^- \in \text{Rad}(A)$, that is, $a \odot b^\sim \in \text{Rad}(A)^*$. On the other hand, since $\text{Rad}(A)$ is a filter of A and $a, b^\sim \in \text{Rad}(A)$ we have $a \odot b^\sim \in \text{Rad}(A)$, a contradiction. We conclude that $a \to b \in \text{Rad}(A)^*$, so $s(a \to b) = 0$.

Similarly, $s(a \rightsquigarrow b) = 0$. Since $a \wedge b \le b$, we have $a \wedge b \in \text{Rad}(A)^*$, so $s(a \wedge b) = 0$.

Thus condition (iii) from Proposition 8.1 and condition (ii) from Proposition 8.2 are verified.

(4) $a \in \text{Rad}(A)^*$ and $b \in \text{Rad}(A)$.

Obviously, $s(a) = 0$ and $s(b) = 1$.

Since $b \le a \to b$ it follows that $a \to b \in \text{Rad}(A)$, so $s(a \to b) = 1$.

Similarly, $s(a \rightsquigarrow b) = 1$. From $b \le a \vee b$ we get $a \vee b \in \text{Rad}(A)$, so $s(a \vee b) = 1$.

Thus condition (iv) from Proposition 8.1 and condition (iv) from Proposition 8.2 are verified.

We conclude that s is a type I and type II state.

Since we have proved that $s(a \to b) = s(a \rightsquigarrow b)$ for all the above cases, it follows that s is strong.   □

Remark 8.5 The generalized states look similar to internal states which have been defined and investigated for MV-algebras ([119]), BL-algebras ([72]), bounded Rℓ-monoids ([114]) and bounded pseudo-hoops ([64]). It has been proved that if A is one of these structures and $\sigma : A \longrightarrow A$ is an internal state on A, then $\sigma(A)$ is a subalgebra of A.

This property does not hold for FL$_w$-algebras. Indeed, consider the FL$_w$-algebra A and the type I and type II state s_4 from Remark 8.1. We can see that $s_4(b) \vee s_4(c) = b \vee c = a \notin s_4(A) = \{0, b, c, 1\}$, hence $s_4(A)$ is not closed under $\vee$.

Proposition 8.11 *Let A be a linearly ordered FL$_w$-algebra and s be an order-preserving type I state or a type II state on A. Then $s(A)$ is linearly ordered.*

Proof According to Proposition 8.6(a), any type II state is order-preserving.

Let $x, y \in s(A)$, thus there exist $a, b \in A$ such that $x = s(a)$ and $y = s(b)$. Since A is linearly ordered, we have $a \leq b$ or $b \leq a$. Taking into consideration that s is order-preserving, it follows that $x = s(a) \leq s(b) = y$ or $y = s(b) \leq s(a) = x$. Hence $s(A)$ is linearly ordered. $\qquad\qquad\square$

8.2 Generalized State-Morphisms

We introduce the notion of generalized state-morphism and we prove that any generalized state morphism is an order-preserving type I state. We also prove that, in certain particular conditions, an order-preserving type I state is a generalized state-morphism. The notion of a strong perfect FL$_w$-algebra is introduced, and it is proved that any strong perfect FL$_w$-algebra admits a generalized state-morphism.

Throughout this section, A and L are FL$_w$-algebras.

Consider the arbitrary function $s : A \longrightarrow L$ and the properties:

(sm_0) $s(0) = 0$ and $s(1) = 1$;
(sm_1) $s(a \vee b) = s(a) \vee s(b)$ for all $a, b \in A$;
(sm_2) $s(a \wedge b) = s(a) \wedge s(b)$ for all $a, b \in A$;
(sm_3) $s(a \rightarrow b) = s(a) \rightarrow s(b)$ and $s(a \rightsquigarrow b) = s(a) \rightsquigarrow s(b)$ for all $a, b \in A$;
(sm_4) $s(a \odot b) = s(a) \odot s(b)$ for all $a, b \in A$.

Lemma 8.1 *If $s : A \longrightarrow L$ is an order-preserving type I state, then:*

(1) (sm_1) *implies* (sm_3);
(2) (sm_2) *implies* (sm_3).

Proof

(1) Applying Proposition 8.1(iv) we have:

$$s(a \rightarrow b) = s(a \vee b) \rightarrow s(b) = \big(s(a) \vee s(b)\big) \rightarrow s(b)$$
$$= \big(s(a) \rightarrow s(b)\big) \wedge \big(s(b) \rightarrow s(b)\big) = s(a) \rightarrow s(b) \quad \text{and}$$

$$s(a \rightsquigarrow b) = s(a \vee b) \rightsquigarrow s(b) = \big(s(a) \vee s(b)\big) \rightsquigarrow s(b)$$
$$= \big(s(a) \rightsquigarrow s(b)\big) \wedge \big(s(b) \rightsquigarrow s(b)\big) = s(a) \rightsquigarrow s(b).$$

(2) Applying Proposition 8.1(iii) we get:

$$s(a \rightarrow b) = s(a) \rightarrow s(a \wedge b) = s(a) \rightarrow \big(s(a) \wedge s(b)\big)$$
$$= \big(s(a) \rightarrow s(a)\big) \wedge \big(s(a) \rightarrow s(b)\big) = s(a) \rightarrow s(b) \quad \text{and}$$
$$s(a \rightsquigarrow b) = s(a) \rightsquigarrow s(a \wedge b) = s(a) \rightsquigarrow \big(s(a) \wedge s(b)\big)$$
$$= \big(s(a) \rightsquigarrow s(a)\big) \wedge \big(s(a) \rightsquigarrow s(b)\big) = s(a) \rightsquigarrow s(b). \qquad \square$$

Lemma 8.2 *Let L be an FL_w-algebra satisfying the* (pDN) *condition and $s : A \longrightarrow L$ be an order-preserving type I state. Then* (sm_2) *implies* (sm_1).

Proof According to Proposition 8.4(1) and $(psbck\text{-}c_{41})$ we have:

$$s(a \vee b)^- = s\big((a \vee b)^-\big) = s\big(a^- \wedge b^-\big) = s\big(a^-\big) \wedge s\big(b^-\big)$$
$$= s(a)^- \wedge s(b)^- = \big(s(a) \vee s(b)\big)^-.$$

Hence $s(a \vee b)^{-\sim} = (s(a) \vee s(b))^{-\sim}$.
Since L satisfies the (pDN) condition, it follows that $s(a \vee b) = s(a) \vee s(b)$. $\quad \square$

Lemma 8.3 *Let $s : A \longrightarrow L$ be an order-preserving type I state. Then:*

(1) (sm_3) *implies* $s(a \odot b)^- = (s(a) \odot s(b))^-$ *and* $s(a \odot b)^\sim = (s(a) \odot s(b))^\sim$ *for all $a, b \in A$;*
(2) *If L satisfies the* (pDN) *condition, then* (sm_3) *implies* (sm_4).

Proof

(1) Applying Proposition 8.4(1) and $(psbck\text{-}c_{37})$ we have:

$$s(a \odot b)^- = s\big((a \odot b)^-\big) = s\big(a \rightarrow b^-\big) = s(a) \rightarrow s\big(b^-\big) = s(a) \rightarrow s(b)^-$$
$$= \big(s(a) \odot s(b)\big)^-.$$

Similarly,

$$s(a \odot b)^\sim = s\big((a \odot b)^\sim\big) = s\big(b \rightsquigarrow a^\sim\big) = s(b) \rightsquigarrow s\big(a^\sim\big) = s(b) \rightsquigarrow s(a)^\sim$$
$$= \big(s(a) \odot s(b)\big)^\sim.$$

(2) This follows by (1) and the (pDN) condition. $\qquad \square$

Definition 8.2 Let A and L be FL_w-algebras. A function $s : A \longrightarrow L$ is a *generalized state-morphism* if it satisfies conditions (sm_0)–(sm_3).

Example 8.6 Let A be the FL_w-algebra from Example 4.4 and L an arbitrary FL_w-algebra. Consider the function $s : A \longrightarrow L$, $s(0) := 0$, $s(a) = s(b) = s(c) = s(1) := 1$. One can easily check that s is a generalized state-morphism.

Proposition 8.12 *Any generalized state-morphism is an order-preserving type I state.*

Proof According to Proposition 8.1(ii), a generalized state-morphism is a type I state. By (sm_1)–(sm_3) the generalized state-morphism is a lattice morphism, so it is an order-preserving function.   □

Proposition 8.13 *Let $s : A \longrightarrow L$ be an order-preserving type I state. If $A/\operatorname{Ker}(s)$ is totally ordered, then s is a generalized state-morphism.*

Proof We check conditions (sm_0)–(sm_3).

(sm_0) This follows from the definition of a type I state.
(sm_1)–(sm_2) Let $a, b \in A$. Since $A/\operatorname{Ker}(s)$ is totally ordered, it follows that $a/\operatorname{Ker}(s) \leq b/\operatorname{Ker}(s)$ or $b/\operatorname{Ker}(s) \leq a/\operatorname{Ker}(s)$.
Suppose $a/\operatorname{Ker}(s) \leq b/\operatorname{Ker}(s)$, thus, by Proposition 8.10, $s(a \rightarrow b) = 1$ and $s(a \rightsquigarrow b) = 1$. By Proposition 8.1(iii), (iv) we have:

$$1 = s(a \rightarrow b) = s(a) \rightarrow s(a \wedge b) = s(a \vee b) \rightarrow s(b).$$

Hence $s(a) \leq s(a \wedge b) \leq s(a \vee b) \leq s(b)$, so $s(a \vee b) = s(a) \vee s(b)$ and $s(a \wedge b) = s(a) \wedge s(b)$.
(sm_3) By Lemma 8.1, (sm_1) implies (sm_3).
Thus s is a generalized state-morphism.   □

Proposition 8.14 *Let $s : A \longrightarrow L$ be a generalized state-morphism and L be totally ordered. Then $A/\operatorname{Ker}(s)$ is totally ordered.*

Proof Let $a, b \in A$. Since L is totally ordered, it follows that $s(a) \leq s(b)$ or $s(b) \leq s(a)$, that is, $s(a \rightarrow b) = s(a) \rightarrow s(b) = 1$ or $s(b \rightarrow a) = s(b) \rightarrow s(a) = 1$.
Hence $a/\operatorname{Ker}(s) \leq b/\operatorname{Ker}(s)$ or $b/\operatorname{Ker}(s) \leq a/\operatorname{Ker}(s)$.
Thus $A/\operatorname{Ker}(s)$ is totally ordered.   □

Proposition 8.15 *Let $s : A \longrightarrow L$ be an order-preserving type I state and L be totally ordered. Then s is a generalized state-morphism iff $A/\operatorname{Ker}(s)$ is totally ordered.*

Proof This follows from Propositions 8.13 and 8.14.   □

Definition 8.3 A perfect FL_w-algebra is said to be *strong perfect* if $\operatorname{Rad}(A)^*$ is closed under $\vee$.

Theorem 8.5 *Any strong perfect FL_w-algebra admits a generalized state-morphism.*

Proof Let A be a strong perfect FL_w-algebra, so $A = \mathrm{Rad}(A) \cup \mathrm{Rad}(A)^*$.
Consider the map $s : A \to L$ defined by

$$s(x) := \begin{cases} 1 & \text{if } x \in \mathrm{Rad}(A) \\ 0 & \text{if } x \in \mathrm{Rad}(A)^*. \end{cases}$$

Obviously, $s(1) = 1$ and $s(0) = 0$, so condition (sm_0) is verified.
Similarly as in Theorem 8.4 we will consider the following cases:

(1) $a, b \in \mathrm{Rad}(A)$.
   Since $a, b, a \wedge b, a \vee b, a \to b, a \rightsquigarrow b \in \mathrm{Rad}(A)$, we have

$$s(a) = s(b) = s(a \wedge b) = s(a \vee b) = s(a \to b) = s(a \rightsquigarrow b) = 1$$

   and the conditions (sm_1)–(sm_3) are verified.
(2) $a, b \in \mathrm{Rad}(A)^*$.
   It follows that $a, b, a \wedge b \in \mathrm{Rad}(A)^*$. Since $\mathrm{Rad}(A)^*$ is closed under $\vee$, we
   also have $a \vee b \in \mathrm{Rad}(A)^*$, so $s(a) = s(b) = s(a \wedge b) = s(a \vee b) = 0$.
   We proved in Theorem 8.4 that $a \to b, a \rightsquigarrow b \in \mathrm{Rad}(A)$, thus $s(a \to b) = s(a \rightsquigarrow b) = 1$.
   It is easy to see that conditions (sm_1)–(sm_3) are satisfied.
(3) $a \in \mathrm{Rad}(A), b \in \mathrm{Rad}(A)^*$.
   Obviously, $s(a) = 1$ and $s(b) = 0$. From $a \wedge b \le b \in \mathrm{Rad}(A)^*$ it follows that
   $a \wedge b \in \mathrm{Rad}(A)^*$, so $s(a \wedge b) = 0$. Similarly, since $a \le a \vee b$ and $a \in \mathrm{Rad}(A)$,
   we get $a \vee b \in \mathrm{Rad}(A)$, thus $s(a \vee b) = 1$. It was proved in Theorem 8.4 that
   $a \to b, a \rightsquigarrow b \in \mathrm{Rad}(A)^*$, hence $s(a \to b) = s(a \rightsquigarrow b) = 0$.
   Conditions (sm_1)–(sm_3) are again satisfied.
(4) $a \in \mathrm{Rad}(A)^*, b \in \mathrm{Rad}(A)$.
   Obviously, $s(a) = 0$ and $s(b) = 1$. From $a \wedge b \le a \in \mathrm{Rad}(A)^*$ it follows that
   $a \wedge b \in \mathrm{Rad}(A)^*$, so $s(a \wedge b) = 0$. Similarly, since $b \le a \vee b$ and $b \in \mathrm{Rad}(A)$,
   we get $a \vee b \in \mathrm{Rad}(A)$, thus $s(a \vee b) = 1$. It was proved in Theorem 8.4 that
   $a \to b, a \rightsquigarrow b \in \mathrm{Rad}(A)$, hence $s(a \to b) = s(a \rightsquigarrow b) = 1$.
   In this case conditions (sm_1)–(sm_3) are also satisfied. Thus s is a generalized
   state-morphism on A. □

Remark 8.6 The FL_w-algebra A from Example 4.4 is perfect with $\mathrm{Rad}(A) = \{a, b, c, 1\}$ and $\mathrm{Rad}(A)^* = \{0\}$. Obviously, $\mathrm{Rad}(A)^*$ is closed under $\vee$, so A is a strong perfect FL_w-algebra. Let L be an arbitrary FL_w-algebra. According to Theorem 8.5 the function $s : A \to L$ defined by

$$s(x) := \begin{cases} 1 & \text{if } x \in \{a, b, c, 1\} \\ 0 & \text{if } x = 0 \end{cases}$$

is a generalized state-morphism on A.

As we can see, it is the generalized state-morphism from Example 8.6.

8.3 Generalized Riečan States

We introduce the notion of a generalized Riečan state and we show that any order-preserving type I state is a generalized Riečan state. Special conditions are given which force the two notions to coincide. Some of the main results are proved using the notion of the Glivenko property defined for the non-commutative case.

In this section A and L will be considered to be good FL_w-algebras.

We recall that if $a, b \in A$, then a is orthogonal to b, denoted by $a \perp b$, if $b^{-\sim} \le a^-$.

Definition 8.4 The function $m : A \longrightarrow L$ is said to be *orthogonal-preserving* if $m(a) \perp m(b)$ whenever $a \perp b$.

Definition 8.5 An orthogonal-preserving function $m : A \longrightarrow L$ is called a *generalized Riečan state* iff the following conditions are satisfied for all $a, b \in A$:

(GR_1) $m(1) = 1$;
(GR_2) if $a \perp b$, then $m(a \oplus b) = m(a) \oplus m(b)$.

Example 8.7 Consider again the FL_w-algebra A from Example 4.4.

One can easily check that $x^{-\sim} = x^{\sim -}$ for any $x \in A$, so A is a good FL_w-algebra.

We claim that the function $m : A \to A$ defined by: $m(0) := 0$, $m(a) = m(b) = m(c) = m(1) := 1$ is a generalized Riečan state on A.

Indeed, the elements $x, y \in A$ with $x \perp y$ are those given in the table below:

x	y	x^-	$y^{-\sim}$	$x \oplus y$	$m(x \oplus y)$	$m(x) \oplus m(y)$
0	0	1	0	0	0	0
0	a	1	1	1	1	1
0	b	1	1	1	1	1
0	c	1	1	1	1	1
0	1	1	1	1	1	1
a	0	0	0	1	1	1
b	0	0	0	1	1	1
c	0	0	0	1	1	1
1	0	0	0	1	1	1

One can easily check that m is a generalized Riečan state.

Proposition 8.16 *Let* $m : A \longrightarrow L$ *be a generalized Riečan state. Then, for all* $a, b \in A$ *the following hold*:

(1) $m(a) \oplus m(a^-) = m(a^\sim) \oplus m(a) = 1$;
(2) $m(a^-)^{-\sim} = m(a)^-$ *and* $m(a^\sim)^{\sim -} = m(a)^\sim$;

(3) *if L satisfies the* (pDN) *condition, then* $m(a^-) = m(a)^-$, $m(a^\sim) = m(a)^\sim$ *and*
   $m(a^{-\sim}) = m(a)$;
(4) $m(0) = 0$;
(5) *if* $b \le a$, *then* $m(a)^- \le m(b)^-$ *and* $m(a)^\sim \le m(b)^\sim$;
(6) *if L satisfies the* (pDN) *condition, then* $b \le a$ *implies* $m(b) \le m(a)$.

Proof

(1) Since $a \perp a^-$ and $a^\sim \perp a$, applying $(psbck\text{-}c_{36})$ we get:

$$m(a) \oplus m(a^-) = m(a \oplus a^-) = m\big((a^{-\sim} \odot a^\sim)^-\big)$$
$$= m\big((a^{\sim -} \odot a^\sim)^-\big) = m(0^-) = m(1) = 1,$$
$$m(a^\sim) \oplus m(a) = m(a^\sim \oplus a) = m\big((a^- \odot a^{\sim-})^\sim\big)$$
$$= m\big((a^- \odot a^{\sim-})^\sim\big) = m(0^\sim) = m(1) = 1.$$

(2) From $a \perp a^-$ it follows that $m(a) \perp m(a^-)$, so $m(a^-)^{-\sim} \le m(a)^-$.
   On the other hand,

$$1 = m(1) = m(a \oplus a^-) = m(a) \oplus m(a^-) = m(a)^- \rightsquigarrow m(a^-)^{-\sim},$$

hence $m(a)^- \le m(a^-)^{-\sim}$.
   Thus $m(a^-)^{-\sim} = m(a)^-$ and similarly, $m(a^\sim)^{\sim-} = m(a)^\sim$.
(3) $m(a^-) = m(a)^-$ and $m(a^\sim) = m(a)^\sim$ follow from (2).
   The identity $m(a^{-\sim}) = m(a)$ follows from $m(a^-) = m(a)^-$ replacing a with
   $a^\sim$ and applying the (pDN) condition for L.
(4) Putting $a = 0$ in (2) we get $1^{-\sim} = m(0)^-$, so $m(0)^- = 1$, that is, $m(0) \to 0 = 1$.
   It follows that $m(0) \le 0$, thus $m(0) = 0$.
(5) From Lemma 1.6(6), $b \le a$ implies $b \perp a^-$, so $m(b) \perp m(a^-)$.
   Applying (2) we get $m(a)^- = m(a^-)^{-\sim} \le m(b)^-$. Similarly, $m(a)^\sim \le m(b)^\sim$.
(6) This follows by (5). $\square$

Theorem 8.6 *Any order-preserving type I state is a generalized Riečan state.*

Proof Let $s : A \longrightarrow L$ be an order-preserving type I state and $a, b \in A$ such that
$a \perp b$. It follows that $a^{-\sim} \le b^\sim$.
   Since s is order-preserving, we get $s(a^{-\sim}) \le s(b^\sim)$.
   Applying Proposition 8.4(1), (2) we get $s(a)^{-\sim} \le s(b)^\sim$.
   Hence $s(a) \perp s(b)$, thus s is orthogonal-preserving.
   By Proposition 8.1(ii) we have:

$$s(a \oplus b) = s(b^\sim \to a^{\sim-}) = s(b^\sim) \to s(a^{\sim-}) = s(b)^\sim \to s(a)^{\sim-} = s(a) \oplus s(b).$$

We conclude that s is a generalized Riečan state. $\square$

Remark 8.7 There exist generalized Riečan states which are not type I or type II states. Indeed, let A be the FL_w-algebra from Example 4.1. The generalized Riečan states $m : A \longrightarrow A$ are the following:

	x				
	0	a	b	c	1
$m_1(x)$	0	a	b	c	1
$m_2(x)$	0	a	b	1	1
$m_3(x)$	0	b	b	c	1
$m_4(x)$	0	b	b	1	1

m_1, m_2, m_3, m_4 are generalized Riečan states on A, but m_1 is the only type I state.

Theorem 8.7 *If A satisfies the (pDN) condition and $m : A \longrightarrow L$ is a generalized Riečan state such that for all $a \in A$, $m(a^-) = m(a)^-$ and $m(a^\sim) = m(a)^\sim$, then m is an order-preserving type I state.*

Proof From $m(a^-) = m(a)^-$ and $m(a^\sim) = m(a)^\sim$ we get $m(a^{-\sim}) = m(a)^{-\sim}$. Let $a, b \in A$ with $b \leq a$, so $b^{-\sim} \leq a^{-\sim}$.

It follows that $b \perp a^-$, so $m(b) \perp m(a^-)$, that is, $m(b) \perp m(a)^-$.
Similarly we have $a^\sim \perp b$, hence $m(a)^\sim \perp m(b)$.
Applying the (pDN) condition of A we get:

$$b \oplus a^- = a^{-\sim} \to b^{\sim -} = a \to b,$$
$$a^\sim \oplus b = a^{\sim -} \rightsquigarrow b^{-\sim} = a \rightsquigarrow b.$$

It follows that:

$$m(a \to b) = m(b \oplus a^-) = m(b) \oplus m(a^-) = m(b) \oplus m(a)^-$$
$$= m(a)^{-\sim} \to m(b)^{-\sim} = m(a^{-\sim}) \to m(b^{-\sim}) = m(a) \to m(b);$$
$$m(a \rightsquigarrow b) = m(a^\sim \oplus b) = m(a^\sim) \oplus m(b) = m(a)^\sim \oplus m(b)$$
$$= m(a)^{-\sim} \rightsquigarrow m(b)^{-\sim} = m(a^{-\sim}) \rightsquigarrow m(b^{-\sim}) = m(a) \rightsquigarrow m(b).$$

By Proposition 8.1(ii), m is a type I state.
From $m(b) \perp m(a^-)$ it follows that $m(b)^{-\sim} \leq m(a)^{-\sim}$, that is, $m(b^{-\sim}) \leq m(a^{-\sim})$.
Thus $m(b) \leq m(a)$, so m is order-preserving. $\square$

Corollary 8.1 *If A and L satisfy the (pDN) condition and $m : A \longrightarrow L$ is a generalized Riečan state, then m is an order-preserving type I state.*

Proof This follows from Theorem 8.7 and Proposition 8.16(3). $\square$

Proposition 8.17 *If A and L are good FL_w-algebras and $m : \mathrm{Reg}(A) \longrightarrow L$ is a generalized Riečan state on $\mathrm{Reg}(A)$, then $M : A \longrightarrow L$ defined by:*

$$M(x) := m(x^{-\sim})$$

is a generalized Riečan state on A such that $M_{|\mathrm{Reg}(A)} = m$.

Proof Obviously, $M(1) = 1$.

Consider $a, b \in A$ such that $a \perp b$, that is, $a^{-\sim} \leq b^{\sim}$.

It follows that $a^{-\sim-\sim} \leq b^{-\sim\sim}$. Hence $a^{-\sim} \perp b^{-\sim}$.

Thus $a \oplus b$ exists iff $a^{-\sim} \oplus b^{-\sim}$ exists.

Since, by Proposition 1.24(5), $(a \oplus b)^{-\sim} = a \oplus b$, it follows that $a \oplus b \in \mathrm{Reg}(A)$.

Moreover, $a^{-\sim} \oplus b^{-\sim} = (b^{-\sim\sim} \odot a^{-\sim\sim})^{-} = (b^{\sim} \odot a^{\sim})^{-} = a \oplus b$.

Hence $M(a \oplus b) = m(a^{-\sim} \oplus b^{-\sim}) = m(a^{-\sim}) \oplus m(b^{-\sim}) = M(a) \oplus M(b)$.

We conclude that M is a generalized Riečan state on A. $\qquad\square$

Example 8.8 Let A be the FL_w-algebra from Example 4.1.

Then $\mathrm{Reg}(A) = \{0, b, 1\}$ and $m : \mathrm{Reg}(A) \longrightarrow A$ is a generalized Riečan state on $\mathrm{Reg}(A)$. One can easily check that $M : A \longrightarrow A$ defined by $M(x) := m(x^{-\sim})$ is the generalized Riečan state m_4 on A from Remark 8.7.

Theorem 8.8 *If A has the Glivenko property and L satisfies the* (pDN) *condition, then any generalized Riečan state $m : A \longrightarrow L$ is an order-preserving type I state.*

Proof Let $m : A \longrightarrow L$ be a generalized Riečan state and $a, b \in A$ such that $b \leq a$. By Lemma 1.6(6) it follows that $b \perp a^{-}$ and $a^{\sim} \perp b$.

Taking into consideration the definition of the operation $\oplus$ and (psbck-c_{19}) we have: $b \oplus a^{-} = a^{-\sim} \to b^{\sim-} = a \to b^{-\sim}$ and $a^{\sim} \oplus b = a^{\sim-} \rightsquigarrow b^{-\sim} = a \rightsquigarrow b^{-\sim}$.

Hence, by the Glivenko property we get:

$$m(a \to b) = m(a \to b)^{-\sim} = m((a \to b)^{-\sim}) = m(a \to b^{-\sim}) = m(b \oplus a^{-})$$

$$= m(b) \oplus m(a^{-}) = m(a)^{-\sim} \to m(b)^{\sim-} = m(a) \to m(b),$$

$$m(a \rightsquigarrow b) = m(a \rightsquigarrow b)^{-\sim} = m((a \rightsquigarrow b)^{-\sim}) = m(a \rightsquigarrow b^{-\sim}) = m(a^{\sim} \oplus b)$$

$$= m(a^{\sim}) \oplus m(b) = m(a)^{-\sim} \rightsquigarrow m(b)^{\sim-} = m(a) \rightsquigarrow m(b)$$

(since L satisfies the (pDN) condition).

According to Proposition 8.1(ii) and Proposition 8.16(6), m is an order-preserving type I state. $\qquad\square$

Corollary 8.2 *If A is a good $R\ell$-monoid and L satisfies the* (pDN) *condition, then any generalized Riečan state $m : A \longrightarrow L$ is an order-preserving type I state.*

Proof This follows from Theorem 8.8, taking into consideration that any good $R\ell$-monoid satisfies the Glivenko property (Remark 4.16). $\qquad\square$

Corollary 8.3 *If A has the Glivenko property, L satisfies the* (pDN) *condition and* $m : A \longrightarrow L$ *is a generalized Riečan state on A, then* $\mathrm{Ker}(m)$ *is a proper and normal filter of A.*

Proof This follows from Theorem 8.8 and Proposition 8.9.   □

Lemma 8.4 *If A has the Glivenko property, L satisfies the* (pDN) *condition and* $m :$ $A \longrightarrow L$ *is a generalized Riečan state on A, then* $a \equiv_{\mathrm{Ker}(m)} b$ *implies* $m(a) = m(b)$.

Proof We know that $a \equiv_{\mathrm{Ker}(m)} b$ is equivalent to $m(a \to b) = m(b \to a) = 1$.

From $m(a \to b) = 1$ and Proposition 8.1(iii) we get $m(a) \leq m(a \wedge b)$.

Since $a \wedge b \leq a$, according to Proposition 8.16 we have $m(a \wedge b) \leq m(a)$.

Thus $m(a) = m(a \wedge b)$. Similarly, from $m(b \to a) = 1$ and Proposition 8.1(iii) we get $m(b) = m(a \wedge b)$.

Hence $m(a) = m(b)$.   □

Theorem 8.9 *If A has the Glivenko property, L satisfies the* (pDN) *condition and m is a generalized Riečan state on A, then the function* $\hat{m} : A/\mathrm{Ker}(m) \to L$ *defined by* $\hat{m}(x/\mathrm{Ker}(m)) := m(x)$ *is a generalized Riečan state on* $A/\mathrm{Ker}(m)$.

Proof First we prove that $\hat{m}$ is well-defined.

Indeed, if $a/\mathrm{Ker}(m) = b/\mathrm{Ker}(m)$, then by Proposition 8.10(2) it follows that $m(a \wedge b) = m(a \vee b)$. By Theorem 8.8, m is an order-preserving type I state and applying Proposition 8.1(i) it follows that $m(d_1(a, b)) = 1$ and $m(d_2(a, b)) = 1$.

Hence $d_1(a, b) \in \mathrm{Ker}(m)$ and $d_2(a, b) \in \mathrm{Ker}(m)$. Thus $a \equiv_{\mathrm{Ker}(m)} b$.

We prove now that $\hat{m}$ is a generalized Riečan state on $A/\mathrm{Ker}(m)$.

We remark that if $a/\mathrm{Ker}(m) \leq b/\mathrm{Ker}(m)$, then there is an element $a_1 \in a/\mathrm{Ker}(m)$ such that $a_1 \leq b$. Indeed, it suffices to take $a_1 = a \wedge b$.

Assume that $a/\mathrm{Ker}(m) \perp b/\mathrm{Ker}(m)$, that is, $(a/\mathrm{Ker}(m))^{-\sim} \leq (b/\mathrm{Ker}(m))^{\sim}$.

It follows that $a^{-\sim}/\mathrm{Ker}(m) \leq b^{\sim}/\mathrm{Ker}(m)$.

Take $a_1 \in a^{-\sim}/\mathrm{Ker}(m)$ such that $a_1^{-\sim} \leq b^{\sim}$, that is, $a_1 \perp b$ and $a_1 \perp b^{-\sim}$.

Therefore, applying Lemma 8.4, Proposition 1.24(5) and Proposition 8.16(3) we get:

$$\hat{m}\big(a/\mathrm{Ker}(m) \oplus b/\mathrm{Ker}(m)\big) = \hat{m}\big(\big((b/\mathrm{Ker}(m))^{\sim} \odot (a/\mathrm{Ker}(m))^{\sim}\big)^{-}\big)$$

$$= \hat{m}\big((b^{\sim} \odot a^{\sim})^{-}/\mathrm{Ker}(m)\big) = m\big((b^{\sim} \odot a^{\sim})^{-}\big)$$

$$= m(a \oplus b) = m\big(a^{-\sim} \oplus b^{-\sim}\big)$$

$$= \hat{m}\big((a/\mathrm{Ker}(m))^{-\sim} \oplus (b/\mathrm{Ker}(m))^{-\sim}\big)$$

$$= \hat{m}\big(a_1/\mathrm{Ker}(m) \oplus (b/\mathrm{Ker}(m))^{-\sim}\big)$$

$$= m\big(a_1 \oplus b^{-\sim}\big) = m(a_1) \oplus m\big(b^{-\sim}\big) = m(a) \oplus m(b)$$

$$= \hat{m}\big(a/\mathrm{Ker}(m)\big) \oplus \hat{m}\big(b/\mathrm{Ker}(m)\big). \qquad \Box$$

8.4 Generalized Local States on Perfect Pseudo-MTL Algebras

In this section we introduce the notion of a generalized local state on a perfect pseudo-MTL algebra A and we prove an extension theorem for this type of state. More precisely we prove that, if A is relatively free of zero divisors, then any generalized local state on A can be extended to a generalized Riečan state on A.

In this section A and L will be perfect pseudo-MTL algebras.

Definition 8.6 A *generalized local state* on A is an orthogonal-preserving function $s : \mathrm{Rad}(A)^* \longrightarrow L$ satisfying the conditions:

(GLS_1) $s(x \oplus y) = s(x) \oplus s(y)$ for all $x, y \in \mathrm{Rad}(A)^*$;
(GLS_2) $s(0) = 0$.

Remark 8.8 For all $x, y \in \mathrm{Rad}(A)^*$ we have $x \oplus y = (y^\sim \odot x^\sim)^-$.

Since $x^\sim, y^\sim \in \mathrm{Rad}(A)$ and $\mathrm{Rad}(A)$ is a filter of A, it follows that $y^\sim \odot x^\sim \in \mathrm{Rad}(A)$, that is, $(y^\sim \odot x^\sim)^- \in \mathrm{Rad}(A)^*$. Thus the function s is well defined.

Example 8.9

(1) *The function $s : \mathrm{Rad}(A)^* \longrightarrow L$, $s(x) := 0$ for all $x \in \mathrm{Rad}(A)^*$ is a generalized local state on the perfect pseudo-MTL algebra A.*
(2) *If m is a generalized Riečan state on the perfect pseudo-MTL algebra A, then $s := m_{|\mathrm{Rad}(A)^*}$ is a generalized local state on A.*

Proposition 8.18 *If s is a generalized local state on A, then the following hold for all $x, y \in \mathrm{Rad}(A)^*$:*

(1) $s(x^{-\sim}) = s(x)^{-\sim}$;
(2) $s(x) \oplus s(y^{-\sim}) = s(x^{-\sim}) \oplus s(y)$;
(3) $s(x) \leq s((x^\sim \odot x^\sim)^-)$.

Proof

(1) Since $x \perp 0$, we have $x \oplus 0 = x^{-\sim}$.
    On the other hand, $s(x \oplus 0) = s(x) \oplus s(0) = s(x) \oplus 0$, thus $s(x^{-\sim}) = s(x)^{-\sim}$.
(2) Since $x, x^{-\sim}, y^{-\sim} \in \mathrm{Rad}(A)^*$, we have $x \perp y^{-\sim}$ and $x^{-\sim} \perp y$.
    Hence $s(x) \oplus s(y^{-\sim}) = s(x \oplus y^{-\sim}) = s((y^{-\sim\sim} \odot x^\sim)^-) = s((y^\sim \odot x^\sim)^-)$
    and $s(x^{-\sim}) \oplus s(y) = s(x^{-\sim} \oplus y) = s((y^\sim \odot x^{-\sim\sim})^-) = s((y^\sim \odot x^\sim)^-)$.
    It follows that $s(x) \oplus s(y^{-\sim}) = s(x^{-\sim}) \oplus s(y)$.
(3) Replacing y with x in the identity $s(x) \oplus s(y^{-\sim}) = s((y^\sim \odot x^\sim)^-)$, we get $s(x) \oplus s(x^{-\sim}) = s((x^\sim \odot x^\sim)^-)$.
    Since $s(x) \leq s(x) \oplus s(x^{-\sim})$, we have $s(x) \leq s((x^\sim \odot x^\sim)^-)$. □

Theorem 8.10 (Extension theorem) *Let A be a perfect pseudo-MTL algebra which is relatively free of zero divisors. Then every generalized local state on A can be extended to a generalized Riečan state on A.*

Proof Let $s : \mathrm{Rad}(A)^* \longrightarrow L$ be a generalized local state on A and the function $m : A \longrightarrow L$ defined by

$$m(x) := \begin{cases} 1 & \text{if } x \in \mathrm{Rad}(A) \\ s(x) & \text{if } x \in \mathrm{Rad}(A)^*. \end{cases}$$

We prove that m is a generalized Riečan state on A. More precisely, since it is obvious that $m(1) = 1$, we must prove that for all $x, y \in A$ such that $x \perp y$, we have $m(x \oplus y) = m(x) \oplus m(y)$.

Consider the following cases:

(1) $x, y \in \mathrm{Rad}(A)$. According to Lemma 5.4, $x \not\perp y$.
(2) $x \in \mathrm{Rad}(A)$ and $y \in \mathrm{Rad}(A)^*$. Applying Lemma 5.3 we get $x \perp y$ iff $y = 0$. We have $m(x \oplus y) = m(x \oplus 0) = m(x^{-\sim}) = 1$ (since $x^{-\sim} \in \mathrm{Rad}(A)$). On the other hand, $m(x) \oplus m(y) = m(x) \oplus m(0) = 1 \oplus s(0) = 1 \oplus 0 = 1$.
    Thus $m(x \oplus y) = m(x) \oplus m(y)$.
(3) $x \in \mathrm{Rad}(A)^*$ and $y \in \mathrm{Rad}(A)$. Applying Lemma 5.3 we get $x \perp y$ iff $x = 0$. We have:

$$m(x \oplus y) = m(0 \oplus y) = m\left(y^{-\sim}\right) = 1 \quad \text{and}$$
$$m(x) \oplus m(y) = m(0) \oplus m(y) = s(0) \oplus 1 = 0 \oplus 1 = 1.$$

    Thus $m(x \oplus y) = m(x) \oplus m(y)$.
(4) $x, y \in \mathrm{Rad}(A)^*$. By Proposition 5.13 it follows that $x \perp y$ and by the definition of a generalized local state we have

$$m(x \oplus y) = s(x \oplus y) = s(x) \oplus s(y) = m(x) \oplus m(y).$$

Thus m is a generalized Riečan state on A. $\square$

We call m the *extension* of the generalized local state s.

Corollary 8.4 *Let A be a perfect pseudo-MTL algebra relatively free of zero divisors. If A and L satisfy the (pDN) condition, then any generalized local state $s : \mathrm{Rad}(A)^* \longrightarrow L$ can be extended to an order-preserving type I state.*

Proof This follows by applying Theorem 8.10 and Corollary 8.1. $\square$

Corollary 8.5 *Let A be a perfect pseudo-MTL algebra relatively free of zero divisors. If A has the Glivenko property and L satisfies the (pDN) condition, then any generalized local state $s : \mathrm{Rad}(A)^* \longrightarrow L$ can be extended to an order-preserving type I state.*

Proof This follows from Theorems 8.10 and 8.8. $\square$

Example 8.10 If s is the generalized local state from Example 8.9(1), then the function $m : A \longrightarrow L$ defined by $m(x) := 1$ for all $x \in \mathrm{Rad}(A)$ and $m(x) := 0$ for all $x \in \mathrm{Rad}(A)^*$ is an extension of s.

8.5 Extension of Generalized States

In this section we discuss possible generalizations of the Horn-Tarski extension theorem ([166]) for the case of generalized states. Such extension theorems will be useful when proving certain completeness theorems of Gaifman type ([123]) in the context of the theory of probabilistic models for fuzzy logics.

We recall that an FL_w-algebra is complete if it is complete as a lattice, that is, all its subsets have both a supremum (join) and an infimum (meet).

Definition 8.7 Let A and L be FL_w-algebras, L complete. Let B be an FL_w-subalgebra of A and $s : B \longrightarrow L$ be a generalized Riečan state (or type I state, or type II state). Consider the functions $s_* : A \to L$ and $s^* : A \longrightarrow L$ defined by:

$$s_*(a) := \sup\{s(x) \mid x \in B, x \leq a\} \quad \text{for any } a \in A;$$

$$s^*(a) := \inf\{s(x) \mid x \in B, a \leq x\} \quad \text{for any } a \in A.$$

s_* is called the *interior state* and s^* the *exterior state* of s.

Lemma 8.5 *For any* $a \in A$, $s_*(a) \leq s^*(a)$.

Definition 8.8 The pair (A, L) satisfies the *extension property* of states (EP) if for any FL_w-subalgebra B of A and for any generalized Riečan state $s : B \to L$ there exists a generalized Riečan state $s' : A \longrightarrow L$ with $s'|_B = s$.

Definition 8.9 The pair (A, L) satisfies the *Horn-Tarski property* (HT) if for any FL_w-subalgebra B of A, for any generalized Riečan state $s : B \longrightarrow L$, and for any $a \in A, r \in [0, 1]$, the following are equivalent:

(i) s can be extended to a generalized Riečan state $s' : A \to L$ such that $s'(a) = r$;
(ii) $s_*(a) \leq r \leq s^*(a)$.

According to Lemma 8.5, (HT) implies (EP).

Remark 8.9 According to the Horn-Tarski theorem ([166]), any pair $(A, [0, 1]_Ł)$ where A is a Boolean algebra and $[0, 1]_Ł$ is the standard MV-algebra, satisfies property (HT).

Remark 8.10 According to Kroupa's theorem ([202]), any pair $(A, [0, 1]_Ł)$, where A is an MV-algebra, satisfies property (EP).

Theorem 8.11 *Let A and L be good FL_w-algebras with L complete satisfying the (pDN) condition. If $(\text{Reg}(A), L)$ has property (HT) (or (EP)), then (A, L) has property (HT) (or (EP), respectively).*

Proof Assume that $(\text{Reg}(A), L)$ has property (HT). Let B be an FL_w-subalgebra of A, $s : B \to L$ be a generalized Riečan state, $a \in A$ and $r \in [0, 1]$. We will show the equivalence of conditions (i), (ii) in Definition 8.9.

(i) $\Rightarrow$ (ii) Assume that there exists a generalized Riečan state $s' : A \to L$ which extends s and $s'(a) = r$. For any $x, y \in A$ with $x \le a \le y$, we have $s'(x) \le s'(a) \le s'(y)$, so $s(x) \le r \le s(y)$. It follows that $s_*(a) \le r \le s^*(a)$.

(ii) $\Rightarrow$ (i) Assume $s_*(a) \le r \le s^*(a)$. Consider $\mathrm{Reg}(B)$ and $\mathrm{Reg}(A)$ and define $m = s|_{\mathrm{Reg}(B)}$. $\mathrm{Reg}(B)$ is an FL_w-subalgebra of $\mathrm{Reg}(A)$ and m is a generalized Riečan state. Consider the interior state $m_* : \mathrm{Reg}(A) \to L$ and the exterior state $m^* : \mathrm{Reg}(A) \to L$ associated with m. For any $x \in \mathrm{Reg}(A)$:

$$m_*(x) = \sup\{m(u) \mid u \in \mathrm{Reg}(B), u \le x\} \le s_*(x);$$
$$m^*(x) = \inf\{m(v) \mid v \in \mathrm{Reg}(B), x \le v\} \ge s^*(x).$$

Consequently, for any $x \in \mathrm{Reg}(A)$ we have $m_*(x) \le s_*(x) \le r \le s^*(x) \le m^*(x)$. Consider $a^{-\sim} \in \mathrm{Reg}(A)$. As $(\mathrm{Reg}(A), L)$ has property (HT), there exists a generalized Riečan state $m' : \mathrm{Reg}(A) \to L$ such that $m'|_{\mathrm{Reg}(B)} = m$ and $m'(a^{-\sim}) = r$. Consider the function $s' : A \longrightarrow L$ defined by $s'(x) = m'(x^{-\sim})$ for any $x \in A$. According to Proposition 8.17, s' is a generalized Riečan state on A. For any $x \in A$, according to Proposition 8.16 we have: $s'(x) = m'(x^{-\sim}) = m(x^{-\sim}) = s(x^{-\sim}) = s(x)$. Also, $s'(a) = m'(a^{-\sim}) = r$.

The other assertion can be proved similarly. $\square$

Corollary 8.6 *Let A be a Heyting algebra. Then $(A, [0, 1]_{\mathrm{Ł}})$ has property* (HT).

Proof $\mathrm{Reg}(A)$ is a Boolean algebra (according to Glivenko's Theorem) and $(\mathrm{Reg}(A), [0, 1]_{\mathrm{Ł}})$ has property (HT) (according to Remark 8.9). $\square$

Corollary 8.7 *Let A be a BL-algebra. Then $(A, [0, 1]_{\mathrm{Ł}})$ satisfies* (EP).

Proof $\mathrm{Reg}(A)$ is an MV-algebra and $(\mathrm{Reg}(A), [0, 1]_{\mathrm{Ł}})$ satisfies (EP) (according to Remark 8.10). $\square$

8.6 Logical Aspects of Generalized States

The non-commutative propositional logics *psMTL* and *psMTL^r* were studied by Hájek in [160] and [161]. The language of both logical systems is based on the binary connectors $\wedge, \vee, \&, \to, \rightsquigarrow$ and the truth constant $\perp$. We will consider the axiomatization of psMTL and psMTL^r from [160, 161] and [162]. In [197] the standard completeness theorem of psMTL^r is proved. The algebraic structures of psMTL are the psMTL-algebras, and those of psMTL^r are the representable psMTL-algebras, characterized by Kühr's axioms [205]. The predicate calculi psMTL$\forall$ and psMTL$^r\forall$, associated with the propositional systems psMTL and psMTL^r, were introduced in [162] (see also [90]). The standard completeness theorem of psMTL$^r\forall$ was proved in [90].

Let C be a schematic extension of psMTL^r (in the sense of [118]). The predicate calculus $C_\forall$ associated with C preserves the axioms of C and, moreover, has the following axioms for the quantifiers:

(Ax.1) $\forall x \varphi(x) \to \varphi(t)$ where the term t can be substituted for x in $\varphi(x)$;
(Ax.2) $\varphi(t) \to \exists x \varphi(x)$ where the term t can be substituted for x in $\varphi(x)$;
(Ax.3) $\forall x (\varphi \to \psi) \to (\varphi \to \forall x \psi)$ (x is not free in φ);
(Ax.4) $\forall x (\varphi \rightsquigarrow \psi) \rightsquigarrow (\varphi \rightsquigarrow \forall x \psi)$ (x is not free in ψ);
(Ax.5) $\forall x (\varphi \to \psi) \to (\exists x \varphi \to \psi)$ (x is not free in ψ);
(Ax.6) $\forall x (\varphi \rightsquigarrow \psi) \rightsquigarrow (\exists x \varphi \rightsquigarrow \psi)$ (x is not free in ψ);
(Ax.7) $\forall x (\varphi \lor v) \to (\forall x \varphi \lor v)$ (x is not free in v).

The deduction rules of $C_\forall$ are those of C plus the generalization:

$$\frac{\varphi}{\forall x \varphi}(G).$$

We fix a schematic extension C of psMTL^r. We denote by E the set of the sentences of $C_\forall$ and by $E/\sim = \{\hat{\varphi} | \varphi \in E\}$ the Lindenbaum-Tarski algebra of $C_\forall$. $E/\sim$ is a psMTL^r-algebra which satisfies the algebraic form of $C_\forall$.

In the following we present a starting point for the development of a theory of probabilistic models for the non-commutative fuzzy logics described above.

Let $D \subseteq E$ such that:

• D contains all the formal theorems of $C_\forall$;
• D is closed under the connectors $\lor, \land, \&, \to, \rightsquigarrow$;
• $\bot \in D$.

Then $D/\sim = \{\hat{\varphi} | \varphi \in D\}$ is a subalgebra of $E/\sim$.
We fix an FL$_w$-algebra L.

Definition 8.10 A function $\mu : D \to L$ is called a *logical probability of type I* on D if for any $\varphi, \psi \in D$:

(P_1) if $\vdash \varphi$ then $\mu(\varphi) = 1$;
(P_2) $\mu(\varphi \to \psi) \to (\mu(\varphi) \to \mu(\psi)) = 1$;
(P_3) $\mu(\varphi \rightsquigarrow \psi) \rightsquigarrow (\mu(\varphi) \rightsquigarrow \mu(\psi)) = 1$;
(P_4) $\mu(\varphi \to \psi) = \mu(\varphi) \to \mu(\varphi \land \psi)$;
(P_5) $\mu(\varphi \rightsquigarrow \psi) = \mu(\varphi) \rightsquigarrow \mu(\varphi \land \psi)$.

Similarly, we define the logical probabilities of type II and of Riečan type.
In the sequel we will only use logical probabilities of type I which we will simply call *logical probabilities*.

Lemma 8.6 *Let* $\mu : D \to L$ *be a logical probability and* $\varphi, \psi \in D$. *Then:*

(i)   *if* $\vdash \varphi \to \psi$ *then* $\mu(\varphi) \le \mu(\psi)$;
(ii) *if* $\vdash \varphi \rightsquigarrow \psi$ *then* $\mu(\varphi) \le \mu(\psi)$;
(iii) *if* $\vdash \varphi \leftrightarrow \psi$ *then* $\mu(\varphi) = \mu(\psi)$;
(iv) *if* $\vdash \varphi \leftrightsquigarrow \psi$ *then* $\mu(\varphi) = \mu(\psi)$.

According to Lemma 8.6, we can define a function $\tilde{\mu} : D/\sim \longrightarrow L$ by $\tilde{\mu}(\hat{\varphi}) := \mu(\varphi)$ for any $\varphi \in D$. Then $\tilde{\mu}$ is a monotone type I state. Based on this fact, the properties of type I states (for example those described in Proposition 8.4) can be transferred to logical probabilities.

Let U be a set of new constants and $\mathcal{C}_\forall(U)$ the language obtained from $\mathcal{C}_\forall$ by adding the constants from U too. Denote by $E(U)$ the set of sentences from $\mathcal{C}_\forall(U)$ and by $E(U)/\sim$ the Lindenbaum-Tarski algebra associated to $\mathcal{C}_\forall(U)$. It is obvious that $E/\sim$ is an FL_w-subalgebra of $E(U)/\sim$.

In the sequel, we assume that the FL_w-algebra L is complete.

We fix a logical probability $\mu : D \to L$ and U a set of new constants. We define the functions $\mu_* : E(U) \to L$ and $\mu^* : E(U) \to L$ as follows, for any $\varphi \in E(U)$:

$$\mu_*(\varphi) := \sup\{\mu(\psi) \mid \psi \in D, \vdash \psi \to \varphi\};$$
$$\mu^*(\varphi) := \inf\{\mu(\psi) \mid \psi \in D, \vdash \varphi \to \psi\}.$$

We consider the monotone type I state $\tilde{\mu} : D/\sim \longrightarrow L$ associated to μ, the interior state $(\tilde{\mu})_* : E(U)/\sim \longrightarrow L$ and the exterior state $(\tilde{\mu})^* : E(U)/\sim \longrightarrow L$ (defined in the previous section).

Lemma 8.7 *For any sentence* $\varphi \in E(U)$, $\mu_*(\varphi) = (\tilde{\mu})_*(\varphi/\sim)$ *and* $\mu^*(\varphi) = (\tilde{\mu})^*(\varphi/\sim)$.

Definition 8.11 We say that the language $\mathcal{C}_\forall$ has the property (HT) if for any set U of new constants, the Lindenbaum-Tarski algebra $E(U)/\sim$ has the property (HT).

Lemma 8.8 *Assume that* $\mathcal{C}_\forall$ *has the property* (HT). *Let* $\mu : D \longrightarrow L$ *be a logical probability,* U *a set of new constants,* $\theta \in E(U)$ *and* $r \in [0, 1]$. *The following are equivalent:*

(a) *There exists a logical probability* $v : E(U) \longrightarrow L$ *such that* $v|_D = \mu$ *and* $v(D) = r$;
(b) $\mu_*(\theta) \leq r \leq \mu^*(\theta)$.

Proof We apply Lemma 8.7.   □

Let U be a set of new constants and $m : E(U) \longrightarrow L$ a logical probability. We introduce the following conditions for the pair (U, m):

(G∃) For any new formula $\phi(x)$ of $\mathcal{C}_\forall(U)$,

$$m\big(\exists x \phi(x)\big) = \sup\left\{ m\left(\bigvee_{i=1}^n \phi(a_i)\right) \,\middle|\, n \in \omega, a_1, \ldots, a_n \in U \right\}.$$

(G∀) For any new formula $\phi(x)$ of $\mathcal{C}_\forall(U)$,

$$m\big(\forall x \phi(x)\big) = \inf\left\{ m\left(\bigwedge_{i=1}^n \phi(a_i)\right) \,\middle|\, n \in \omega, a_1, \ldots, a_n \in U \right\}.$$

Properties $(G\exists)$ and $(G\forall)$ were inspired by Gaifman's conditions on the probability models of the first-order logics [123]. The pair (U, m) is called a *probabilistic structure* on $C_\forall$ if it satisfies the conditions $(G\exists)$ and $(G\forall)$. The probabilistic structure (U, m) is a *model* of the logical probability $\mu : D \longrightarrow L$ if $m|_D = \mu$.

Definition 8.12 We say that the system of the logic $C_\forall$ satisfies *Gaifman's completeness theorem* (GC Th) if any logical probability $\mu : D \longrightarrow L$ admits a model.

Remark 8.11 According to [123], the classical predicate calculus satisfies (GC Th). In [133], it was proved that the infinite-valued Łukasiewicz predicate logic Ł$\forall$ satisfies (GC Th).

Conditions (HT) and (GC Th) are important in the development of a theory of probabilistic models for the logical system $C_\forall$. In the sequel, we will try to present a few ideas in this direction.

Let (U, u) and (V, v) be two probabilistic structures for $C_\forall$. We say that (V, v) is an *elementary extension* of (U, u) and we write $(U, u) \preceq (V, v)$ if $U \subseteq V$ and for any $\varphi \in E(U)$, $u(\varphi) = v(\varphi)$. The probabilistic structures (U, u) and (V, v) are *elementarily equivalent* if for any $\varphi \in E$, $u(\varphi) = v(\varphi)$. In this case we write $(U, u) \equiv (V, v)$.

Remark 8.12 Suppose $C_\forall$ is the classical first order logic. Let U be a set of new constants and $E_0(U)$ the set of quantifier-free sentences of $C_\forall$. In [123] the following result has been proved: any logical probability $\mu : E_0(U) \longrightarrow [0, 1]$ extends in a unique way to a logical probability $\hat{\mu} : E(U) \longrightarrow [0, 1]$ which satisfies Gaifman's conditions $(G\forall)$ and $(G\exists)$. An important open problem is to prove some similar results for various systems $C_\forall$. This would allow us to define a notion of probabilistic substructure and to develop a good part of the probabilistic theory models (see [130] for probabilistic models of classical finite order logic).

Consider a family $(U_\alpha, u_\alpha)_{\alpha < \lambda}$ indexed by the ordinals $\alpha < \lambda$ such that:

- for any $\alpha < \lambda$, U_α is a nonempty set;
- $U_\alpha \subseteq U_\beta$ for any $\alpha \leq \beta < \lambda$;
- for any $\alpha < \lambda$, $u_\alpha : E(U_\alpha) \to L$ is a logical probability;
- $u_\beta|_{E(U_\alpha)} = u_\alpha$, for any $\alpha \leq \beta < \lambda$.

Let $U = \bigcup_{\alpha < \lambda} U_\alpha$ and $u : E(U) \longrightarrow L$ be the function defined by $u(\phi) = u_\alpha(\phi)$, if $\phi \in E(U_\alpha)$. It is obvious that u is well defined. Furthermore, $u|_{E(U_\alpha)} = u_\alpha$, for any $\alpha < \lambda$. We will denote the pair (U, u) by $\bigcup_{\alpha < \lambda}(U_\alpha, u_\alpha)$.

Proposition 8.19

(i) *If (U_α, u_α) satisfies $(G\forall)$ for any $\alpha < \lambda$, then (U, u) satisfies $(G\forall)$.*

(ii) *If (U_α, u_α) satisfies $(G\exists)$ for any $\alpha < \lambda$, then (U, u) satisfies $(G\exists)$.*

Proof

(i) Consider the sentence $\forall x \phi(x) \in E(U)$, so there exists an ordinal $\alpha < \lambda$ with $\forall x \phi(x) \in E(U_\alpha)$. Let $a_1, \ldots, a_n \in U$, so there exists a $\beta < \lambda$ such that $\alpha \leq \beta$ and $a_1, \ldots, a_n \in U_\beta$. Then $\vdash \forall x \phi(x) \to \bigwedge_{i=1}^n \varphi(a_i)$ in the language $C_\forall(U_\beta)$ so, according to Lemma 8.6(i), we have $u(\forall x \varphi(x)) \leq u(\bigwedge_{i=1}^n \varphi(a_i))$. In this way we obtain the inequality $u(\forall x \varphi(x)) \leq \inf\{u(\bigwedge_{i=1}^n \varphi(a_i)) \mid n \in \omega, a_1, \ldots, a_n \in U\}$.

The converse inequality follows:

$$u\big(\forall x \varphi(x)\big) = u_\alpha\big(\forall x \varphi(x)\big) = \inf\left\{ u_\beta\left(\bigwedge_{i=1}^n \varphi(a_i) \right) \,\middle|\, n \in \omega, a_1, \ldots, a_n \in U_\beta \right\}$$

$$= \inf\left\{ u\left(\bigwedge_{i=1}^n \varphi(a_i) \right) \,\middle|\, n \in \omega, a_1, \ldots, a_n \in U_\beta \right\}$$

$$\geq \inf\left\{ u\left(\bigwedge_{i=1}^n \varphi(a_i) \right) \,\middle|\, n \in \omega, a_1, \ldots, a_n \in U \right\}.$$

(ii) Similar to (i).   □

A sequence of probabilistic structures $(U_\alpha, u_\alpha)_{\alpha < \lambda}$ is called *elementary* if $(U_\alpha, u_\alpha) \preceq (U_\beta, u_\beta)$ for any $\alpha \leq \beta < \lambda$. If $(U_\alpha, u_\alpha)_{\alpha < \lambda}$ is an elementary sequence of probabilistic structures then, according to Proposition 8.19, $(U, u) = \bigcup_{\alpha < \lambda}(U_\alpha, u_\alpha)$ is another probabilistic structure and $(U_\alpha, u_\alpha) \preceq (U, u)$ for any $\alpha < \lambda$.

Theorem 8.12 *Assume that $C_\forall$ satisfies* (HT) *and* (GC Th).

Let (U, u) and (V, v) be two elementarily equivalent probabilistic structures. Then there exists a probabilistic structure (W, w) such that $(U, u) \preceq (W, w)$ and $(V, v) \preceq (W, w)$.

Proof Consider the sentences $\varphi(\vec{a}) \in E(U)$, $\psi(\vec{b}) \in E(V)$ with $\vec{a} = (a_1, \ldots, a_n)$ in U^n and $\vec{b} = (b_1, \ldots, b_m)$ in V^m. Assume that $\vdash \varphi(\vec{a}) \leftrightarrow \psi(\vec{b})$ in $C_\forall(U \cup V)$. Let $\vec{x} = (x_1, \ldots, x_n)$ and $\vec{y} = (y_1, \ldots, y_m)$ with $\{x_1, \ldots, x_n\} \cap \{y_1, \ldots, y_m\} = \emptyset$. We notice that $\vdash \varphi(\vec{a}) \leftrightarrow \forall \vec{x}\, \varphi(\vec{x})$ in $C_\forall(U)$, $\vdash \psi(\vec{b}) \leftrightarrow \forall \vec{y}\, \psi(\vec{y})$ in $C_\forall(V)$, so $\vdash \forall \vec{x}\, \varphi(\vec{x}) \leftrightarrow \vdash \forall \vec{y}\, \psi(\vec{y})$ in $C_\forall(U \cup V)$. But $\forall \vec{x}\, (\varphi\, \vec{x})$ and $\forall \vec{y}\, \psi(\vec{y})$ belong to E, so $\vdash \forall \vec{x}\, \varphi(\vec{x}) \leftrightarrow \forall \vec{y}\, \psi(\vec{y})$ in $C_\forall$. Because $(U, u) \equiv (V, v)$, it follows that $u(\varphi(\vec{a})) = u(\forall \vec{x}\, \varphi(\vec{x})) = v(\forall \vec{y}\, \psi(\vec{y})) = v(\psi(\vec{b}))$.

Let $\varphi = \varphi(\vec{a}) \in E(V) \setminus E(U)$, where $\vec{a} = (a_1, \ldots, a_n)$ has components in V. We will show that there exists a logical probability $\mu : E(U \cup V) \to L$ such that $\mu|_{E(U)} = u$ and $\mu(\varphi) = v(\varphi)$. Since $C_\forall$ satisfies (HT) and using Lemma 8.8, it is enough to prove the inequality $u_*(\varphi) \leq v(\varphi) \leq u^*(\varphi)$, where $u_* : E(U \cup V) \longrightarrow L$ and $u^* : E(U \cup V) \longrightarrow L$ are the functions defined before Lemma 8.7.

Let $\vdash \psi \rightarrow \varphi(\overrightarrow{a})$ in $C_\forall(U \cup V)$ with $\psi \in E(U)$. Because $\varphi(\overrightarrow{a}) \leftrightarrow \forall \overrightarrow{x} \; \varphi(\overrightarrow{x})$ and $\forall \overrightarrow{x} \; \varphi(\overrightarrow{x}) \in E$ we have $u(\psi) \leq u(\forall \overrightarrow{x} \; \varphi(\overrightarrow{x})) = v(\varphi(\overrightarrow{x})) = v(\varphi)$, by applying Lemma 8.6 and the fact that $(U, u) \equiv (V, v)$. It follows that $u_*(\varphi) \leq v(\varphi)$. The inequality $v(\varphi) \leq u^*(\varphi)$ can be proved similarly.

Using the above, by transfinite induction, we construct a logical probability $v : E(U \cup V) \longrightarrow L$ such that $v|_{E(U)} = u$ and $v|_{E(V)} = v$. Applying property (GC Th) to v, we get that there exists a probability structure (W, w) such that $U \cup V \subseteq W$ and $w|_{E(U)} = v|_{E(U)} = u$ and $w|_{E(V)} = v|_{E(V)} = v$. So $(U, u) \preceq (W, w)$ and $(V, v) \preceq (W, w)$. $\qquad\square$

The notions and results from this section suggest the possibility of developing a probabilistic theory of models for $C_\forall$ logics. Such investigations should start from the literature of the probabilistic theory of models for the classical predicate calculus (in particular the works [123, 253]).

The following open problems could be investigated:

1. Robinson's consistency theorem (p. 114 in [251]) is a result of the classical theory of models that can be expressed in terms of pure semantics. The goal is to be able to easily give a similar statement for the case of $C_\forall$ logics. For example, consistency theories from Robinson's theorem will be replaced with logical probabilities. The proof of this assertion will require the existence of a Gaifman type completeness theorem ([123]) (which holds, for example, in the case of the probabilistic models of Łukasiewicz logics [133] and it can use the basic probabilistic structures). A consistency theorem of such type has not been proved even for the case of classical probabilistic models.

2. The omitting types theorem is another important result from the classical theory of models which could be expressed in terms of logical probabilities and of the above mentioned probability structures. We mention that a version of the omitting types theorem was proved in the case of some multiple-valued logics in [231]. However, a probabilistic version of the omitting types theorem has not been proved even in the case of classical predicate logics.

3. The Lowenheim-Skolem type theorems (in the case of probabilistic models for classical logic, this topic has been investigated in [123, 253]).

4. Ultraproducts of probabilistic models of $C_\forall$ logics (the case of ultraproducts of classical probabilistic models has been studied in [253]).

5. Model-completion and model-companion of a logical probability and existential complete probabilistic structures (in the case of classical probabilistic models, this subject has been studied in [133]).

Chapter 9
Pseudo-hoops with Internal States

Flaminio and Montagna ([119]) endowed the MV-algebras with a unary operation called an internal state, or a state operator, satisfying some basic properties of states and the new structures are called state MV-algebras. Di Nola and Dvurečenskij presented the notion of a state-morphism MV-algebra which is a stronger variation of a state MV-algebra ([78]). Subdirectly irreducible state-morphism MV-algebras have been characterized in [80], and some classes of state-morphism MV-algebras were presented in [79]. The notion of a state operator was extended by Rachůnek and Šalounová in [243] to the case of GMV-algebras (pseudo-MV algebras). State operators and state-morphism operators on BL-algebras were introduced and investigated in [72] and subdirectly irreducible state-morphism BL-algebras were studied in [101]. Recently, Dvurečenskij, Rachůnek and Šalounová presented state Rℓ-monoids and state-morphism Rℓ-monoids ([114, 115]). A general approach of state-morphism algebras was presented in [18].

In this chapter we study these concepts for the more general fuzzy structures, namely the pseudo-hoops, and we present state pseudo-hoops and state-morphism pseudo-hoops. We define the notions of a state operator, strong state operator, state-morphism operator and weak state-morphism operator and we study their properties. We prove that every strong state pseudo-hoop is a state pseudo-hoop and any state operator on an idempotent pseudo-hoop is a weak state-morphism operator. It is proved that for an idempotent pseudo-hoop A, a state operator on $\text{Reg}(A)$ can be extended to a state operator on A. One of the main results of the chapter consists of proving that every perfect pseudo-hoop admits a state operator. Other results compare the state operators with states and generalized states on a pseudo-hoop. Some conditions are given for a state operator to be a generalized state and for a generalized state to be a state operator.

L.C. Ciungu, *Non-commutative Multiple-Valued Logic Algebras*,
Springer Monographs in Mathematics, DOI 10.1007/978-3-319-01589-7_9,
© Springer International Publishing Switzerland 2014

9.1 State Pseudo-hoops

In what follows $(A, \odot, \rightarrow, \rightsquigarrow, 0, 1)$ will be a bounded pseudo-hoop.

Definition 9.1 A *state pseudo-hoop* is a pair (A, σ) where A is a bounded pseudo-hoop and $\sigma : A \longrightarrow A$ is a mapping, called a *state operator*, such that for any $x, y \in A$ the following conditions are satisfied:

(S_1) $\sigma(0) = 0$;
(S_2) $\sigma(x \rightarrow y) = \sigma(x) \rightarrow \sigma(x \wedge y)$ and $\sigma(x \rightsquigarrow y) = \sigma(x) \rightsquigarrow \sigma(x \wedge y)$;
(S_3) $\sigma(x \odot y) = \sigma(x) \odot \sigma(x \rightsquigarrow x \odot y) = \sigma(y \rightarrow x \odot y) \odot \sigma(y)$;
(S_4) $\sigma(\sigma(x) \odot \sigma(y)) = \sigma(x) \odot \sigma(y)$;
(S_5) $\sigma(\sigma(x) \rightarrow \sigma(y)) = \sigma(x) \rightarrow \sigma(y)$ and $\sigma(\sigma(x) \rightsquigarrow \sigma(y)) = \sigma(x) \rightsquigarrow \sigma(y)$.

$\mathrm{Ker}(\sigma) := \{x \in A \mid \sigma(x) = 1\}$ is called the *kernel* of σ.
A state operator is called *faithful* if $\mathrm{Ker}(\sigma) = 1$.

Proposition 9.1 *If (A, σ) is a state pseudo-hoop, then for all $x, y \in A$ the following hold:*

(1) $\sigma(1) = 1$;
(2) $\sigma(x^-) = \sigma(x)^-$ *and* $\sigma(x^\sim) = \sigma(x)^\sim$;
(3) $x \leq y$ *implies* $\sigma(x) \leq \sigma(y)$;
(4) $\sigma(x \odot y) \geq \sigma(x) \odot \sigma(y)$; *if* $x \odot y = 0$, *then* $\sigma(x \odot y) = \sigma(x) \odot \sigma(y)$; *if A is good and $y \perp x$, then* $\sigma(x \odot y) = \sigma(x) \odot \sigma(y)$;
(5) $\sigma(x \odot y^\sim) \geq \sigma(x) \odot \sigma(y)^\sim$ *and* $\sigma(y^- \odot x) \geq \sigma(y)^- \odot \sigma(x)$; *if $x \leq y$, then* $\sigma(x \odot y^\sim) = \sigma(x) \odot \sigma(y)^\sim$ *and* $\sigma(y^- \odot x) = \sigma(y)^- \odot \sigma(x)$;
(6) $\sigma(x \wedge y) = \sigma(x) \odot \sigma(x \rightsquigarrow y) = \sigma(y \rightarrow x) \odot \sigma(y)$;
(7) $\sigma(x \rightarrow y) \leq \sigma(x) \rightarrow \sigma(y)$ *and* $\sigma(x \rightsquigarrow y) \leq \sigma(x) \rightsquigarrow \sigma(y)$; *if x and y are comparable, then* $\sigma(x \rightarrow y) = \sigma(x) \rightarrow \sigma(y)$ *and* $\sigma(x \rightsquigarrow y) = \sigma(x) \rightsquigarrow \sigma(y)$;
(8) $\sigma(x \rightarrow y) \odot \sigma(y \rightarrow x) \leq d_1(\sigma(x), \sigma(y))$ *and* $\sigma(x \rightsquigarrow y) \odot \sigma(y \rightsquigarrow x) \leq d_2(\sigma(x), \sigma(y))$;
(9) $\sigma^2(x) = \sigma(x)$;
(10) *if A is good, then:* $\sigma(x \oplus y) \leq \sigma(x) \oplus \sigma(y)$; $\sigma(\sigma(x) \oplus \sigma(y)) = \sigma(x) \oplus \sigma(y)$; *if $x \perp_{no} y$, then* $\sigma(x \oplus y) = \sigma(x) \oplus \sigma(y)$; $\sigma(x \oplus x^-) = \sigma(x^\sim \oplus x) = 1$;
(11) $\sigma(A) = \{x \in A \mid \sigma(x) = x\}$;
(12) $\sigma(x \rightarrow y) = \sigma(x) \rightarrow \sigma(y)$ *iff* $\sigma(y \rightarrow x) = \sigma(y) \rightarrow \sigma(x)$ *iff* $\sigma(x \rightsquigarrow y) = \sigma(x) \rightsquigarrow \sigma(y)$ *iff* $\sigma(y \rightsquigarrow x) = \sigma(y) \rightsquigarrow \sigma(x)$;
(13) $\sigma(\sigma(x) \wedge \sigma(y)) = \sigma(x) \wedge \sigma(y)$;
(14) $\sigma(\sigma(x) \vee_1 \sigma(y)) = \sigma(x) \vee_1 \sigma(y)$ *and* $\sigma(\sigma(x) \vee_2 \sigma(y)) = \sigma(x) \vee_2 \sigma(y)$; $\sigma(x \vee_1 y) \leq \sigma(x) \vee_1 \sigma(y)$ *and* $\sigma(x \vee_2 y) \leq \sigma(x) \vee_2 \sigma(y)$; *if x and y are comparable, then* $\sigma(x \vee_1 y) = \sigma(x) \vee_1 \sigma(y)$ *and* $\sigma(x \vee_2 y) = \sigma(x) \vee_2 \sigma(y)$;
(15) *if σ is faithful, then $x < y$ implies $\sigma(x) < \sigma(y)$*;
(16) *if σ is faithful, then either $\sigma(x) = x$ or $\sigma(x)$ and x are not comparable*;
(17) *if A is linearly ordered and σ is faithful, then $\sigma(x) = x$ for all $x \in A$.*

Proof

(1) $\sigma(1) = \sigma(0 \to 0) = \sigma(0) \to \sigma(0 \wedge 0) = 1$.
(2) Applying (S_2) we get:

$$\sigma\left(x^-\right) = \sigma(x \to 0) = \sigma(x) \to \sigma(x \wedge 0) = \sigma(x) \to \sigma(0)$$

$$= \sigma(x) \to 0 = \sigma(x)^-.$$

Similarly for $\sigma(x^\sim) = \sigma(x)^\sim$.
(3) By *(pshoop-c₃)* we get $x = y \odot (y \rightsquigarrow x)$, so $\sigma(x) = \sigma(y \odot (y \rightsquigarrow x)) = \sigma(y) \odot \sigma(y \rightsquigarrow y \odot (y \rightsquigarrow x)) \leq \sigma(y)$.
(4) From $x \odot y \leq x \odot y$ we get $y \leq x \rightsquigarrow x \odot y$, so by (3) we have $\sigma(y) \leq \sigma(x \rightsquigarrow x \odot y)$.

Applying (S_3) we get: $\sigma(x \odot y) = \sigma(x) \odot \sigma(x \rightsquigarrow x \odot y) \geq \sigma(x) \odot \sigma(y)$.
If $x \odot y = 0$, then $\sigma(x \odot y) = 0$, so $\sigma(x \odot y) = \sigma(x) \odot \sigma(y) = 0$.
If A is good and $y \perp x$, by Proposition 1.28 we have $x \odot y = 0$, hence $\sigma(x \odot y) = \sigma(x) \odot \sigma(y) = 0$.
(5) $\sigma(x \odot y^\sim) \geq \sigma(x) \odot \sigma(y)^\sim$ and $\sigma(y^- \odot x) \geq \sigma(y)^- \odot \sigma(x)$ follow from (4) and (2).

If $x \leq y$ we have $y^\sim \leq x^\sim$, $y^- \leq x^-$, so $x \odot y^\sim \leq x \odot x^\sim = 0$ and $y^- \odot x \leq x^- \odot x = 0$. It follows that $\sigma(x \odot y^\sim) = \sigma(y^- \odot x) = 0$, hence $\sigma(x \odot y^\sim) = \sigma(x) \odot \sigma(y)^\sim = 0$ and $\sigma(y^- \odot x) = \sigma(y)^- \odot \sigma(x) = 0$.
(6)

$$\sigma(x \wedge y) = \sigma\left(x \odot (x \rightsquigarrow y)\right) = \sigma(x) \odot \sigma\left(x \rightsquigarrow \left(x \odot (x \rightsquigarrow y)\right)\right)$$

$$= \sigma(x) \odot \sigma(x \rightsquigarrow x \wedge y) = \sigma(x) \odot \sigma(x \rightsquigarrow y) \quad \text{and}$$

$$\sigma(x \wedge y) = \sigma\left((y \to x) \odot y\right) = \sigma\left(y \to \left((y \to x) \odot y\right)\right) \odot \sigma(y)$$

$$= \sigma(y \to x \wedge y) \odot \sigma(y) = \sigma(y \to x) \odot \sigma(y).$$

(7) By (S_2) and *(psbck-c₁₀)* we have:

$$\sigma(x \to y) = \sigma(x) \to \sigma(x \wedge y) \leq \sigma(x) \to \sigma(y) \quad \text{and}$$

$$\sigma(x \rightsquigarrow y) = \sigma(x) \rightsquigarrow \sigma(x \wedge y) \leq \sigma(x) \rightsquigarrow \sigma(y).$$

If $x \leq y$, then $\sigma(x) \leq \sigma(y)$ and $\sigma(x \to y) = \sigma(x) \to \sigma(x \wedge y) = \sigma(x) \to \sigma(x) = 1$.
We also have $\sigma(x) \to \sigma(y) = 1$, thus $\sigma(x \to y) = \sigma(x) \to \sigma(y)$.
Similarly, $\sigma(x \rightsquigarrow y) = \sigma(x) \rightsquigarrow \sigma(y)$.
If $y \leq x$, then $x \wedge y = y$ and the equalities follow from (S_2).
(8) By (7) we have $\sigma(x \to y) \leq \sigma(x) \to \sigma(y)$ and $\sigma(y \to x) \leq \sigma(y) \to \sigma(x)$, hence $\sigma(x \to y) \odot \sigma(y \to x) \leq d_1(\sigma(x), \sigma(y))$.
Similarly, $\sigma(x \rightsquigarrow y) \odot \sigma(y \rightsquigarrow x) \leq d_2(\sigma(x), \sigma(y))$.
(9) Applying (1) and (S_4) we have:

$$\sigma^2(x) = \sigma\left(\sigma(x)\right) = \sigma\left(\sigma(x) \odot \sigma(1)\right) = \sigma(x) \odot \sigma(1) = \sigma(x).$$

(10) From $\sigma(y^- \odot x^-) \geq \sigma(y^-) \odot \sigma(x^-)$ we get $(\sigma(y^-) \odot \sigma(x^-))^\sim \geq (\sigma(y^- \odot x^-))^\sim$.

Applying (2) it follows that $(\sigma(y)^- \odot \sigma(x)^-)^\sim \geq \sigma((y^- \odot x^-)^\sim)$.
Thus $\sigma(x \oplus y) \leq \sigma(x) \oplus \sigma(y)$.
By (2) and (9) we get:

$$\sigma\big(\sigma(x) \oplus \sigma(y)\big) = \sigma\big((\big(\sigma(y)^- \odot \sigma(x)^-\big)^\sim\big) = \big(\sigma\big(\sigma(y^-) \odot \sigma(x^-)\big)\big)^\sim$$
$$= \big(\sigma(y^-) \odot \sigma(x^-)\big)^\sim = \big(\sigma(y)^- \odot \sigma(x)^-\big)^\sim$$
$$= \sigma(x) \oplus \sigma(y).$$

Obviously, $\sigma(x \oplus 0) = \sigma(x) \oplus \sigma(0)$ and $\sigma(0 \oplus x) = \sigma(0) \oplus \sigma(x)$.
Since $x \perp_{no} y$, we have $x^- \perp y^-$, so by (4) and (2) we have

$$\sigma\big(y^- \odot x^-\big) = \sigma\big(y^-\big) \odot \sigma\big(x^-\big) = \sigma(y)^- \odot \sigma(x)^-.$$

Hence

$$\sigma(x \oplus y) = \sigma\big((y^- \odot x^-)^\sim\big) = \big(\sigma(y^- \odot x^-)\big)^\sim = \big(\sigma(y)^- \odot \sigma(x)^-\big)^\sim$$
$$= \sigma(x) \oplus \sigma(y).$$

For the last assertion we have:

$$\sigma\big(x \oplus x^-\big) = \big(\sigma\big(x^{-\sim} \odot x^\sim\big)^-\big) = \big(\sigma\big(x^{\sim-} \odot x^\sim\big)\big)^- = \big(\sigma(0)\big)^- = 1 \quad \text{and}$$
$$\sigma\big(x^\sim \oplus x\big) = \big(\sigma\big(x^- \odot x^{\sim-}\big)^\sim\big) = \big(\sigma\big(x^- \odot x^{-\sim}\big)\big)^\sim = \big(\sigma(0)\big)^\sim = 1.$$

(11) Let $y \in \sigma(A)$, so there exists an $x \in A$ such that $y = \sigma(x)$.
Hence $\sigma(y) = \sigma^2(x) = \sigma(x) = y$. It follows that $y \in \{x \in A \mid \sigma(x) = x\}$.
Conversely, if $y \in \{x \in A \mid \sigma(x) = x\}$ it follows that $y \in \sigma(A)$.

(12) Suppose $\sigma(x \to y) = \sigma(x) \to \sigma(y)$. Applying (S_2), (6) and $(pshoop\text{-}c_3)$ we get:

$$\sigma(y \to x) = \sigma(y) \to \sigma(y \wedge x) = \sigma(y) \to \sigma(x \to y) \odot \sigma(x)$$
$$= \sigma(y) \to \big(\sigma(x) \to \sigma(y)\big) \odot \sigma(x) = \sigma(y) \to \sigma(x) \wedge \sigma(y)$$
$$= \sigma(y) \to \sigma(x).$$

Similarly, if $\sigma(y \to x) = \sigma(y) \to \sigma(x)$, then $\sigma(x \to y) = \sigma(x) \to \sigma(y)$.
Suppose again that $\sigma(x \to y) = \sigma(x) \to \sigma(y)$, so $\sigma(y \to x) = \sigma(y) \to \sigma(x)$.
Then we have:

$$\sigma(x \rightsquigarrow y) = \sigma(x) \rightsquigarrow \sigma(x \wedge y) = \sigma(x) \rightsquigarrow \big(\sigma(y \to x) \odot \sigma(y)\big)$$
$$= \sigma(x) \rightsquigarrow \big((\sigma(y) \to \sigma(x)) \odot \sigma(y)\big) = \sigma(x) \rightsquigarrow \sigma(x) \wedge \sigma(y)$$
$$= \sigma(x) \rightsquigarrow \sigma(y).$$

Similarly, if $\sigma(x \rightsquigarrow y) = \sigma(x) \rightsquigarrow \sigma(y)$, then $\sigma(x \to y) = \sigma(x) \to \sigma(y)$.

Finally, we can prove in the same manner that $\sigma(x \rightsquigarrow y) = \sigma(x) \rightsquigarrow \sigma(y)$ implies $\sigma(y \rightsquigarrow x) = \sigma(y) \rightsquigarrow \sigma(x)$.

(13) Applying (6), (9), (S_5) and (*pshoop-c3*) we get:

$$\sigma\big(\sigma(x) \wedge \sigma(y)\big) = \sigma^2(x) \odot \sigma\big(\sigma(x) \rightsquigarrow \sigma(y)\big) = \sigma(x) \odot \big(\sigma(x) \rightsquigarrow \sigma(y)\big)$$
$$= \sigma(x) \wedge \sigma(y).$$

(14) Applying (S_5) and (9) we get:

$$\sigma\big(\sigma(x) \vee_1 \sigma(y)\big) = \sigma\big((\sigma(x) \rightarrow \sigma(y)) \rightsquigarrow \sigma(y)\big) = \sigma\big(\sigma(x) \rightarrow \sigma(y)\big) \rightsquigarrow \sigma(y)$$
$$= \big(\sigma(x) \rightarrow \sigma(y)\big) \rightsquigarrow \sigma(y) = \sigma(x) \vee_1 \sigma(y).$$

Similarly, $\sigma(\sigma(x) \vee_2 \sigma(y)) = \sigma(x) \vee_2 \sigma(y)$.
The second part follows applying (7) twice.

(15) By (3) $x < y$ implies $\sigma(x) \le \sigma(y)$. Suppose $\sigma(x) = \sigma(y)$. From (S_2) it follows that $\sigma(y \rightarrow x) = \sigma(y) \rightarrow \sigma(x) = 1$, that is, $y \rightarrow x \in \mathrm{Ker}(\sigma) = \{1\}$. Thus $y \rightarrow x = 1$, hence $y \le x$, a contradiction. It follows that $\sigma(x) < \sigma(y)$.

(16) Consider $x \in A$ such that $\sigma(x) \ne x$ and let x and $\sigma(x)$ be comparable. We have $x < \sigma(x)$ or $\sigma(x) < x$, so $\sigma(x) < \sigma(x)$, a contradiction. It follows that either $\sigma(x) = x$ or $\sigma(x)$ and x are not comparable.

(17) Since A is linearly ordered it follows that x and $\sigma(x)$ are comparable. Hence by (16), $\sigma(x) = x$.   □

Corollary 9.1 *Let* (A, σ) *be a linearly ordered state pseudo-hoop. Then for all* $x, y \in A$ *the following hold*:

(1) $\sigma(x \rightarrow y) = \sigma(x) \rightarrow \sigma(y)$ *and* $\sigma(x \rightsquigarrow y) = \sigma(x) \rightsquigarrow \sigma(y)$;
(2) $\sigma(x \vee_1 y) = \sigma(x) \vee_1 \sigma(y)$ *and* $\sigma(x \vee_2 y) = \sigma(x) \vee_2 \sigma(y)$;
(3) *if* A *has* (SO) *property, then*:

$$\sigma\big(x \oplus y^-\big) = \sigma(x) \oplus \sigma\big(y^-\big) \quad and \quad \sigma\big(y^\sim \oplus x\big) = \sigma\big(y^\sim\big) \oplus \sigma(x) \quad or$$
$$\sigma\big(y \oplus x^-\big) = \sigma(y) \oplus \sigma\big(x^-\big) \quad and \quad \sigma\big(x^\sim \oplus y\big) = \sigma\big(x^\sim\big) \oplus \sigma(y).$$

Proof

(1) This follows from Proposition 9.1(7).
(2) This follows from Proposition 9.1(14).
(3) Consider the following cases:

(a) If $x = 0$, then applying Proposition 1.24(3) and Proposition 9.1(2) we have:

$$\sigma\big(0 \oplus y^-\big) = \sigma\big(y^{-\sim-}\big) = \sigma\big(y^-\big) = \sigma(y)^- \quad \text{and}$$
$$\sigma(0) \oplus \sigma\big(y^-\big) = 0 \oplus \sigma\big(y^-\big) = \sigma\big(y^-\big)^{\sim-} = \sigma(y)^-.$$

Thus $\sigma(x \oplus y^-) = \sigma(x) \oplus \sigma(y^-)$ and similarly $\sigma(y^\sim \oplus x) = \sigma(y^\sim) \oplus \sigma(x)$.

(b) If $y = 0$, then according to Proposition 1.24(4) we have:

$$\sigma\left(x \oplus 0^-\right) = \sigma(x \oplus 1) = \sigma(1) = 1 \quad \text{and} \quad \sigma(1) \oplus \sigma\left(y^-\right) = 1 \oplus \sigma\left(y^-\right) = 1.$$

Thus $\sigma(x \oplus y^-) = \sigma(x) \oplus \sigma(y^-)$ and similarly $\sigma(y^\sim \oplus x) = \sigma(y^\sim) \oplus \sigma(x)$.

(c) Assume $x \neq 0$, $y \neq 0$ and $x \leq y$.

According to Proposition 1.6(6), $x \perp y^-$ and $y^\sim \perp x$.

Applying Proposition 9.1(10) we have

$$\sigma\left(x \oplus y^-\right) = \sigma(x) \oplus \sigma\left(y^-\right) \quad \text{and} \quad \sigma\left(y^\sim \oplus x\right) = \sigma\left(y^\sim\right) \oplus \sigma(x).$$

Similarly, if $x \neq 0$, $y \neq 0$ and $y \leq x$ we get

$$\sigma\left(y \oplus x^-\right) = \sigma(y) \oplus \sigma\left(x^-\right) \quad \text{and} \quad \sigma\left(x^\sim \oplus y\right) = \sigma\left(x^\sim\right) \oplus \sigma(y). \qquad \square$$

Proposition 9.2 *Let (A, σ) be a state pseudo-hoop. Consider the properties*:

(a) $\sigma(x \rightarrow y) = \sigma(x) \rightarrow \sigma(y)$ *or* $\sigma(x \rightsquigarrow y) = \sigma(x) \rightsquigarrow \sigma(y)$ *for all* $x, y \in A$;
(b) $\sigma(x \wedge y) = \sigma(x) \wedge \sigma(y)$ *for all* $x, y \in A$;
(c) $\sigma(x \odot y) = \sigma(x) \odot \sigma(y)$ *for all* $x, y \in A$;
(d) $\sigma(x \vee_1 y) = \sigma(x) \vee_1 \sigma(y)$ *and* $\sigma(x \vee_2 y) = \sigma(x) \vee_2 \sigma(y)$ *for all* $x, y \in A$.

Then:

(a) *is equivalent to* (b);
(a) *implies* (c) *and* (d).

Proof According to Proposition 9.1(12), σ preserves $\rightarrow$ iff it preserves $\rightsquigarrow$.

(a) $\Rightarrow$ (b) By Proposition 9.1(6) and (*pshoop-c3*) we have:

$$\sigma(x \wedge y) = \sigma(x) \odot \sigma(x \rightsquigarrow y) = \sigma(x) \odot \left(\sigma(x) \rightsquigarrow \sigma(y)\right) = \sigma(x) \wedge \sigma(y).$$

(b) $\Rightarrow$ (a) Applying (S_2) we get:

$$\sigma(x \rightarrow y) = \sigma(x) \rightarrow \sigma(x \wedge y) = \sigma(x) \rightarrow \left(\sigma(x) \wedge \sigma(y)\right) = \sigma(x) \rightarrow \sigma(y).$$

Similarly, $\sigma(x \rightsquigarrow y) = \sigma(x) \rightsquigarrow \sigma(y)$.

(a) $\Rightarrow$ (c) By (*psHOOP$_3$*) we have:

$$\sigma(x \odot y) \rightarrow \sigma(z) = \sigma(x \odot y \rightarrow z) = \sigma\left(x \rightarrow (y \rightarrow z)\right)$$

$$= \sigma(x) \rightarrow \left(\sigma(y) \rightarrow \left(\sigma(z)\right)\right) = \left(\sigma(x) \odot \sigma(y)\right) \rightarrow \sigma(z).$$

Taking $z = \sigma(x) \odot \sigma(y)$ we get:

$$\sigma(x \odot y) \rightarrow \sigma\left(\sigma(x) \odot \sigma(y)\right) = \left(\sigma(x) \odot \sigma(y)\right) \rightarrow \sigma\left(\sigma(x) \odot \sigma(y)\right)$$

$$= \left(\sigma(x) \odot \sigma(y)\right) \rightarrow \left(\sigma(x) \odot \sigma(y)\right) = 1.$$

Thus $\sigma(x \odot y) \leq \sigma(\sigma(x) \odot \sigma(y)) = \sigma(x) \odot \sigma(y)$.

Applying Proposition 9.1(4), we get $\sigma(x \odot y) = \sigma(x) \odot \sigma(y)$.

(a) $\Rightarrow$ (d) This follows by the definitions of $\vee_1$ and $\vee_2$, applying (a).   □

Let A be a bounded pseudo-hoop and $\sigma : A \longrightarrow A$ be a mapping such that for all $x, y \in A$:

(S_3') $\sigma(x \odot y) = \sigma(y^- \vee_1 x) \odot \sigma(y) = \sigma(x) \odot \sigma(x^\sim \vee_2 y)$.

Definition 9.2 A mapping $\sigma : A \longrightarrow A$ is called a *strong state operator* on A if σ satisfies conditions $(S_1), (S_2), (S_3'), (S_4), (S_5)$.

A pair (A, σ) such that A is a bounded pseudo-hoop and σ is a strong state operator on A is called a *strong state pseudo-hoop*.

A state operator σ is called a *C-state operator* if it satisfies the following condition:

(C) $\sigma(x \vee_1 y) = \sigma(y \vee_1 x)$ and $\sigma(x \vee_2 y) = \sigma(y \vee_2 x)$.

A pair (A, σ) such that A is a bounded pseudo-hoop and σ is a C-state operator on A is called a *C-state pseudo-hoop*.

If a C-state operator is strong, then we call it a *C-strong state operator*.

Remark 9.1 Every state Wajsberg pseudo-hoop is a C-state Wajsberg pseudo-hoop.

Proposition 9.3 *Let A be a bounded pseudo-hoop. If $\sigma : A \longrightarrow A$ is an order-preserving mapping satisfying condition* (C), *then* $\sigma(x \vee_1 y) = \sigma(x \vee_2 y)$ *for all $x, y \in A$.*

Proof First we prove the equality for $y \le x$.

Applying Proposition 1.5(4) and condition (C) we get:

$$\sigma(x \vee_1 y) = \sigma(y \vee_1 x) = \sigma(x) \quad \text{and} \quad \sigma(x \vee_2 y) = \sigma(x),$$

i.e., $\sigma(x \vee_1 y) = \sigma(x \vee_2 y)$.

Assume now that x and y are arbitrary elements of A. Using again Proposition 1.5(4), condition (C) and the first part of the proof, we get:

$$\sigma(x \vee_1 y) = \sigma\big(x \vee_1 (x \vee_1 y)\big) = \sigma\big((x \vee_1 y) \vee_1 x\big)$$
$$= \sigma\big((x \vee_1 y) \vee_2 x\big) \ge \sigma(y \vee_2 x)$$
$$= \sigma\big(x \vee_2 (y \vee_2 x)\big) \ge \sigma(x \vee_2 y)$$
$$= \sigma\big(y \vee_2 (x \vee_2 y)\big) = \sigma\big((x \vee_2 y) \vee_2 y\big)$$
$$\ge \sigma(x \vee_1 y).$$

Thus $\sigma(x \vee_1 y) = \sigma(x \vee_2 y)$.   □

Corollary 9.2 *If σ is a C-state operator, then $\sigma(x \vee_1 y) = \sigma(x \vee_2 y)$.*

Theorem 9.1 *Every strong state pseudo-hoop is a state pseudo-hoop.*

Proof Consider the strong state pseudo-hoop (A, σ) and $x, y \in A$.

Taking into consideration that $y^- \leq y \rightarrow x$ and $x^\sim \leq x \rightsquigarrow y$ we get:

$$y^- \vee_1 (y \rightarrow x) = y \rightarrow x \quad \text{and} \quad x^\sim \vee_2 (x \rightsquigarrow y) = x \rightsquigarrow y.$$

Then applying (S_3') we have:

$$\sigma(x \wedge y) = \sigma\big(x \odot (x \rightsquigarrow y)\big) = \sigma(x) \odot \sigma\big(x^\sim \vee_2 (x \rightsquigarrow y)\big) = \sigma(x) \odot \sigma(x \rightsquigarrow y).$$

It follows that $\sigma(x \odot y) = \sigma(x \wedge (x \odot y)) = \sigma(x) \odot \sigma(x \rightsquigarrow (x \odot y))$.

Similarly, $\sigma(x \wedge y) = \sigma((y \rightarrow x) \odot y) = \sigma(y^- \vee_1 (y \rightarrow x)) \odot \sigma(y) = \sigma(y \rightarrow x) \odot \sigma(y)$, so $\sigma(x \odot y) = \sigma((x \odot y) \wedge y) = \sigma(y \rightarrow (x \odot y)) \odot \sigma(y)$.

Thus condition (S_3') implies condition (S_3), hence σ is a state operator on A. $\square$

Proposition 9.4 *If σ is a strong state operator on a bounded pseudo-hoop A such that $x^\sim \leq y$ or $y^- \leq x$ for some $x, y \in A$, then $\sigma(x \odot y) = \sigma(x) \odot \sigma(y)$.*

Proof Since σ is a strong state operator, it satisfies the condition

$$\sigma(x \odot y) = \sigma\big(y^- \vee_1 x\big) \odot \sigma(y) = \sigma(x) \odot \sigma\big(x^\sim \vee_2 y\big).$$

According to Proposition 1.5(4), $y^- \leq x$ implies $\sigma(y^- \vee_1 x) = \sigma(x)$ and $x^\sim \leq y$ implies $\sigma(x^\sim \vee_2 y) = \sigma(y)$. Thus $\sigma(x \odot y) = \sigma(x) \odot \sigma(y)$. $\square$

Proposition 9.5 *If σ is a state operator on a linearly ordered bounded pseudo-hoop A, then σ is a pseudo-hoop endomorphism such that $\sigma^2 = \sigma$.*

Proof Since (A, σ) is a linearly ordered state pseudo-hoop, according to Corollary 9.1, σ preserves $\rightarrow$ and $\rightsquigarrow$. Applying Proposition 9.2, it follows that σ preserves $\odot$. Taking into consideration that σ also preserves the constants 0 (by (S_1)) and 1 (by Proposition 9.1(1)), we conclude that σ is an endomorphism. Condition $\sigma^2 = \sigma$ follows from Proposition 9.1(9). $\square$

Proposition 9.6 *If (A, σ) is a state pseudo-hoop, then $\sigma(A)$ is a subalgebra of A.*

Proof By (S_4), (S_5), (S_1) and Proposition 9.1(1), $\sigma(A)$ is closed under the operations $\odot$, $\rightarrow$, $\rightsquigarrow$, 0, 1. Thus $\sigma(A)$ is a subalgebra of A. $\square$

Definition 9.3

(1) A state operator σ on a bounded pseudo-hoop A is called a *weak state-morphism operator* on A if for all $x, y \in A$:

(S_6) $\sigma(x \odot y) = \sigma(x) \odot \sigma(y)$.

In this case (A, σ) is called a *weak state-morphism pseudo-hoop*.

(2) A bounded pseudo-hoop endomorphism $\sigma : A \longrightarrow A$ is said to be a *state-morphism operator* if $\sigma^2 = \sigma$.

Obviously, a state-morphism operator is always a weak state-morphism operator.

Example 9.1 ([114])

(1) If A is a bounded pseudo-hoop, then the identity id_M is a state operator on A.
(2) Let A be a bounded pseudo-hoop and $B = A \times A$. Then the mappings $\sigma_1, \sigma_2 :$ $B \longrightarrow B$ such that $\sigma_1(x_1, x_2) := (x_1, x_1)$, $\sigma_2(x_1, x_2) := (x_2, x_2)$ are state-morphism operators on the bounded pseudo-hoop B.

Remark 9.2 From Propositions 9.2 and 9.4 it follows that:

(1) If σ is a state operator on A preserving $\rightarrow$ or preserving $\rightsquigarrow$, then σ is a weak state-morphism operator and a state-morphism operator.
(2) If σ is a strong state operator on a bounded pseudo-hoop A such that $x^\sim \leq y$ or $y^- \leq x$ for all $x, y \in A$, then σ is a weak state-morphism operator.

Proposition 9.7 *If A is a bounded cancellative pseudo-hoop, then any state operator σ on A is a weak state-morphism operator.*

Proof According to (S_3) and taking into consideration that in a cancellative pseudo-hoop $y \rightarrow x \odot y = x$, we get:

$$\sigma(x \odot y) = \sigma(y \rightarrow x \odot y) \odot \sigma(y) = \sigma(x) \odot \sigma(y).$$

Thus σ is a weak state-morphism operator on A.
(The proof for the case $x \rightsquigarrow x \odot y = y$ is similar.) □

Theorem 9.2 *If A is a bounded idempotent pseudo-hoop, then any state operator σ on A is a weak state-morphism operator and a state-morphism operator.*

Proof Consider $x, y \in A$. Since they are idempotent, so are $\sigma(x)$ and $\sigma(y)$.
Hence, applying the property of idempotent elements and Proposition 9.1(4) we get:

$$\sigma(x \wedge y) = \sigma(x \odot y) \geq \sigma(x) \odot \sigma(y) = \sigma(x) \wedge \sigma(y).$$

On the other hand, $\sigma(x \wedge y) \leq \sigma(x) \wedge \sigma(y) = \sigma(x) \odot \sigma(y)$.
Thus $\sigma(x \wedge y) = \sigma(x \odot y) = \sigma(x) \odot \sigma(y) = \sigma(x) \wedge \sigma(y)$.
Hence σ is a weak state-morphism operator on A.
Since $\wedge$ is preserved, according to Proposition 9.2((a) $\Leftrightarrow$ (b)), one of $\rightarrow$, $\rightsquigarrow$ is preserved as well. It then follows from Proposition 9.1(12) that the other arrow is also preserved. The constants 0 and 1 are preserved by (S_1) and Proposition 9.1(1), respectively. Thus σ is an endomorphism on A.
Since from Proposition 9.1(9) we have $\sigma^2 = \sigma$, it follows that σ is also a state-morphism operator on A. □

Proposition 9.8 *If σ is a state operator on a bounded pseudo-hoop A, then* $\mathrm{Ker}(\sigma)$
is a normal filter of A.

Proof Since $\sigma(1) = 1$, it follows that $1 \in \mathrm{Ker}(\sigma)$.

Consider $x, y \in \mathrm{Ker}(\sigma)$. Then $\sigma(x \odot y) \geq \sigma(x) \odot \sigma(y) = 1$, hence $\sigma(x \odot y) = 1$,
that is, $x \odot y \in \mathrm{Ker}(\sigma)$. If $x \in \mathrm{Ker}(\sigma)$ and $y \in A$ are such that $x \leq y$, then we have
$1 = \sigma(x) \leq \sigma(y)$, that is, $y \in \mathrm{Ker}(\sigma)$.

Thus $\mathrm{Ker}(\sigma)$ is a filter of A.

Let $x, y \in A$ such that $x \to y \in \mathrm{Ker}(\sigma)$, so $\sigma(x \to y) = 1$.

Applying (S_2) we get $\sigma(x) \to \sigma(x \wedge y) = 1$, hence $\sigma(x) \leq \sigma(x \wedge y)$.

Therefore $\sigma(x) = \sigma(x \wedge y)$ and applying again (S_2) we have $\sigma(x \rightsquigarrow y) = \sigma(x) \rightsquigarrow \sigma(x \wedge y) = 1$. Thus $x \rightsquigarrow y \in \mathrm{Ker}(\sigma)$.

Similarly, $x \rightsquigarrow y \in \mathrm{Ker}(\sigma)$ implies $x \to y \in \mathrm{Ker}(\sigma)$ and we conclude that $\mathrm{Ker}(\sigma)$
is a normal filter of A. $\square$

9.2 On the Existence of State Operators on Pseudo-hoops

In this section we investigate the existence of state operators, proving that every
perfect pseudo-hoop admits a nontrivial state operator.

In what follows A will be a bounded pseudo-hoop.

We recall that any bounded idempotent pseudo-hoop A is good (Remark 2.3).

Proposition 9.9 *Let (A, σ) by an idempotent state pseudo-hoop. Then:*

(1) $\sigma(x \wedge y) = \sigma(x \odot y) = \sigma(x) \odot \sigma(y) = \sigma(x) \wedge \sigma(y)$ *for all* $x, y \in A$;
(2) $\sigma(x \to y) = \sigma(x) \to \sigma(y)$ *and* $\sigma(x \rightsquigarrow y) = \sigma(x) \rightsquigarrow \sigma(y)$ *for all* $x, y \in A$;
(3) $(x \odot y)^{-\sim} = (x \wedge y)^{-\sim} = x^{-\sim} \wedge y^{-\sim} = x^{-\sim} \odot y^{-\sim}$ *for all* $x, y \in A$.

Proof

(1) This follows from the proof of Theorem 9.2.
(2) By Proposition 9.2 it follows that $\sigma(x \to y) = \sigma(x) \to \sigma(y)$ or $\sigma(x \rightsquigarrow y) = \sigma(x) \rightsquigarrow \sigma(y)$ for all $x, y \in A$, which are equivalent according to Proposition 9.1(12).
(3) Since A is idempotent, $x \odot y = x \wedge y$ for all $x, y \in A$.
    Applying $(pshoop\text{-}c_{10})$ we get:

$$(x \odot y)^{-\sim} = (x \wedge y)^{-\sim} = x^{-\sim} \wedge y^{-\sim} = x^{-\sim} \odot y^{-\sim}. \qquad \square$$

Proposition 9.10 *Let (A, σ) be a state pseudo-hoop and $x, y \in \mathrm{Reg}(A)$. Then:*

(1) $\sigma(x) \in \mathrm{Reg}(A)$;
(2) *if A is good, then* $x \oplus y, x \wedge y, x \to y, x \rightsquigarrow y, x \vee_1 y, x \vee_2 y \in \mathrm{Reg}(A)$;
(3) *if A is idempotent, then* $x \odot y \in \mathrm{Reg}(A)$.

Proof

(1) By Proposition 9.1(2) we have:

$$\sigma(x)^{-\sim} = \sigma\left(x^{-\sim}\right) = \sigma(x) \quad \text{and} \quad \sigma(x)^{\sim-} = \sigma\left(x^{\sim-}\right) = \sigma(x),$$

thus $\sigma(x) \in \text{Reg}(A)$.

(2) Applying Proposition 1.25 we get:

$$(x \oplus y)^{-\sim} = \left(\left(y^{-} \odot x^{-}\right)^{\sim}\right)^{-\sim} = \left(y^{-} \odot x^{-}\right)^{\sim} = x \oplus y.$$

From the goodness property we have $(x \oplus y)^{\sim-} = (x \oplus y)^{-\sim} = x \oplus y$.
Thus $x \oplus y \in \text{Reg}(A)$.
Applying (*pshoop-c₁₀*) we have $(x \wedge y)^{-\sim} = x^{-\sim} \wedge y^{-\sim} = x \wedge y$, so $x \wedge y \in \text{Reg}(A)$.

By (*pshoop-c₉*) we have:

$$(x \to y)^{-\sim} = x^{-\sim} \to y^{-\sim} = x \to y \quad \text{and}$$

$$(x \rightsquigarrow y)^{-\sim} = x^{-\sim} \rightsquigarrow y^{-\sim} = x \rightsquigarrow y.$$

Thus $x \to y, x \rightsquigarrow y \in \text{Reg}(A)$.
As a consequence, it follows that $x \vee_1 y, x \vee_2 y \in \text{Reg}(A)$.

(3) From Proposition 9.9 we have $(x \odot y)^{-\sim} = x^{-\sim} \odot y^{-\sim} = x \odot y$, so $x \odot y \in \text{Reg}(A)$. □

Lemma 9.1 *Any state operator σ on a locally finite pseudo-hoop is faithful.*

Proof Assume that there exists an x with $0 < x < 1$ such that $\sigma(x) = 1$. Then there is an integer $n \geq 1$ such that $x^n = 0$, hence $0 = \sigma(0) = \sigma(x^n) \geq \sigma(x)^n = 1$, a contradiction. Thus σ is faithful. □

Proposition 9.11 *If A is a strongly simple locally finite basic pseudo-hoop, then the identity is the unique state operator on A.*

Proof Let σ be a state operator on A. By Lemma 9.1 it follows that σ is faithful. Since every strongly simple basic pseudo-hoop is linearly ordered (Corollary 2.2), applying Proposition 9.1(17), we get $\sigma(x) = x$ for all $x \in A$. □

Theorem 9.3 *Let A be an idempotent pseudo-hoop and $\sigma : \text{Reg}(A) \to \text{Reg}(A)$ be a state operator on $\text{Reg}(A)$. Then the mapping $\tilde{\sigma} : A \to A$ defined by*

$$\tilde{\sigma}(x) := \sigma\left(x^{-\sim}\right)$$

is a state operator on A such that $\tilde{\sigma}_{|\text{Reg}(A)} = \sigma$.

Proof Obviously, $\tilde{\sigma}(0) = \sigma(0) = 0$, so condition (S_1) is verified.

Applying ($pshoop$-c_9) and Proposition 9.9 we get:

$$\tilde{\sigma}(x \to y) = \sigma\big((x \to y)^{-\sim}\big) = \sigma\big(x^{-\sim} \to y^{-\sim}\big) = \sigma\big(x^{-\sim}\big) \to \sigma\big(x^{-\sim} \wedge y^{-\sim}\big)$$
$$= \sigma\big(x^{-\sim}\big) \to \sigma\big((x \wedge y)^{-\sim}\big) = \tilde{\sigma}(x) \to \tilde{\sigma}(x \wedge y).$$

Similarly, $\tilde{\sigma}(x \rightsquigarrow y) = \tilde{\sigma}(x) \rightsquigarrow \tilde{\sigma}(x \wedge y)$, so $\tilde{\sigma}$ satisfies (S_2).

By Proposition 9.9 we also have:

$$\tilde{\sigma}(x \odot y) = \sigma\big((x \odot y)^{-\sim}\big) = \sigma\big(x^{-\sim} \odot y^{-\sim}\big) = \sigma\big(x^{-\sim}\big) \odot \sigma\big(x^{-\sim} \rightsquigarrow x^{-\sim} \odot y^{-\sim}\big)$$
$$= \sigma\big(x^{-\sim}\big) \odot \sigma\big(x^{-\sim} \rightsquigarrow (x \odot y)^{-\sim}\big) = \sigma\big(x^{-\sim}\big) \odot \sigma\big((x \rightsquigarrow x \odot y)^{-\sim}\big)$$
$$= \tilde{\sigma}(x) \odot \tilde{\sigma}(x \rightsquigarrow x \odot y).$$

Similarly, $\tilde{\sigma}(x \odot y) = \tilde{\sigma}(y \to x \odot y) \odot \tilde{\sigma}(y)$, hence $\tilde{\sigma}$ satisfies (S_3).

For condition (S_4) we have:

$$\tilde{\sigma}\big(\tilde{\sigma}(x) \odot \tilde{\sigma}(y)\big) = \sigma\big(\big((\sigma\big(x^{-\sim}\big) \odot \sigma\big(y^{-\sim}\big))\big)^{-\sim}\big) = \sigma\big(\sigma\big(x^{-\sim}\big)^{-\sim} \odot \sigma\big(y^{-\sim}\big)^{-\sim}\big)$$
$$= \sigma\big(\sigma\big(x^{-\sim}\big) \odot \sigma\big(y^{-\sim}\big)\big) = \sigma\big(x^{-\sim}\big) \odot \sigma\big(y^{-\sim}\big) = \tilde{\sigma}(x) \odot \tilde{\sigma}(y)$$

thus it is verified too.

Finally we have:

$$\tilde{\sigma}\big(\tilde{\sigma}(x) \to \tilde{\sigma}(y)\big) = \sigma\big(\big((\sigma\big(x^{-\sim}\big) \to \sigma\big(y^{-\sim}\big))\big)^{-\sim}\big) = \sigma\big(\sigma\big(x^{-\sim}\big)^{-\sim} \to \sigma\big(y^{-\sim}\big)^{-\sim}\big)$$
$$= \sigma\big(\sigma\big(x^{-\sim}\big) \to \sigma\big(y^{-\sim}\big)\big) = \sigma\big(x^{-\sim}\big) \to \sigma\big(y^{-\sim}\big)$$
$$= \tilde{\sigma}(x) \to \tilde{\sigma}(y)$$

and similarly $\tilde{\sigma}\big(\tilde{\sigma}(x) \rightsquigarrow \tilde{\sigma}(y)\big) = \tilde{\sigma}(x) \rightsquigarrow \tilde{\sigma}(y)$, that is, condition (S_5) for $\tilde{\sigma}$.

We conclude that $\tilde{\sigma}$ is a state operator on A.

If $x \in \mathrm{Reg}(A)$, then $\tilde{\sigma}(x) = \sigma(x^{-\sim}) = \sigma(x)$, so $\tilde{\sigma}_{|\mathrm{Reg}(A)} = \sigma$. □

Corollary 9.3 *If A is an idempotent pseudo-hoop, then any state operator on $\mathrm{Reg}(A)$ can be extended to a state operator on A.*

Theorem 9.4 *Every perfect pseudo-hoop admits a nontrivial state operator.*

Proof Let A be a perfect pseudo-hoop, so $A = \mathrm{Rad}(A) \cup \mathrm{Rad}(A)^*$. We will prove that the map $\sigma : A \to A$ defined by

$$\sigma(x) := \begin{cases} 1 & \text{if } x \in \mathrm{Rad}(A) \\ 0 & \text{if } x \in \mathrm{Rad}(A)^* \end{cases}$$

is a state operator on A.

Obviously, $\sigma(0) = 0$, hence (S_1) is satisfied.

We consider the following cases:

(1) $a, b \in \text{Rad}(A)$.

   Obviously, $\sigma(a) = \sigma(b) = 1$. Since $\text{Rad}(A)$ is a filter of A and $b \leq a \rightarrow b$, it follows that $a \rightarrow b \in \text{Rad}(A)$. Hence $\sigma(a \rightarrow b) = 1$. Similarly, $\sigma(a \rightsquigarrow b) = 1$.

   From the definition of a filter we have $a \odot b, a \wedge b \in \text{Rad}(A)$.

   Thus $\sigma(a) = \sigma(b) = \sigma(a \rightarrow b) = \sigma(a \rightsquigarrow b) = \sigma(a \wedge b) = \sigma(a \odot b) = 1$.

   Since $a \odot b \in \text{Rad}(A)$ and $a \odot b \leq a \rightsquigarrow a \odot b, a \odot b \leq b \rightarrow a \odot b$ it follows that $a \rightsquigarrow a \odot b, b \rightarrow a \odot b \in \text{Rad}(A)$, so $\sigma(a \rightsquigarrow a \odot b) = \sigma(b \rightarrow a \odot b) = 1$.

   One can easily check that conditions (S_2)–(S_5) are satisfied.

(2) $a, b \in \text{Rad}(A)^*$.

   In this case, $\sigma(a) = \sigma(b) = 0$ and we will prove that $a \rightarrow b, a \rightsquigarrow b \in \text{Rad}(A)$. Indeed, suppose that $a \rightarrow b \in \text{Rad}(A)^*$. Since $a \leq a^{-\sim}$, it follows that $a^{-\sim} \rightarrow b \leq a \rightarrow b$, so $a^{-\sim} \rightarrow b \in \text{Rad}(A)^*$. But $a^- \leq a^{-\sim} \rightarrow b$, hence $a^- \in \text{Rad}(A)^*$, contradicting condition (ii) in the definition of a perfect pseudo-hoop ($a \in \text{Rad}(A)^* = D(A)^*$ iff $a^- \in \text{Rad}(A) = D(A)$). It follows that $a \rightarrow b \in \text{Rad}(A)$ and similarly $a \rightsquigarrow b \in \text{Rad}(A)$. Hence $\sigma(a \rightarrow b) = \sigma(a \rightsquigarrow b) = 1$.

   From $a \wedge b \leq b, a \odot b \leq b$, we get $a \wedge b, a \odot b \in \text{Rad}(A)^*$, thus $\sigma(a \wedge b) = \sigma(a \odot b) = 0$.

   We can see that conditions (S_2)–(S_5) are also verified.

(3) $a \in \text{Rad}(A), b \in \text{Rad}(A)^*$.

   Obviously, $\sigma(a) = 1$ and $\sigma(b) = 0$.

   We show that $a \rightarrow b \in \text{Rad}(A)^*$. Indeed, suppose that $a \rightarrow b \in \text{Rad}(A)$.

   Because $b \leq b^{-\sim}$, we have $a \rightarrow b \leq a \rightarrow b^{-\sim}$, so $a \rightarrow b^{-\sim} \in \text{Rad}(A)$. This means that $(a \odot b^\sim)^- \in \text{Rad}(A)$, that is, $a \odot b^\sim \in \text{Rad}(A)^*$. On the other hand, since $\text{Rad}(A)$ is a filter of A and $a, b^\sim \in \text{Rad}(A)$ we have $a \odot b^\sim \in \text{Rad}(A)$, a contradiction. We conclude that $a \rightarrow b \in \text{Rad}(A)^*$, so $\sigma(a \rightarrow b) = 0$.

   Similarly, $a \rightsquigarrow b \in \text{Rad}(A)^*$, so $\sigma(a \rightsquigarrow b) = 0$.

   Since $a \wedge b \leq b, a \odot b \leq b$, we have $a \wedge b, a \odot b \in \text{Rad}(A)^*$, so $\sigma(a \wedge b) = \sigma(a \odot b) = 0$.

   Moreover, $a \in \text{Rad}(A)$ and $a \odot b \in \text{Rad}(A)^*$ implies $a \rightsquigarrow a \odot b \in \text{Rad}(A)^*$, hence $\sigma(a \rightsquigarrow a \odot b) = 0$.

   It is easy to see that conditions (S_2)–(S_5) are satisfied.

(4) $a \in \text{Rad}(A)^*$ and $b \in \text{Rad}(A)$.

   Taking into consideration that $b \leq a \rightarrow b, b \leq a \rightsquigarrow b$ we have $a \rightarrow b, a \rightsquigarrow b \in \text{Rad}(A)$. From $a \wedge b, a \odot b \leq a$ we get $a \wedge b, a \odot b \in \text{Rad}(A)^*$.

   Hence $\sigma(a) = 0$, $\sigma(b) = 1$, $\sigma(a \wedge b) = \sigma(a \odot b) = 0$, $\sigma(a \rightarrow b) = \sigma(a \rightsquigarrow b) = 1$.

   Applying case (3), $b \in \text{Rad}(A)$ and $a \odot b \in \text{Rad}(A)^*$ implies $b \rightarrow a \odot b \in \text{Rad}(A)^*$, so $\sigma(b \rightarrow a \odot b) = 0$.

   Thus conditions (S_2)–(S_5) are also satisfied.

   We conclude that σ is a state operator on A, that is, (A, σ) is a state pseudo-hoop.   □

Remark 9.3 The state operator σ defined in Theorem 9.4 is a C-state operator. Indeed, in cases (1), (3) and (4) in the proof of Theorem 9.4 we have $a \vee_1 b, b \vee_1 a, a \vee_2 b, b \vee_2 a \in \text{Rad}(A)$, so $\sigma(a \vee_1 b) = \sigma(b \vee_1 a) = 1$ and

$\sigma(a \vee_2 b) = \sigma(b \vee_2 a) = 1$. In case (2), $a \vee_1 b, b \vee_1 a, a \vee_2 b, b \vee_2 a \in \mathrm{Rad}(A)^*$, hence $\sigma(a \vee_1 b) = \sigma(b \vee_1 a) = 0$ and $\sigma(a \vee_2 b) = \sigma(b \vee_2 a) = 0$.

Thus σ is a C-state operator.

9.3 State Operators and States on Pseudo-hoops

States on bounded pseudo-hoops have been investigated in [56]. In this section we show that there is a close connection between state operators and states on pseudo-hoops.

Theorem 9.5 *Let σ be a state operator on the bounded pseudo-hoop A preserving $\rightarrow$ or $\rightsquigarrow$. If s is a Bosbach state on A, then the mapping $s_\sigma : A \rightarrow [0, 1]$ defined by $s_\sigma(x) := s(\sigma(x))$ is a Bosbach state on A.*

Proof Obviously, $s_\sigma(0) = 0$ and $s_\sigma(1) = 1$, so (B_3) is verified.

It is sufficient to assume that just one of the arrows $\rightarrow$, $\rightsquigarrow$ is preserved, the preservation of the other one is then implied by Proposition 9.1(12).

Assume σ preserves $\rightarrow$. It follows that:

$$s_\sigma(x) + s_\sigma(x \rightarrow y) = s(\sigma(x)) + s(\sigma(x \rightarrow y)) = s(\sigma(x)) + s(\sigma(x) \rightarrow \sigma(y))$$
$$= s(\sigma(y)) + s(\sigma(y) \rightarrow \sigma(x)) = s(\sigma(y)) + s(\sigma(y \rightarrow x))$$
$$= s_\sigma(y) + s_\sigma(y \rightarrow x).$$

Thus s_σ satisfies (B_1) and similarly s_σ satisfies (B_2).

It follows that s_σ is a Bosbach state on A. $\square$

Corollary 9.4 *Let (A, σ) be a linearly ordered state pseudo-hoop and s be a Bosbach state on A. Then the mapping $s_\sigma : A \rightarrow [0, 1]$ defined by $s_\sigma(x) := s(\sigma(x))$ is a Bosbach state on A.*

Proof According to Corollary 9.1, σ preserves $\rightarrow$ and $\rightsquigarrow$, hence by Theorem 9.5, s_σ is a Bosbach state on A. $\square$

Corollary 9.5 *Let (A, σ) be an idempotent state pseudo-hoop and s be a Bosbach state on A. Then the mapping $s_\sigma : A \rightarrow [0, 1]$ defined by $s_\sigma(x) := s(\sigma(x))$ is a Bosbach state on A.*

Proof By Proposition 9.9, σ preserves $\rightarrow$ and $\rightsquigarrow$, hence by Theorem 9.5, s_σ is a Bosbach state on A. $\square$

Theorem 9.6 *Let A be a bounded pseudo-hoop and σ be a state operator on A preserving $\rightarrow$ or $\rightsquigarrow$. If s is a Bosbach state on $\mathrm{Reg}(A)$, then the mapping $\tilde{s}_\sigma : A \rightarrow [0, 1]$ defined by $\tilde{s}_\sigma(x) := s(\sigma(x^{-\sim}))$ is a Bosbach state on A.*

Proof Obviously, $s_\sigma(0) = 0$ and $s_\sigma(1) = 1$, so (B_3) is verified.

If σ preserves one of the arrows $\to$, $\rightsquigarrow$, then by Proposition 9.1(12) the second one is also preserved. Assume σ preserves $\to$.

Applying (*pshoop-c9*), we have:

$$
\begin{aligned}
\tilde{s}_\sigma(x) + \tilde{s}_\sigma(x \to y) &= s\big(\sigma\big(x^{-\sim}\big)\big) + s\big(\sigma\big((x \to y)^{-\sim}\big)\big) \\
&= s\big(\sigma\big(x^{-\sim}\big)\big) + s\big(\sigma\big(x^{-\sim} \to y^{-\sim}\big)\big) \\
&= s\big(\sigma\big(x^{-\sim}\big)\big) + s\big(\sigma\big(x^{-\sim}\big) \to \sigma\big(y^{-\sim}\big)\big) \\
&= s\big(\sigma\big(y^{-\sim}\big)\big) + s\big(\sigma\big(y^{-\sim}\big) \to \sigma\big(x^{-\sim}\big)\big) \\
&= s\big(\sigma\big(y^{-\sim}\big)\big) + s\big(\sigma\big(y^{-\sim} \to x^{-\sim}\big)\big) \\
&= s\big(\sigma\big(y^{-\sim}\big)\big) + s\big(\sigma\big((y \to x)^{-\sim}\big)\big) \\
&= \tilde{s}_\sigma(y) + \tilde{s}_\sigma(y \to x).
\end{aligned}
$$

Thus $\tilde{s}_\sigma$ satisfies condition (B_1).

Similarly, $\tilde{s}_\sigma(x) + \tilde{s}_\sigma(x \rightsquigarrow y) = \tilde{s}_\sigma(y) + \tilde{s}_\sigma(y \rightsquigarrow x)$, so condition (B_2) is also satisfied. It follows that $\tilde{s}_\sigma$ is a Bosbach state on A. $\square$

Theorem 9.7 *Let A be a good pseudo-hoop satisfying the* (SO) *property, σ be a state operator and s be a Riečan state on A. Then the mapping $s_\sigma : A \to [0,1]$ defined by $s_\sigma(x) := s(\sigma(x))$ is a Riečan state on A.*

Proof Obviously, $s_\sigma(1) = 1$.

It is easy to check that $s_\sigma(x \oplus 0) = s_\sigma(x) + s_\sigma(0)$ and $s_\sigma(0 \oplus x) = s_\sigma(0) + s_\sigma(x)$.

Consider $x, y \in A$ such that $x \neq 0$, $y \neq 0$ and $x \perp y$. It follows that $\sigma(x) \perp \sigma(y)$.

By the (SO) property we have $x \perp_{no} y$ and applying Proposition 9.1(10) we get $\sigma(x \oplus y) = \sigma(x) \oplus \sigma(y)$. Hence:

$$
s_\sigma(x \oplus y) = s\big(\sigma(x \oplus y)\big) = s\big(\sigma(x) \oplus \sigma(y)\big) = s\big(\sigma(x)\big) + s\big(\sigma(y)\big) = s_\sigma(x) + s_\sigma(y).
$$

Thus s_σ is a Riečan state on A. $\square$

Theorem 9.8 *Let A be a good pseudo-hoop with the* (SO) *property and τ be a state operator on A. If s is a Riečan state on $\mathrm{Reg}(A)$, then the mapping $\tilde{s}_\tau : A \to [0,1]$ defined by $\tilde{s}_\tau(x) := s(\tau(x^{-\sim}))$ is a Riečan state on A.*

Proof Since s is a Riečan state on $\mathrm{Reg}(A)$, according to Proposition 6.15(3), $s(x^{-\sim}) = s(x)$ for all $x \in \mathrm{Reg}(A)$.

Obviously, $\tilde{s}_\tau(1) = 1$, $\tilde{s}_\tau(x \oplus 0) = \tilde{s}_\tau(x) + \tilde{s}_\tau(0)$ and $\tilde{s}_\tau(0 \oplus x) = \tilde{s}_\tau(0) + \tilde{s}_\tau(x)$.

Consider $x, y \in A$ such that $x \neq 0$, $y \neq 0$ and $x \perp y$.

It follows that $x^{-\sim} \perp y^{-\sim}$ (Proposition 1.6(7)).

Hence by the (SO) property, $x^{-\sim} \perp_{no} y^{-\sim}$.

Since $(x \oplus y)^{-\sim} = x^{-\sim} \oplus y^{-\sim}$ (Proposition 1.24(5)), applying Proposition 9.1(10) we get:

$$\tilde{s}_{\tau}(x \oplus y) = s\left(\tau\left((x \oplus y)^{-\sim}\right)\right) = s\left(\tau\left(x^{-\sim} \oplus y^{-\sim}\right)\right) = s\left(\tau\left(x^{-\sim}\right) \oplus \tau\left(y^{-\sim}\right)\right)$$
$$= s\left(\tau\left(x^{-\sim}\right)\right) \oplus s\left(\tau\left(y^{-\sim}\right)\right) = \tilde{s}_{\tau}(x) + \tilde{s}_{\tau}(y)$$

(since s is a Riečan state and from $x^{-\sim} \perp y^{-\sim}$ it follows that $\tau(x^{-\sim}) \perp \tau(y^{-\sim})$).
Thus $\tilde{s}_{\tau}$ is a Riečan state on A. $\qquad\qquad\qquad\qquad\qquad\qquad\qquad\qquad\qquad\qquad\quad\square$

9.4 State Operators and Generalized States on Pseudo-hoops

The concept of generalized states has been extended to the case of pseudo-BCK algebras and pseudo-hoops ([68]). In this section we investigate the connection between state operators and generalized states on a pseudo-hoop. Some conditions are given for a state operator to be a generalized state and for a generalized state to be a state operator.

Let A be a bounded pseudo-hoop and $s : A \longrightarrow A$ be an arbitrary function such that $s(0) = 0$ and $s(x \vee_1 y) = s(y \vee_2 x)$ for all $x, y \in A$. Then s is said to be a *generalized Bosbach state of type I* or a *type I state* if it satisfies one of the following equivalent conditions:

(bsI_1) for all $x, y \in A$ with $x \geq y$, $s(x \rightarrow y) = s(x) \rightarrow s(y)$ and $s(x \rightsquigarrow y) = s(x) \rightsquigarrow s(y)$;

(bsI_2) for all $x, y \in A$, $s(x \vee_1 y) = s(x \rightarrow y) \rightsquigarrow s(y)$ and $s(x \vee_2 y) = s(x \rightsquigarrow y) \rightarrow s(y)$;

(bsI_3) for all $x, y \in A$, $s(x \rightarrow y) \rightsquigarrow s(y) = s(y \rightsquigarrow x) \rightarrow s(x)$ and $s(1) = 1$;

(bsI_4) for all $x, y \in A$ with $x \geq y$, $s(x) = s(x \rightarrow y) \rightsquigarrow s(y) = s(x \rightsquigarrow y) \rightarrow s(y)$;

(bsI_5) for all $x, y \in A$, $s(x \rightarrow y) = s(x \vee_1 y) \rightarrow s(y)$ and $s(x \rightsquigarrow y) = s(x \vee_2 y) \rightsquigarrow s(y)$;

(bsI_6) for all $x, y \in A$, $s(x \rightarrow y) = s(x) \rightarrow s(x \wedge y)$ and $s(x \rightsquigarrow y) = s(x) \rightsquigarrow s(x \wedge y)$.

Proposition 9.12 *Let A be a bounded pseudo-hoop and $s : A \longrightarrow A$ be an order-preserving type I state on A. Then the following hold for all $a, b \in A$:*

(1) $s(a \odot b) = s(b \rightarrow a \odot b) \odot s(b) = s(a) \odot s(a \rightsquigarrow a \odot b)$;
(2) $s(a) \odot s(b) \leq s(a \odot b)$.

Proof

(1) Since $a \odot b \leq a, b$, applying (bsI_1) we have:

$$s(b \rightarrow a \odot b) \odot s(b) = \left(s(b) \rightarrow s(a \odot b)\right) \odot s(b)$$
$$= s(b) \wedge s(a \odot b) = s(a \odot b) \quad \text{and}$$

$$s(a) \odot s(a \rightsquigarrow a \odot b) = s(a) \odot \left(s(a) \rightsquigarrow s(a \odot b) \right)$$

$$= s(a) \wedge s(a \odot b) = s(a \odot b).$$

(2) From $a \le b \to a \odot b$ we have $s(a) \le s(b \to a \odot b)$.

Applying (1) we get: $s(a) \odot s(b) \le s(b \to a \odot b) \odot s(b) = s(a \odot b)$.   □

Let A be a bounded pseudo-hoop and $s : A \longrightarrow A$ be an arbitrary function such that $s(0) = 0$ and $s(x \vee_1 y) = s(y \vee_2 x)$ for all $x, y \in A$. The function s is said to be a *generalized Bosbach state of type II* or a *type II state* if it satisfies one of the following equivalent conditions:

($bsII_1$) for all $x, y \in A$ with $x \ge y$, $s(x \to y) = s(x) \rightsquigarrow s(y)$ and $s(x \rightsquigarrow y) = s(x) \to s(y)$;

($bsII_2$) for all $x, y \in A$, $s(x \vee_1 y) = s(x \to y) \to s(y)$ and $s(x \vee_2 y) = s(x \rightsquigarrow y) \rightsquigarrow s(y)$;

($bsII_3$) for all $x, y \in A$, $s(x \to y) \to s(y) = s(y \rightsquigarrow x) \rightsquigarrow s(x)$ and $s(1) = 1$;

($bsII_4$) for all $x, y \in A$ with $x \ge y$, $s(x) = s(x \to y) \to s(y) = s(x \rightsquigarrow y) \rightsquigarrow s(y)$;

($bsII_5$) for all $x, y \in A$, $s(x \to y) = s(x \vee_1 y) \rightsquigarrow s(y)$ and $s(x \rightsquigarrow y) = s(x \vee_2 y) \to s(y)$;

($bsII_6$) for all $x, y \in A$, $s(x \to y) = s(x) \rightsquigarrow s(x \wedge y)$ and $s(x \rightsquigarrow y) = s(x) \to s(x \wedge y)$.

Let A be a bounded Wajsberg pseudo-hoop and $s : A \longrightarrow A$ be a mapping satisfying $s(0) = 0$, $s(1) = 1$ and $s(x \vee_1 y) = s(y \vee_2 x)$. Then:

(1) s is a type I state iff $s(d_1(x, y)) = s(x \vee y) \to s(x \wedge y)$ and $s(d_2(x, y)) = s(x \vee y) \rightsquigarrow s(x \wedge y)$;

(2) s is a type II state iff $s(d_1(x, y)) = s(x \vee y) \rightsquigarrow s(x \wedge y)$ and $s(d_2(x, y)) = s(x \vee y) \to s(x \wedge y)$.

Let A be a bounded pseudo-hoop. An endomorphism $h : A \longrightarrow A$ satisfying the condition $h(x \vee_1 y) = h(y \vee_2 x)$ for all $x, y \in A$ is called a *generalized state-morphism*. If, moreover, $h(x \to y) = h(x \rightsquigarrow y)$ for all $x, y \in A$, then h is a *strong generalized state-morphism*.

A mapping $m : A \longrightarrow A$ is called a *generalized Riečan state* iff the following conditions are satisfied for all $x, y \in A$:

(rs_1) $m(1) = 1$;

(rs_2) for all $x, y \in A$, if $x \perp y$, then $m(x) \perp m(y)$ and $m(x \oplus y) = m(x) \oplus m(y)$.

Proposition 9.13 *Every C-state operator on a bounded pseudo-hoop is a type I state.*

Proof Let σ be a state operator on a bounded pseudo-hoop A.

From (S_1) we have $\sigma(0) = 0$.

By condition (C) and Proposition 9.3 we get $\sigma(x \vee_1 y) = \sigma(y \vee_2 x)$.

Since condition (S_2) in the definition of a state operator is condition (bsI_6), it follows that σ is a type I state on A.   □

Corollary 9.6 *Every perfect pseudo-hoop admits a type I state.*

Proof According to Theorem 9.4 and Remark 9.3, every perfect pseudo-hoop has a C-state operator, hence by Proposition 9.13 every perfect pseudo-hoop admits a type I state.   □

Proposition 9.14 *Let (A, σ) be an idempotent state pseudo-hoop such that $\sigma(x \vee_1 y) = \sigma(y \vee_2 x)$. Then σ is a generalized state-morphism on A.*

Proof This follows by Propositions 9.9 and Proposition 9.2.   □

Proposition 9.15 *Let (A, σ) be a linearly ordered state pseudo-hoop such that $\sigma(x \vee_1 y) = \sigma(y \vee_2 x)$. Then σ is a generalized state-morphism on A.*

Proof This follows by Corollary 9.1 and Proposition 9.2.   □

Proposition 9.16 *If (A, σ) is a good state pseudo-hoop satisfying the (SO) property, then σ is a generalized Riečan state on A.*

Proof From Proposition 9.1(1) we have $\sigma(1) = 1$, that is, (rs_1).
It is easy to check that $\sigma(x \oplus 0) = \sigma(x) \oplus \sigma(0)$ and $\sigma(0 \oplus x) = \sigma(0) \oplus \sigma(x)$.
Consider $x, y \in A$ such that $x \neq 0$, $y \neq 0$ and $x \perp y$.
From the (SO) property we have $x \perp_{no} y$ and applying Proposition 9.1(10), we get $\sigma(x \oplus y) = \sigma(x) \oplus \sigma(y)$, so (rs_2) is verified too.
Thus σ is a generalized Riečan state on A.   □

Theorem 9.9 *If A is a linearly ordered bounded pseudo-hoop and $s : A \longrightarrow A$ is an order-preserving type I state such that $s^2(x) = s(x) \leq x$ for all $x \in A$, then s is a state operator on A.*

Proof Applying the hypothesis and the definition of a type I state, we will verify axioms (S_1)–(S_5) from the definition of a state operator.

(S_1) $s(0) = 0$:
    This follows from the definition of a type I state.
(S_2) $s(a \rightarrow b) = s(a) \rightarrow s(a \wedge b)$ and $s(a \rightsquigarrow b) = s(a) \rightsquigarrow s(a \wedge b)$:
    This is the condition (bsI_6).
(S_3) $s(a \odot b) = s(a) \odot s(a \rightsquigarrow a \odot b) = s(b \rightarrow a \odot b) \odot s(b)$:
    This follows from Proposition 9.12(1).
(S_4) $s(s(a) \odot s(b)) = s(a) \odot s(b)$:
    Since $s(x) \leq x$ for all $x \in A$ we have $s(s(a) \odot s(b)) \leq s(a) \odot s(b)$.
    On the other hand, from Proposition 9.12(2), replacing a with $s(a)$ and b with $s(b)$ we get $s^2(a) \odot s^2(b) \leq s(s(a) \odot s(b))$, that is, $s(a) \odot s(b) \leq s(s(a) \odot s(b))$.
    Thus $s(s(a) \odot s(b)) = s(a) \odot s(b)$.

(S_5) $s(s(a) \to s(b)) = s(a) \to s(b)$ and $s(s(a) \rightsquigarrow s(b)) = s(a) \rightsquigarrow s(b)$:
Since A is linearly ordered we consider the cases:

(a) $b \le a$, so $s(b) \le s(a)$. According to condition (bsI_1) we get $s(s(a) \to s(b)) = s^2(a) \to s^2(b) = s(a) \to s(b)$.

(b) $a \le b$, so $s(a) \le s(b)$. It follows that $s(a) \to s(b) = 1$, thus $s(s(a) \to s(b)) = s(a) \to s(b) = s(1) = 1$.

Similarly, $s(s(a) \rightsquigarrow s(b)) = s(a) \rightsquigarrow s(b)$.
We conclude that s is a state operator on A. $\square$

Theorem 9.10 *If A is a linearly ordered bounded pseudo-hoop and $s : A \longrightarrow A$ is an order-preserving type I state such that $s^2 = s$ and $s(x \odot y) = s(x) \odot s(y)$ for all $x, y \in A$, then s is a weak state-morphism operator on A.*

Proof The axioms (S_1), (S_2), (S_3) and (S_5) are verified in a similar way as in Theorem 9.9.

For axiom (S_4) we have: $s(s(a) \odot s(b)) = s^2(a) \odot s^2(b) = s(a) \odot s(b)$.

Since $s(x \odot y) = s(x) \odot s(y)$ for all $x, y \in A$, it follows that s is a weak state-morphism operator on A. $\square$

Theorem 9.11 *Let σ be a C-state operator on the bounded pseudo-hoop A preserving $\to$ or $\rightsquigarrow$. If $s : A \longrightarrow A$ is a type I (type II) state on A, then $s_\sigma : A \longrightarrow A$ defined by $s_\sigma(x) := s(\sigma(x))$ is a type I (type II) state on A.*

Proof Obviously, $s_\sigma(0) = s(\sigma(0)) = s(0) = 0$.

We remark again that, if σ preserves one of the arrows $\to$, $\rightsquigarrow$, then by Proposition 9.1(12) the second one is also preserved.

Since σ is a C-state operator on A, applying Corollary 9.2 we have:

$$s_\sigma(x \vee_1 y) = s\big(\sigma(x \vee_1 y)\big) = s\big(\sigma(x \vee_2 y)\big) = s_\sigma(x \vee_2 y).$$

On the other hand, from $\sigma(x \vee_2 y) = \sigma(y \vee_2 x)$, we get $s_\sigma(x \vee_2 y) = s_\sigma(y \vee_2 x)$. Hence $s_\sigma(x \vee_1 y) = s_\sigma(y \vee_2 x)$.

Let s be a type I state on A, so it satisfies (bsI_1).

Consider $y \le x$. It follows that $\sigma(y) \le \sigma(x)$ and taking into consideration that σ preserves $\to$, we get:

$$s_\sigma(x \to y) = s\big(\sigma(x \to y)\big) = s\big(\sigma(x) \to \sigma(y)\big) = s\big(\sigma(x)\big) \to s\big(\sigma(y)\big)$$
$$= s_\sigma(x) \to s_\sigma(y).$$

Similarly, $s_\sigma(x \rightsquigarrow y) = s_\sigma(x) \rightsquigarrow s_\sigma(y)$ for all $x, y \in A$.
Hence s_σ satisfies (bsI_1), thus it is a type I state on A.
Suppose s is a type II state on A, so it satisfies $(bsII_1)$.
Assume $y \le x$, so $\sigma(y) \le \sigma(x)$. Since σ preserves $\to$, we get:

$$s_\sigma(x \to y) = s\big(\sigma(x \to y)\big) = s\big(\sigma(x) \to \sigma(y)\big) = s\big(\sigma(x)\big) \rightsquigarrow s\big(\sigma(y)\big)$$
$$= s_\sigma(x) \rightsquigarrow s_\sigma(y).$$

Similarly, $s_\sigma(x \rightsquigarrow y) = s_\sigma(x) \to s_\sigma(y)$ for all $x, y \in A$.
Thus s_σ satisfies $(bsII_1)$, hence it is a type II state on A. □

Corollary 9.7 *Let (A, σ) be a linearly ordered C-state pseudo-hoop and s be a type I (type II) state on A. Then the mapping $s_\sigma : A \longrightarrow A$ defined by $s_\sigma(x) := s(\sigma(x))$ is a type I (type II) state on A.*

Proof According to Corollary 9.1, σ preserves $\to$ and $\rightsquigarrow$, hence by Theorem 9.11, s_σ is a type I (type II) state on A. □

Corollary 9.8 *Let (A, σ) be an idempotent C-state pseudo-hoop and s be a Bosbach state on A. Then the mapping $s_\sigma : A \to A$ defined by $s_\sigma(x) := s(\sigma(x))$ is a type I (type II) state on A.*

Proof By Proposition 9.9 we have $\sigma(x \wedge y) = \sigma(x) \wedge \sigma(y)$. According to Proposition 9.2, σ preserves $\to$ or $\rightsquigarrow$, hence by Theorem 9.11, s_σ is a type I (type II) state on A. □

Remark 9.4 The state operator σ from Corollaries 9.7 and 9.8 is an endomorphism satisfying condition (C). Moreover, $\sigma(A)$ is a Wajsberg sub-pseudo-hoop of A.

References

1. Aglianò, P., Ferreirim, I.M.A., Montagna, F.: Basic hoops: an algebraic study of continuous t-norms. Stud. Log. **87**, 73–98 (2007)
2. Anderson, M., Feil, T.: Lattice-Ordered Groups: An Introduction. Reidel, Dordrecht (1988)
3. Bahls, P., Cole, J., Galatos, N., Jipsen, P., Tsinakis, C.: Cancellative residuated lattices. Algebra Univers. **50**, 83–106 (2003)
4. Balbes, R., Dwinger, P.: Distributive Lattices. University of Missouri Press, Columbia (1974)
5. Belluce, L.P.: Semisimple algebras of infinite valued logic and bold fuzzy set theory. Can. J. Math. **38**, 1356–1379 (1986)
6. Belluce, L.P.: Semisimple and complete MV-algebras. Algebra Univers. **29**, 1–9 (1992)
7. Belluce, L.P., Di Nola, A.: Yosida type representation for perfect MV-algebras. Math. Log. Q. **42**, 551–563 (1996)
8. Belluce, L.P., Di Nola, A., Lettieri, A.: Local MV-algebras. Rend. Circ. Mat. Palermo **42**, 347–361 (1993)
9. Birkhoff, G.: Lattice Theory. Am. Math. Soc., Providence (1967)
10. Blok, W.J., Ferreirim, I.M.A.: On the structure of hoops. Algebra Univers. **43**, 233–257 (2000)
11. Blount, K., Tsinakis, C.: The structure of residuated lattices. Int. J. Algebra Comput. **13**, 437–461 (2003)
12. Blyth, T.S.: Lattices and Ordered Algebraic Structures. Springer, Berlin (2005)
13. Boicescu, V.: Contributions to the study of Łukasiewicz algebras. Ph.D. thesis, University of Bucharest (1984) (in Romanian)
14. Boicescu, V., Filipoiu, A., Georgescu, G., Rudeanu, S.: Łukasiewicz-Moisil Algebras. North-Holland, Amsterdam (1991)
15. Bosbach, B.: Komplementäre Halbgruppen. Axiomatik und Aritmetik. Fundam. Math. **64**, 257–287 (1969)
16. Bosbach, B.: Komplementäre Halbgruppen. Kongruenzen und Quotienten. Fundam. Math. **69**, 1–14 (1970)
17. Bosbach, B.: Residuation groupoids. Results Math. **5**, 107–122 (1982)
18. Botur, M., Dvurečenskij, A.: State-morphism algebras—general approach. Fuzzy Sets Syst. **218**, 90–102 (2013)
19. Burris, S., Sankappanavar, H.P.: A Course in Universal Algebra. Springer, New York (1981)
20. Buşneag, C.: States on Hilbert algebras. Stud. Log. **94**, 177–188 (2010)
21. Buşneag, C.: State-morphisms on Hilbert algebras. An. Univ. Craiova, Ser. Mat. Inform. **37**, 58–64 (2010)
22. Buşneag, D., Piciu, D.: On the lattice of ideals of an MV-algebra. Sci. Math. Jpn. **56**, 362–367 (2002)

23. Buşneag, D., Piciu, D.: On the lattice of deductive systems of a BL-algebra. Cent. Eur. J. Math. **2**, 221–237 (2003)

24. Buşneag, D., Piciu, D.: Localization of MV-algebras and ℓu-groups. Algebra Univers. **50**, 359–380 (2003)

25. Buşneag, D., Piciu, D.: Boolean BL-algebra of fractions. An. Univ. Craiova, Ser. Mat. Inform. **31**, 1–19 (2004)

26. Buşneag, D., Piciu, D.: MV-algebra of fractions and maximal MV-algebra of quotients. J. Mult.-Valued Log. Soft Comput. **10**, 363–383 (2004)

27. Buşneag, D., Piciu, D.: Boolean MV-algebra of fractions. Math. Rep. (Bucur.) **7**, 265–280 (2005)

28. Buşneag, D., Piciu, D.: Localization of pseudo-BL algebra. Rev. Roum. Math. Pures Appl. **L**, 495–513 (2005)

29. Buşneag, D., Piciu, D.: BL-algebra of fractions and maximal BL-algebra of quotients. Soft Comput. **9**, 544–555 (2005)

30. Buşneag, D., Piciu, D.: On the lattice of filters of a pseudo-BL algebra. J. Mult.-Valued Log. Soft Comput. **12**, 217–248 (2006)

31. Buşneag, D., Piciu, D.: Residuated lattice of fractions relative to a meet-closed system. Bull. Math. Soc. Sci. Math. Roum. **49**, 13–24 (2006)

32. Buşneag, D., Piciu, D.: Pseudo-BL algebra of fractions and maximal pseudo-BL algebra of quotients. Southeast Asian Bull. Math. **31**, 639–665 (2007)

33. Buşneag, D., Piciu, D.: Localization of pseudo-MV algebras and ℓ-groups with strong unit. Int. Rev. Fuzzy Math. **2**, 63–95 (2007)

34. Buşneag, D., Rudeanu, S.: A glimpse of deductive systems in algebra. Cent. Eur. J. Math. **8**, 688–705 (2010)

35. Buşneag, D., Piciu, D., Jeflea, A.: Archimedean residuated lattices. An. ştiinţ. Univ. "Al.I. Cuza" Iaşi, Mat. **LVI**, 227–252 (2010)

36. Ceterchi, R.: Pseudo-Wajsberg algebras. Mult. Valued Log. **6**, 67–88 (2001)

37. Chajda, I., Kühr, J.: A note on interval MV-algebra. Math. Slovaca **56**, 47–52 (2006)

38. Chang, C.C.: Algebraic analysis of many-valued logic. Trans. Am. Math. Soc. **88**, 467–490 (1958)

39. Chang, C.C.: A new proof of the completeness of the Łukasiewicz axioms. Trans. Am. Math. Soc. **93**, 74–80 (1959)

40. Chovanec, F.: States and observables on MV-algebras. Tatra Mt. Math. Publ. **3**, 55–65 (1993)

41. Cignoli, R., D'Ottaviano, I.M.L., Mundici, D.: Algebraic Foundations of Many-Valued Reasoning. Kluwer Academic, Dordrecht (2000)

42. Ciungu, L.C.: L-partitions generated by L-fuzzy sets. An. Univ. Craiova, Ser. Mat. Inform. **31**, 28–34 (2004)

43. Ciungu, L.C.: Classes of residuated lattices. An. Univ. Craiova, Ser. Mat. Inform. **33**, 189–207 (2006)

44. Ciungu, L.C.: Algebraic models for multiple-valued logic algebras. Ph.D. thesis, University of Bucharest (2007)

45. Ciungu, L.C.: Convergences in perfect BL-algebras. Mathw. Soft Comput. **14**, 67–80 (2007)

46. Ciungu, L.C.: Some classes of pseudo-MTL algebras. Bull. Math. Soc. Sci. Math. Roum. **50**, 223–247 (2007)

47. Ciungu, L.C.: States on perfect pseudo-MV algebras. J. Appl. Math. Stat. Inform. **3**, 153–163 (2007)

48. Ciungu, L.C.: On perfect pseudo-BCK algebras with pseudo-product. An. Univ. Craiova, Ser. Mat. Inform. **34**, 29–42 (2007)

49. Ciungu, L.C.: Bosbach and Riečan states on residuated lattices. J. Appl. Funct. Anal. **2**, 175–188 (2008)

50. Ciungu, L.C.: Convergence with a fixed regulator in perfect MV-algebras. Demonstr. Math. **41**, 1–10 (2008)

51. Ciungu, L.C.: Convergences in algebraic nuanced structures. J. Concr. Appl. Math. **3**, 267–277 (2008)

52. Ciungu, L.C.: States on pseudo-BCK algebras. Math. Rep. (Bucur.) **10**, 17–36 (2008)
53. Ciungu, L.C.: Convergence with a fixed regulator in residuated lattices. Ital. J. Pure Appl. Math. **26**, 93–102 (2009)
54. Ciungu, L.C.: On the existence of states on fuzzy structures. Southeast Asian Bull. Math. **33**, 1041–1062 (2009)
55. Ciungu, L.C.: Directly indecomposable residuated lattices. Iran. J. Fuzzy Syst. **6**, 7–18 (2009)
56. Ciungu, L.C.: Algebras on subintervals of pseudo-hoops. Fuzzy Sets Syst. **160**, 1099–1113 (2009)
57. Ciungu, L.C.: The radical of a perfect residuated structure. Inf. Sci. **179**, 2695–2709 (2009)
58. Ciungu, L.C.: On the convergence with fixed regulator in residuated structures. In: Symbolic and Quantitative Approaches to Reasoning with Uncertainty, Proc. 10th European Conference ECSQARU. Lecture Notes in Artificial Intelligence, pp. 899–910 (2009)
59. Ciungu, L.C.: On pseudo-BCK algebras with pseudo-double negation. An. Univ. Craiova, Ser. Mat. Inform. **37**, 19–26 (2010)
60. Ciungu, L.C.: Toward a probability theory on nuanced MV-algebras. J. Mult.-Valued Log. Soft Comput. **16**, 221–246 (2010)
61. Ciungu, L.C.: Local pseudo-BCK algebras with pseudo-product. Math. Slovaca **61**, 127–154 (2011)
62. Ciungu, L.C.: Weight and nonlinearity of Boolean functions. Turk. J. Math. **36**, 520–529 (2012)
63. Ciungu, L.C.: Relative negations in non-commutative fuzzy structures. Soft Comput. (2013). doi:10.1007/s00500-013-1054-2
64. Ciungu, L.C.: Bounded pseudo-hoops with internal states. Math. Slovaca (to appear)
65. Ciungu, L.C.: Submeasures on nuanced MV-algebras (submitted)
66. Ciungu, L.C., Cusick, T.W.: Sum of digits sequences modulo m. Theor. Comput. Sci. **35**, 4738–4741 (2011)
67. Ciungu, L.C., Dvurečenskij, A.: Measures, states and de Finetti maps on pseudo-BCK algebras. Fuzzy Sets Syst. **160**, 1099–1113 (2010)
68. Ciungu, L.C., Kühr, J.: New probabilistic model for pseudo-BCK algebras and pseudo-hoops. J. Mult.-Valued Log. Soft Comput. **20**, 373–400 (2013)
69. Ciungu, L.C., Riečan, B.: General form of probabilities on IF-sets. In: Fuzzy Logic and Applications, Proc. 8th International Workshop WILF. Lecture Notes in Artificial Intelligence, pp. 101–107 (2009)
70. Ciungu, L.C., Riečan, B.: Representation theorem for probabilities on IFS-events. Inf. Sci. **180**, 793–798 (2010)
71. Ciungu, L.C., Riečan, B.: A proof of the inclusion-exclusion principle for IF-states (submitted)
72. Ciungu, L.C., Dvurečenskij, A., Hyčko, M.: State BL-algebras. Soft Comput. **15**, 619–634 (2011)
73. Ciungu, L.C., Kelemenová, J., Riečan, B.: A new point of view to the inclusion-exclusion principle. In: Proc. 2012 IEEE 6th International Conference "Intelligent Systems", September 6–8, Sofia, Bulgaria, pp. 142–144 (2012)
74. Ciungu, L.C., Georgescu, G., Mureşan, C.: Generalized Bosbach states: part I. Arch. Math. Log. **52**, 335–376 (2013)
75. Ciungu, L.C., Georgescu, G., Mureşan, C.: Generalized Bosbach states: part II. Arch. Math. Log. (2013). doi:10.1007/s00153-013-0339-6
76. Darnel, M.R.: Theory of Lattice-Ordered Groups. Dekker, New York (1995)
77. Di Nola, A.: Algebraic analysis of Łukasiewicz logic, ESSLLI. Summer school, Utrecht (1999)
78. Di Nola, A., Dvurečenskij, A.: State-morphism MV-algebras. Ann. Pure Appl. Log. **161**, 161–173 (2009)
79. Di Nola, A., Dvurečenskij, A.: On some classes of state-morphism MV-algebras. Math. Slovaca **59**, 517–534 (2009)

80. Di Nola, A., Dvurečenskij, A.: On varieties of MV-algebras with internal states. Int. J. Approx. Reason. **51**, 680–694 (2010)
81. Di Nola, A., Lettieri, A.: Perfect MV-algebras are categorically equivalent to Abelian ℓ-groups. Stud. Log. **53**, 417–432 (1994)
82. Di Nola, A., Leuştean, L.: Compact representations of BL-algebras. Arch. Math. Log. **42**, 737–761 (2003)
83. Di Nola, A., Georgescu, G., Lettieri, A.: Extending probabilities to states on MV-algebras. Coll. Logicum, Ann. Kurt Gödel Soc. **3**, 31–50 (1999)
84. Di Nola, A., Georgescu, G., Leuştean, I.: States on perfect MV-algebras. In: Novák, V., Perfilieva, I. (eds.) Discovering the World with Fuzzy Logic. Studies in Fuzziness and Soft Computing, vol. 57, pp. 105–125. Physica-Verlag, Heidelberg (2000)
85. Di Nola, A., Georgescu, G., Iorgulescu, A.: Pseudo-BL algebras: part I. Mult. Valued Log. **8**, 673–714 (2002)
86. Di Nola, A., Georgescu, G., Iorgulescu, A.: Pseudo-BL algebras: part II. Mult. Valued Log. **8**, 717–750 (2002)
87. Di Nola, A., Flondor, P., Leuştean, I.: MV-modules. J. Algebra **1**, 21–40 (2003)
88. Di Nola, A., Dvurečenskij, A., Tsinakis, C.: Perfect GMV-algebras. Commun. Algebra **36**, 1221–1249 (2008)
89. Di Nola, A., Georgescu, G., Leuştean, I.: MV-algebras. Manuscript
90. Diaconescu, D.: Kripke-style semantics for non-commutative monoidal t-norm logic. J. Mult.-Valued Log. Soft Comput. **16**, 247–263 (2010)
91. Diaconescu, D., Georgescu, G.: Tense operators on MV-algebras and Łukasiewicz-Moisil algebras. Fundam. Inform. **81**, 379–408 (2007)
92. Diaconescu, D., Georgescu, G.: On the forcing semantics for monoidal t-norm based logic. J. Univers. Comput. Sci. **13**, 1550–1572 (2007)
93. Dilworth, R.P.: Non-commutative residuated lattices. Trans. Am. Math. Soc. **46**, 426–444 (1939)
94. Dvurečenskij, A.: Measures and states on BCK-algebras. Atti Semin. Mat. Fis. Univ. Modena Reggio Emilia **47**, 511–528 (1999)
95. Dvurečenskij, A.: On partial addition in pseudo-MV algebras. In: Proc. 4th International Symposium on Economic Informatics, pp. 952–960. INFOREC Printing House, Bucharest (1999)
96. Dvurečenskij, A.: States on pseudo-MV algebras. Stud. Log. **68**, 301–327 (2001)
97. Dvurečenskij, A.: Pseudo-MV algebras are intervals in ℓ-groups. J. Aust. Math. Soc. **70**, 427–445 (2002)
98. Dvurečenskij, A.: Every linear pseudo-BL algebra admits a state. Soft Comput. **6**, 495–501 (2007)
99. Dvurečenskij, A.: Aglianò-Montagna type decomposition of linear pseudo-hoops and its applications. J. Pure Appl. Algebra **211**, 851–861 (2007)
100. Dvurečenskij, A.: On n-perfect GMV-algebras. J. Algebra **319**, 4921–4946 (2008)
101. Dvurečenskij, A.: Subdirectly irreducible state-morphism BL-algebras. Arch. Math. Log. **50**, 145–160 (2011)
102. Dvurečenskij, A.: States on quantum structures versus integrals. Int. J. Theor. Phys. **50**, 3761–3777 (2011)
103. Dvurečenskij, A., Graziano, M.G.: Commutative BCK-algebras and lattice ordered groups. Math. Jpn. **49**, 159–174 (1999)
104. Dvurečenskij, A., Hyčko, M.: On the existence of states for linear pseudo-BL algebras. Atti Semin. Mat. Fis. Univ. Modena Reggio Emilia **53**, 93–110 (2005)
105. Dvurečenskij, A., Hyčko, M.: Algebras on subintervals of BL-algebras, pseudo-BL algebras and bounded residuated Rℓ-monoids. Math. Slovaca **56**, 125–144 (2006)
106. Dvurečenskij, A., Kowalski, T.: On decomposition of pseudo-BL algebras. Math. Slovaca **61**, 307–326 (2011)
107. Dvurečenskij, A., Kühr, J.: On the structure of linearly ordered pseudo-BCK algebras. Arch. Math. Log. **48**, 771–791 (2009)

108. Dvurečenskij, A., Pulmannova, S.: New Trends in Quantum Structures. Kluwer Academic, Dordrecht (2000)
109. Dvurečenskij, A., Rachůnek, J.: Probabilistic averaging in bounded commutative Rℓ-monoids. Discrete Math. **306**, 1317–1326 (2006)
110. Dvurečenskij, A., Rachůnek, J.: Probabilistic averaging in bounded non-commutative Rℓ-monoids. Semigroup Forum **72**, 190–206 (2006)
111. Dvurečenskij, A., Rachůnek, J.: On Riečan and Bosbach states for bounded non-commutative Rℓ-monoids. Math. Slovaca **56**, 487–500 (2006)
112. Dvurečenskij, A., Vetterlein, T.: Algebras in the positive cone of *po*-groups. Order **19**, 127–146 (2002)
113. Dvurečenskij, A., Giuntini, R., Kowalski, T.: On the structure of pseudo-BL algebras and pseudo-hoops in quantum logics. Found. Phys. **49**, 1519–1542 (2010)
114. Dvurečenskij, A., Rachůnek, J., Šalounová, D.: State operators on generalizations of fuzzy structures. Fuzzy Sets Syst. **187**, 58–76 (2012)
115. Dvurečenskij, A., Rachůnek, J., Šalounová, D.: Erratum to "State operators on generalizations of fuzzy structures" [Fuzzy Sets Syst. 187 (2012) 58–76]. Fuzzy Sets Syst. **194**, 97–99 (2012)
116. Dymek, G., Walendziak, A.: Semisimple, Archimedean and semilocal pseudo-MV algebras. Sci. Math. Jpn. **66**, 217–226 (2007)
117. Esteva, F., Godo, L.: Monoidal t-norm based logic: towards a logic for left-continuous t-norms. Fuzzy Sets Syst. **124**, 271–288 (2001)
118. Flaminio, T.: Strong non-standard completeness for fuzzy logics. Soft Comput. **12**, 321–333 (2008)
119. Flaminio, T., Montagna, F.: MV-algebras with internal states and probabilistic fuzzy logics. Int. J. Approx. Reason. **50**, 138–152 (2009)
120. Flondor, P., Leuştean, I.: Tensor products of MV-algebras. Soft Comput. **7**, 446–457 (2003)
121. Flondor, P., Sularia, M.: On a class of residuated semilattice monoids. Fuzzy Sets Syst. **138**, 149–176 (2003)
122. Flondor, P., Georgescu, G., Iorgulescu, A.: Pseudo-t-norms and pseudo-BL algebras. Soft Comput. **5**, 355–371 (2001)
123. Gaifman, H.: Concerning measures on first-order calculi. Isr. J. Math. **2**, 1–18 (1964)
124. Galatos, N.: Varieties of residuated lattices. Ph.D. thesis, Vanderbilt University, Nashville (2003)
125. Galatos, N.: Minimal varieties of residuated lattices. Algebra Univers. **52**, 215–239 (2005)
126. Galatos, N., Jipsen, P.: A survey of generalized basic logic algebras. In: Cintula, P., Hanikova, Z., Svejdar, V. (eds.) Witnessed Years: Essays in Honour of Petr Hajek, pp. 305–331. College Publications, London (2009)
127. Galatos, N., Ono, H.: Glivenko theorems for substructural logics over FL. J. Symb. Log. **71**, 1353–1384 (2006)
128. Galatos, N., Tsinakis, C.: Generalized MV-algebras. J. Algebra **283**, 254–291 (2005)
129. Galatos, N., Jipsen, P., Kowalski, T., Ono, H.: Residuated Lattices: An Algebraic Glimpse at Substructural Logics. Elsevier, Amsterdam (2007)
130. Georgescu, G.: Some model theory for probability structures. Rep. Math. Log. **35**, 103–113 (2001)
131. Georgescu, G.: Bosbach states on fuzzy structures. Soft Comput. **8**, 217–230 (2004)
132. Georgescu, G.: An extension theorem for submeasures on Łukasiewicz-Moisil algebras. Fuzzy Sets Syst. **158**, 1782–1790 (2007)
133. Georgescu, G.: States on polyadic MV-algebras. Stud. Log. **94**, 231–243 (2010)
134. Georgescu, G.: Probabilistic models for intuitionistic predicate logic. J. Log. Comput. **21**, 1165–1176 (2011)
135. Georgescu, G., Iorgulescu, A.: Pseudo-MV algebras: a non-commutative extension of MV-algebras. In: Proc. 4th International Symposium of Economic Informatics, pp. 961–968. IN-FOREC Printing House, Bucharest (1999)

136. Georgescu, G., Iorgulescu, A.: Pseudo-BL algebras: a non-commutative extension of BL-algebras. In: Abstracts of the 5th International Conference FSTA 2000, Slovakia, pp. 90–92 (2000)
137. Georgescu, G., Iorgulescu, A.: Pseudo-MV algebras. Mult. Valued Log. **6**, 95–135 (2001)
138. Georgescu, G., Iorgulescu, A.: Pseudo-BCK algebras: an extension of BCK-algebras. In: Proc. DMTCS'01: Combinatorics, Computability and Logic, pp. 97–114. Springer, London (2001)
139. Georgescu, G., Leuştean, I.: Convergence in perfect MV-algebras. J. Math. Anal. Appl. **228**, 96–111 (1998)
140. Georgescu, G., Leuştean, I.: Probabilities on Łukasiewicz-Moisil algebras. Int. J. Approx. Reason. **18**, 201–215 (1998)
141. Georgescu, G., Leuştean, I.: A representation theorem for monadic Pavelka algebras. J. Univers. Comput. Sci. **6**, 105–111 (2000)
142. Georgescu, G., Leuştean, I.: Towards a probability theory based on Moisil logic. Soft Comput. **4**, 19–26 (2000)
143. Georgescu, G., Leuştean, L.: Some classes of pseudo-BL algebras. J. Aust. Math. Soc. **73**, 127–153 (2002)
144. Georgescu, G., Popescu, A.: Non-commutative fuzzy structures and pairs of weak negations. Fuzzy Sets Syst. **143**, 129–155 (2004)
145. Georgescu, G., Popescu, A.: Similarity convergence in residuated structures. Log. J. IGPL **13**, 389–413 (2005)
146. Georgescu, G., Liguori, F., Martini, G.: Convergence in MV-algebras. Mathw. Soft Comput. **4**, 41–52 (1997)
147. Georgescu, G., Iorgulescu, A., Leuştean, I.: Monadic and closure MV-algebras. Mult. Valued Log. **3**, 235–257 (1998)
148. Georgescu, G., Leuştean, L., Preoteasa, V.: Pseudo-hoops. J. Mult.-Valued Log. Soft Comput. **11**, 153–184 (2005)
149. Georgescu, G., Iorgulescu, A., Rudeanu, S.: Grigore C. Moisil (1906–1973) and his school in algebraic logic. Int. J. Comput. Commun. Control **1**, 81–99 (2006)
150. Georgescu, G., Leuştean, I., Popescu, A.: Order convergence and distance on Łukasiewicz-Moisil algebras. J. Mult.-Valued Log. Soft Comput. **12**, 33–69 (2006)
151. Georgescu, G., Iorgulescu, A., Rudeanu, S.: Some Romanian researchers in the algebra of logic. In: Grigore C. Moisil and His Followers in the Field of Theoretical Computer Science, pp. 86–120. Romanian Academy Ed., Bucharest (2007)
152. Georgescu, G., Leuştean, L., Mureşan, C.: Maximal residuated lattices with lifting Boolean center. Algebra Univers. **63**, 83–99 (2010)
153. Gerla, B., Leuştean, I.: Similarity MV-algebras. Fundam. Inform. **69**, 287–300 (2006)
154. Glivenko, V.: Sur quelques points de la logique de M. Brouwer. Bull. Cl. Sci., Acad. R. Belg. **15**, 183–188 (1929)
155. Goodearl, K.R.: Partially Ordered Abelian Groups with Interpolation. Mathematical Surveys and Monographs, vol. 20. Am. Math. Soc., Providence (1986)
156. Grätzer, G.: Lattice Theory. First Concepts and Distributive Lattices. A Series of Books in Mathematics. Freeman, San Francisco (1972)
157. Grigolia, R.S.: Algebraic analysis of Łukasiewicz-Tarski's n-valued logical systems. In: Wójcicki, R., Malinkowski, G. (eds.) Selected Papers on Łukasiewicz Sentential Calculi, pp. 81–92. Polish Academy of Sciences, Wroclav (1977)
158. Hájek, P.: Metamathematics of Fuzzy Logic. Kluwer Academic, Dordrecht (1998)
159. Hájek, P.: Basic fuzzy logic and BL-algebras. Soft Comput. **2**, 124–128 (1998)
160. Hájek, P.: Observations on non-commutative fuzzy logic. Soft Comput. **8**, 28–43 (2003)
161. Hájek, P.: Fuzzy logics with non-commutative conjunctions. J. Log. Comput. **13**, 469–479 (2003)
162. Hájek, P., Ševčik, J.: On fuzzy predicate calculi with non-commutative conjuction. In: Proc. East West Fuzzy Colloquium, Zittau, pp. 103–110 (2004)

163. Halaš, R., Kühr, J.: Deductive systems and annihilators of pseudo BCK-algebras. Ital. J. Pure Appl. Math. **25**, 83–94 (2009)
164. Heath, T.L.: The Works of Archimedes. Dover, New York (1953)
165. Höhle, U.: Commutative residuated monoids. In: Höhle, U., Klement, P. (eds.) Non-classical Logics and Their Applications to Fuzzy Subsets, pp. 53–106. Kluwer Academic, Dordrecht (1995)
166. Horn, A., Tarski, A.: Measures in Boolean algebras. Trans. Am. Math. Soc. **64**, 467–497 (1948)
167. Idziak, P.M.: Lattice operations in BCK-algebras. Math. Jpn. **29**, 839–846 (1984)
168. Imai, Y., Iséki, K.: On axiom systems of propositional calculi. XIV. Proc. Jpn. Acad. **42**, 19–22 (1966)
169. Ioniţă, C.: Pseudo-MV algebras as semigroups. In: Proc. 4th International Symposium on Economic Informatics, pp. 983–987. INFOREC Printing House, Bucharest (1999)
170. Iorgulescu, A.: $(1 + \theta)$-valued Łukasiewicz-Moisil algebras. Ph.D. thesis, University of Bucharest (1984) (in Romanian)
171. Iorgulescu, A.: Connections between MV_n-algebras and n-valued Łukasiewicz-Moisil algebras—I. Discrete Math. **181**, 155–177 (1998)
172. Iorgulescu, A.: Connections between MV_n-algebras and n-valued Łukasiewicz-Moisil algebras—II. Discrete Math. **202**, 113–134 (1999)
173. Iorgulescu, A.: Connections between MV_n-algebras and n-valued Łukasiewicz-Moisil algebras—IV. J. Univers. Comput. Sci. **6**, 139–154 (2000)
174. Iorgulescu, A.: Classes of BCK-algebras—part I. Preprint 1, Institute of Mathematics of the Romanian Academy (2004)
175. Iorgulescu, A.: Classes of BCK-algebras—part II. Preprint 2, Institute of Mathematics of the Romanian Academy (2004)
176. Iorgulescu, A.: Classes of BCK-algebras—part III. Preprint 3, Institute of Mathematics of the Romanian Academy (2004)
177. Iorgulescu, A.: Classes of BCK-algebras—part IV. Preprint 4, Institute of Mathematics of the Romanian Academy (2004)
178. Iorgulescu, A.: Classes of BCK-algebras—part V. Preprint 5, Institute of Mathematics of the Romanian Academy (2004)
179. Iorgulescu, A.: On pseudo-BCK algebras and porims. Sci. Math. Jpn. **16**, 293–305 (2004)
180. Iorgulescu, A.: Pseudo-Iséki algebras. Connection with pseudo-BL algebras. J. Mult.-Valued Log. Soft Comput. **11**, 263–308 (2005)
181. Iorgulescu, A.: Classes of pseudo-BCK algebras—part I. J. Mult.-Valued Log. Soft Comput. **12**, 71–130 (2006)
182. Iorgulescu, A.: Classes of pseudo-BCK algebras—part II. J. Mult.-Valued Log. Soft Comput. **12**, 575–629 (2006)
183. Iorgulescu, A.: On BCK algebras—part I.a: an attempt to treat unitarily the algebras of logic. New algebras. J. Univers. Comput. Sci. **13**, 1628–1654 (2007)
184. Iorgulescu, A.: On BCK algebras—part I.b: an attempt to treat unitarily the algebras of logic. New algebras. J. Univers. Comput. Sci. **14**, 3686–3715 (2008)
185. Iorgulescu, A.: On BCK algebras—part II: new algebras. The ordinal sum (product) of two bounded BCK algebras. Soft Comput. **12**, 835–856 (2008)
186. Iorgulescu, A.: Algebras of Logic as BCK-Algebras. ASE Ed., Bucharest (2008)
187. Iorgulescu, A.: Monadic involutive BCK-algebras. Acta Univ. Apulensis, Mat.-Inform. **15**, 159–178 (2008)
188. Iorgulescu, A.: Classes of examples of pseudo-MV algebras, pseudo-BL algebras and divisible bounded non-commutative residuated lattices. Soft Comput. **14**, 313–327 (2010)
189. Iorgulescu, A.: On BCK algebras—part III: classes of examples of proper MV algebras, BL algebras and divisible bounded residuated lattices, with or without condition (WNM). J. Mult.-Valued Log. Soft Comput. **16**, 341–386 (2010)
190. Iorgulescu, A.: The implicative-group—a term equivalent definition of the group coming from algebras of logic—part I. Preprint 11, Institute of Mathematics of the Romanian

Academy (2011)

191. Iorgulescu, A.: Connections between MV_n-algebras and n-valued Łukasiewicz-Moisil algebras—III. Manuscript

192. Iorgulescu, A.: On BCK algebras—part IV: classes of examples of finite proper IMTL algebras, MTL and bounded $\alpha\gamma$ algebras, with or without condition (WNM) (submitted)

193. Iorgulescu, A.: On BCK algebras—part V: classes of examples of finite proper bounded α, β, γ, $\beta\gamma$ algebras and BCK(P) lattices (residuated lattices), with or without conditions (WNM) and (DN) (submitted)

194. Isèki, K.: An algebra related with a propositional calculus. Proc. Jpn. Acad. **42**, 26–29 (1966)

195. Jakubík, J.: On intervals and the dual of a pseudo-MV algebra. Math. Slovaca **56**, 213–221 (2006)

196. Jakubík, J.: On interval subalgebras of generalized MV-algebras. Math. Slovaca **56**, 387–395 (2006)

197. Jenei, S., Montagna, F.: A proof of standard completeness for Esteva and Godo's logic MTL. Stud. Log. **70**, 183–192 (2002)

198. Jipsen, P.: An overview of generalized basic logic algebras. Neural Netw. World **13**, 491–500 (2003)

199. Jipsen, P., Montagna, F.: On the structure of generalized BL-algebras. Algebra Univers. **55**, 227–238 (2006)

200. Jipsen, P., Tsinakis, C.: A survey of residuated lattices. In: Martinez, J. (ed.) Ordered Algebraic Structures, pp. 19–56. Kluwer Academic, Dordrecht (2002)

201. Kowalski, T., Ono, H.: Residuated lattices: an algebraic glimpse at logics without contraction. Monograph (2001)

202. Kroupa, T.: Every state on semisimple MV-algebra is integral. Fuzzy Sets Syst. **157**, 2771–2782 (2006)

203. Kroupa, T.: Representation and extension of states on MV-algebras. Arch. Math. Log. **45**, 381–392 (2006)

204. Krull, W.: Axiomatische Begründung der allgemeinen Idealtheorie. Sitzungsber. Phys.-Med. Soz. Erlangen **56**, 47–63 (1924)

205. Kühr, J.: Pseudo-BL algebras and DRℓ-monoids. Math. Bohem. **128**, 199–208 (2003)

206. Kühr, J.: Pseudo-BCK algebras and residuated lattices. Contrib. Gen. Algebra **16**, 139–144 (2005)

207. Kühr, J.: On a generalization of pseudo-MV algebras. J. Mult.-Valued Log. Soft Comput. **12**, 373–389 (2006)

208. Kühr, J.: Pseudo-BCK algebras and related structures. Univerzita Palackého v Olomouci (2007)

209. Kühr, J.: Pseudo-BCK semilattices. Demonstr. Math. **40**, 495–516 (2007)

210. Kühr, J.: Representable pseudo-BCK algebras and integral residuated lattices. J. Algebra **317**, 354–364 (2007)

211. Kühr, J., Mundici, D.: De Finetti theorem and Borel states in [0, 1]-valued algebraic logic. Int. J. Approx. Reason. **46**, 605–616 (2007)

212. Leuştean, I.: Local pseudo-MV algebras. Soft Comput. **5**, 386–395 (2001)

213. Leuştean, I.: Contributions to the theory of MV-algebras: MV-modules. Ph.D. thesis, Faculty of Mathematics and Computer Science, University of Bucharest (2003)

214. Leuştean, I.: α-convergence and complete distributivity in MV-algebras. J. Mult.-Valued Log. Soft Comput. **12**, 309–319 (2006)

215. Leuştean, I.: Non-commutative Łukasiewicz propositional logic. Arch. Math. Log. **45**, 191–213 (2006)

216. Leuştean, I.: A determination principle for algebras of n-Łukasiewicz logic. J. Algebra **320**, 3694–3719 (2008)

217. Leuştean, I.: Tensor products of probability MV-algebras. J. Mult.-Valued Log. Soft Comput. **16**, 405–419 (2010)

218. Leuştean, I.: Metric completions of MV-algebras with states. An approach to stochastic independence. J. Log. Comput. **21**, 493–508 (2011)

219. Leuştean, I.: Hahn-Banach theorems for MV-algebras. Soft Comput. **16**, 1845–1850 (2012)
220. Leuştean, L.: Representations of many-valued algebras. Ph.D. thesis, Faculty of Mathematics and Computer Science, University of Bucharest (2003)
221. Leuştean, L.: Sheaf representations of BL-algebras. Soft Comput. **9**, 897–909 (2005)
222. Leuştean, L.: Baer extensions of BL-algebras. J. Mult.-Valued Log. Soft Comput. **12**, 321–336 (2006)
223. Łukasiewicz, J.: On three-valued logic. Ruch Filoz. **5**, 169–171 (1920) (in Polish)
224. Łukasiewicz, J., Tarski, A.: Untersuchungen über den Aussagenkalkül. Comp. Rend. Soc. Sci. et Lettres Varsovie Cl. III **23**, 30–50 (1930)
225. Meng, J., Jun, Y.B.: BCK-Algebras. Kyung Moon Sa Co., Seoul (1994)
226. Moisil, Gr.C.: Recherches sur les logiques non-chrysippiennes. Ann. Sci. Univ. Jassy **26**, 431–466 (1940)
227. Moisil, Gr.C.: Notes sur les logiques non-chrysippiennes. Ann. Sci. Univ. Jassy **27**, 86–98 (1941)
228. Moisil, Gr.C.: Essais sur les Logique Non-Chrysippiennes. Romanian Academy Ed., Bucharest (1972)
229. Mundici, D.: Interpretation of AFC*-algebras in Łukasiewicz sentential calculus. J. Funct. Anal. **65**, 15–63 (1986)
230. Mundici, D.: Averaging the truth-value in Łukasiewicz logic. Stud. Log. **55**, 113–127 (1995)
231. Murinová, P., Novák, V.: Omitting types in fuzzy logic with evaluated syntax. Math. Log. Q. **52**, 259–268 (2006)
232. Noguera, C., Esteva, F., Gispert, J.: On some varieties of MTL-algebras. Log. J. IGPL **13**, 443–466 (2005)
233. Noguera, C., Esteva, F., Gispert, J.: Perfect and bipartite IMTL-algebras and disconnected rotations of prelinear semihoops. Arch. Math. Log. **44**, 869–886 (2005)
234. Ono, H.: Substructural logics and residuated lattices: an introduction. In: Hendricks, F.V., Malinowski, J. (eds.) 50 Years of Studia Logica. Trends in Logic, vol. 20, pp. 177–212. Kluwer Academic, Dordrecht (2003)
235. Ono, H.: Completions of algebras and completeness of modal and substructural logics. In: Advances in Modal Logic, vol. 4, pp. 335–353. King's College Publications, London (2003)
236. Ono, H., Komori, Y.: Logics without contraction rule. J. Symb. Log. **50**, 169–201 (1985)
237. Pavelka, J.: On fuzzy logic II. Enriched residuated lattices and semantics of propositional calculi. Z. Math. Log. Grundl. Math. **25**, 119–134 (1979)
238. Piciu, D.: Algebras of Fuzzy Logic. Universitaria Ed., Craiova (2007)
239. Post, E.L.: Introduction to a general theory of elementary propositions. Am. J. Math. **43**, 163–185 (1921)
240. Rachůnek, J.: A duality between algebras of basic logic and bounded representable DRℓ-monoids. Math. Bohem. **126**, 561–569 (2001)
241. Rachůnek, J.: A non-commutative generalization of MV-algebras. Czechoslov. Math. J. **52**, 255–273 (2002)
242. Rachůnek, J., Šalounová, D.: A generalization of local fuzzy structures. Soft Comput. **11**, 565–571 (2007)
243. Rachůnek, J., Šalounová, D.: State operators on GMV-algebras. Soft Comput. **15**, 327–334 (2011)
244. Rachůnek, J., Slezák, V.: Bounded dually residuated lattice ordered monoids as a generalization of fuzzy structures. Math. Slovaca **56**, 223–233 (2006)
245. Riečan, B.: On the probability on BL-algebras. Acta Math. Nitra **4**, 3–13 (2000)
246. Riečan, B.: On the probability theory on MV-algebras. Soft Comput. **4**, 49–57 (2000)
247. Riečan, B.: Almost everywhere convergence in MV-algebras with product. Soft Comput. **5**, 396–399 (2001)
248. Riečan, B.: On the Dobrakov submeasure on fuzzy sets. Fuzzy Sets Syst. **151**, 635–641 (2005)
249. Riečan, B., Mundici, D.: Probability on MV-algebras. In: Pap, E. (ed.) Handbook of Measure Theory, pp. 869–909. North-Holland, Amsterdam (2002)

250. Riečan, B., Neubrunn, T.: Integral, Measure and Ordering. Kluwer Academic, Dordrecht (1997)
251. Robinson, A.: Introduction to Model Theory and the Metamathematics of Algebra. North-Holland, Amsterdam (1974)
252. Rudeanu, S.: Localizations and fractions in algebra of logic. J. Mult.-Valued Log. Soft Comput. **16**, 467–504 (2010)
253. Scott, D., Krauss, P.: Assigning probabilities to logical formulas. In: Hintikka, J., Suppes, P. (eds.) Aspects of Inductive Logic, pp. 219–264. North-Holland, Amsterdam (1966)
254. Torrens, A.: An approach to Glivenko's theorem in algebraizable logics. Stud. Log. **88**, 349–383 (2008)
255. Turunen, E.: BL-algebras of basic fuzzy logic. Mathw. Soft Comput. **6**, 49–61 (1999)
256. Turunen, E.: Mathematics Behind Fuzzy Logic. Advances in Soft Computing. Physica-Verlag, Heidelberg (1999)
257. Turunen, E.: Boolean deductive systems of BL-algebras. Arch. Math. Log. **40**, 467–473 (2001)
258. Turunen, E., Mertanen, J.: States on semi-divisible residuated lattices. Soft Comput. **12**, 353–357 (2008)
259. Turunen, E., Mertanen, J.: States on semi-divisible generalized residuated lattices reduce to states on MV-algebras. Fuzzy Sets Syst. **22**, 3051–3064 (2008)
260. Turunen, E., Sessa, S.: Local BL-algebras. Mult. Valued Log. **6**, 229–249 (2001)
261. Ward, M.: Residuated distributive lattices. Duke Math. J. **6**, 641–651 (1940)
262. Ward, M., Dilworth, R.P.: Residuated lattices. Trans. Am. Math. Soc. **45**, 335–354 (1939)
263. Zhou, H., Zhao, B.: Generalized Bosbach and Riečan states based on relative negations in residuated lattices. Fuzzy Sets Syst. **187**, 33–57 (2012)

Index

L.C. Ciungu, *Non-commutative Multiple-Valued Logic Algebras*,
Springer Monographs in Mathematics, DOI 10.1007/978-3-319-01589-7,
© Springer International Publishing Switzerland 2014

Printed in the United States
By Bookmasters